Prevention of Thermal Cracking in Concrete at Early Ages

Also available from E & FN Spon

Autogenous Shrinkage of Concrete
Edited by E. Tazawa

Concrete under Severe Conditions 2
Edited by O. Gjørv, K. Sakai and N. Banthia

Corrosion of Steel in Concrete
J.P. Broomfield

Durability of Concrete in Cold Climates
M. Pigeon and R. Pleau

Freeze–Thaw Durability of Concrete
Edited by M. Pigeon, M.J. Setzer and J. Marchand

Frost Resistance of Concrete
M.J. Setzer and R. Auberg

High Performance Concrete
P.C. Aïtcin

Integrated Design and Environmental Issues in Concrete Technology
Edited by K. Sakkai

Mechanisms of Chemical Degradation of Cement-based Systems
Edited by K.L. Scrivener and J.F. Young

Optimization Methods for Material Design of Cement-based Composites
Edited by A.M. Brandt

Prevention and Permeability of Concrete
H.W. Reinhardt

Polymers in Concrete
Y. Ohama

Design Aids For Eurocode 2
Edited by The Concrete Societies of the UK, The Netherlands and Germany

Introduction to Eurocode 2
D. Beckett and A. Alexandrou

Repair, Protection and Waterproofing of Concrete Structures, Third edition
P. Perkins

For more information about these and other titles please contact: The Marketing Department, E & FN Spon, 11 New Fetter Lane, London, EC4P 4EE, UK Tel: 0171 842 2180

RILEM REPORT 15

Prevention of Thermal Cracking in Concrete at Early Ages

State-of-the-Art Report prepared by
RILEM Technical Committee 119
Avoidance of Thermal Cracking in Concrete at Early Ages

Edited by
R. Springenschmid
Technical University of Munich
Germany

CRC Press is an imprint of the
Taylor & Francis Group, an **informa** business

CRC Press
Taylor & Francis Group
6000 Broken Sound Parkway NW, Suite 300
Boca Raton, FL 33487-2742

First issued in paperback 2019

ISBN-13: 978-0-419-22310-8 (hbk)
ISBN-13: 978-0-367-44767-0 (pbk)

Publisher's Note
This book has been prepared from camera-ready copy and word-processing discs provided by the individual contributors.

British Library Cataloguing in Publication Data
A catalogue record for this book is available from the British Library.

Visit the Taylor & Francis Web site at
http://www.taylorandfrancis.com

and the CRC Press Web site at
http://www.crcpress.com

Contents

List of Contributors

Stig Bernander
Luleå University of Technology, Division of Structural Engineering,
Luleå, Sweden

Rolf Breitenbücher
Philipp Holzmann AG, Zentrales Baustofflabor,
Frankfurt am Main, Germany

Klaas van Breugel
Delft University of Technology, Faculty of Civil Engineering,
Delft, The Netherlands

Mats Emborg
Luleå University of Technology, Division of Structural Engineering,
Luleå, Sweden

Martin Laube
Technische Universität Braunschweig, IBMB, Brunswick, Germany

Martin Mangold
Barg Betontechnik GmbH & Co., Berlin, Germany

Paolo Morabito
ENEL spa, DSR-CRIS, Milan, Italy

Ferdinand S. Rostásy
Technische Universität Braunschweig, IBMB, Brunswick, Germany

Rupert Springenschmid
Technische Universität München, Baustoffinstitut, Munich, Germany

Tada-aki Tanabe
University of Nagoya, Department of Civil Engineering, Nagoya, Japan

Preface

This book describes the work of RILEM Technical Committee -119 TCE on the "Avoidance of Thermal Cracking in Concrete at Early Ages" which was performed between 1989 and 1997. It also includes important results presented at the International Symposium on Thermal Cracking in Concrete at Early Ages between October 10th and 12th, 1994 in Munich.

The avoidance of cracking of young concrete is one of the main problems of concrete technology. The aim of the work of the members of the RILEM TC 119 and many civil engineers and scientists all over the world was to replace earlier methods, based purely on field experience, by modern concepts for the prediction of stresses in concrete at early ages and the factors influencing them.

The book should be a basis for further research. Furthermore – and this is even more important – it should serve as a basis for the avoidance of thermal cracking by the engineers who have to design and construct mass or medium sized concrete structures and by those who have to select the concrete materials and design the concrete mix.

As Chairman of RILEM TC 119, I would like to thank the authors of the sections of this book and all others who have contributed to this work. They passed on their experience generously. Mr. J.-L. Bostvironnois assisted me in most of this editorial work. I would like to take this opportunity to thank him.

Rupert Springenschmid
Munich, March 1998

Introduction

Cracks in mass concrete structures caused by the hydration heat of the cement are a well-known phenomenon since the beginning of this century. Methods of avoiding such cracks have been developed mainly for large concrete dams and other massive hydraulic engineering structures. In order to reduce the heat development, pozzolanas, and since 1932, low-heat cements have been used. Further progress aimed at reducing the high maximum temperature caused by the hydration heat has been made by the use of very low cement contents, coarse aggregates, cooling the concrete materials, limitation of lift-heights and by pipe cooling.

Although the processes of heat generation and dissipation were familiar to civil engineers, the specification of the maximum permissible temperature in the concrete bulk and the temperature difference between the base and the inside of the concrete block was based solely on experience. The decisive properties of a particular concrete such as its tensile strength or its coefficient of thermal expansion were not considered.

In the past decades, cracks in the structural concrete of foundations, bridges, tunnel linings and other medium-sized concrete elements have become an increasing problem. It has been found that drying shrinkage is often of minor importance. The heat of hydration as well as other temperature changes have been found to be the primary cause of restraint stresses and cracks in unreinforced as well as reinforced concrete.

In the late sixties, the first attempts were made to estimate the size of stresses due to restrained thermal deformations and compare them with the increasing tensile strength of the concrete at early ages. Two points turned out to be extremely problematic:

- The results of thermal stress calculations depend strongly on the evaluation of the stiffness of the concrete as it increases during its transformation from a semi-liquid to a solid state. However, the stiffness is difficult to measure and predict.
- Restraint stresses could not be determined with conventional methods and therefore no data were available to verify the results of stress calculation.

In 1969 the first laboratory equipment, the cracking frame, was developed in order to perform model tests. By measuring the stress response of concrete at early ages to changing temperature we gained a deeper understanding of the changes which occur when the expansion or contraction of a concrete element is prevented and stresses are produced. The Munich Temperature Stress Testing Machine (1984) and similar machines in several other research institutes now allow stress measurement for any degree of restraint.

In recent years much research work has been devoted to the calculation of the early age restraint stresses and to the determination of the risk of cracking. Computer programs have taken account of the materials properties, the hydration heat development, the increase of stiffness and the decrease of relaxation capacity, the increasing tensile strength, the coefficient of thermal expansion and the influence of chemical reactions on the deformation. All these factors depend largely on age, temperature, cement type, cement brand and concrete mix composition. Realistically speaking, it is only possible to assess roughly the effects of these factors.

Promising new methods have been developed in Japan, France and Germany to measure restraint stresses in situ. The comparison of test results both in the laboratory and in the field with the results of calculations is a source of further progress in this area.

In recent years high-strength concrete has proved to be an extremely sensitive material regarding cracking at an early age. This is not only a consequence of the hydration heat. Autogenous shrinkage due to self-desiccation and chemical reactions of the sulphate phase may also be important.

Unexpected cracking of structural or mass concrete cannot always be attributed to the inexperience of the practical engineers. The limited knowledge of many problems in this area has strongly encouraged research all over the world. In 1989 RILEM, the International Union of Testing and Research Laboratories for Materials and Structures, established a Technical Committee, TC 119, on „the Avoidance of Thermal Cracking in Concrete at Early Ages“. The tasks of this committee are as well as the exchange of opinions and experience, the preparation of State of the Art Reports and to make RILEM Recommendations for the following test methods:

- Determination of the semi-adiabatic hydration heat
- Determination of the adiabatic hydration heat
- Restraint stress measurements in the laboratory with the cracking frame
- Restraint stress measurements in situ with the stress-meter.

In the meantime the recommendations have been published in *Materials and Structures/Matériaux et Constructions*, **30**, October 1997, pp 451-464, and are reprinted here in chapter 10.

The International Symposium in Munich between October 10th and 12th, 1994 thus took place in a dynamic period of development. A number of questions regarding the avoidance of early age cracking have already been solved and the results are ready for practical application. To answer other questions we have many suggestions based on test results or from theoretical considerations.

The avoidance of early age cracking is a task which requires theoretical knowledge, sound engineering judgement and extensive experience. Furthermore, dedicated engineers must ensure that all the necessary considerations are carried out in practice.

Rupert Springenschmid
Munich, March 1998

1

Methods to Determine the Heat of Hydration of Concrete

P. Morabito

1.1 INTRODUCTION

In order to minimise the risk of thermal cracking in concrete structures a knowledge of the expected temperature rise during the hydration of cement is desirable. The factors controlling the rate of temperature rise, the maximum temperature and the thus temperature distribution are complex and many and include:

- Rate of heat evolution
- Total amount of heat evolved
- Proportion and composition of cement
- Dimensions of pour
- Environmental conditions - temperature, wind, humidity
- Temperature of concrete at placing
- Mix proportions
- Aggregate type.

Numerous laboratory techniques have been developed to measure the temperature rise in concrete, ranging from sophisticated calorimeters used to monitor the temperature in very small samples of cement to temperature measurements at the centre of a large insulated block of concrete. The testing methods being used can be divided into the following categories: isothermal, adiabatic and semi-adiabatic. Isothermal methods are usually applied to pure cement pastes and make them possible to measure the heat generated by samples kept at constant temperature. The majority

Prevention of Thermal Cracking in Concrete at Early Ages. Edited by R. Springenschmid. RILEM Report 15. Published in 1998 by E & FN Spon, 11 New Fetter Lane, London EC4P 4EE, UK. ISBN 0 419 22310 X

of standard tests [1-4] currently in use rely on this method. Adiabatic methods use calorimeters for measuring the temperature rise in a sample being hydrated with no heat exchanges with the external medium. Semi-adiabatic methods use calorimeters or equipment where the heat exchange with the external environment is limited.

The ideal testing method should be able to reproduce the temperature = f(time) curve of each type of concrete during concreting. Since the isothermal tests do not take account of the change in reactivity of the cement with changing temperature, prediction of temperature rise of concrete from these results can be very difficult; the isothermal conditions of the test do not reflect the conditions in the real structure where the temperature is continually changing.

Adiabatic and semi-adiabatic methods are to be preferred. True adiabatic conditions are impossible to achieve but they do at least more closely resemble conditions at the centre of a large pour than do the isothermal conditions. A more realistic test would perhaps be to use semi-adiabatic methods where some heat is lost from the calorimeter. It should be realised however that even these tests may not truly reflect the conditions in the structure where the rate of heat loss is continually changing and would be difficult if not impossible to accurately reproduce them in the laboratory.

Adiabatic and semi-adiabatic calorimeters have been developed to measure the temperature rise in cement pastes, cement mortars and concrete samples. Although those developed for use with pastes and mortars [5, 6] are often much smaller than those used for concretes, the results obtained do not give a direct measure of the temperature rise in the concrete. The results can only be used to estimate the behaviour of the concrete from a knowledge of the mix proportions and thermal characteristics of the aggregates. Calculations of this kind can be extremely difficult because the kinetics involved are significantly different due largely to the buffering effects of the aggregate [30].

This report is concerned with the characteristics of adiabatic and semi-adiabatic calorimeters designed to measure the temperature rise and/or heat evaluation in concrete samples.

1.2 DEFINITIONS

For the purpose of this report the following definitions shall apply.

Adiabatic calorimeter. - Being impossible to have materials with an infinite thermal resistance, heat losses are prevented by controlling the temperature of the environment surrounding the sample in order to be as close as possible to the temperature of the sample at any time of the hydration. In order to achieve this conditions some form of heat will need to be supplied to the environment in which the test specimen is kept. However no calorimeter can be claimed to be truly adiabatic, even the most sophisticated are prone to some heat loss.

Semi-adiabatic calorimeter. - In the majority of semi-adiabatic calorimeters no attempt is made to heat the environment in which the specimen is placed. Most calorimeters of this type rely only on some form of insulation to slow down the rate of heat loss from the sample.

1.3 REVIEW OF AVAILABLE METHODS

1.3.1 Adiabatic calorimeter

In this report adiabatic calorimeters will be broadly classified by the method used to control the temperature of the environment in which the specimen is kept. For this purpose researchers have adopted one of three basic approaches. In the majority of cases the sample is surrounded by a heated water jacket, in some the air surrounding the specimen is heated and in others the container in which the specimen is kept has been heated. In addition, in the majority of calorimeters some form of insulation is provided between the sample and the control environment.

(a) Adiabatic calorimeters using water as the insulating medium
Work carried out by the Sumitomo Cement Company of Japan [7-10] has led to the development and production of a commercially available adiabatic calorimeter. The original development work involved making comparisons between two different basic types of calorimeters, one using air and the other using water as the insulating medium. By making comparisons with temperature measurements taken at the centre of a large insulated block, they concluded that the temperature rise measured by the apparatus using water (Type A) agreed very well with that from the large block. The results from the apparatus using air (Type B) were lower than that of the core temperature of the block.

The Type A apparatus was also used to examine the effect of specimen size on adiabatic temperature rise. Two cylindrical specimens of different sizes were used, namely: 60 cm diameter x 60 cm and 30 cm diameter x 30 cm high; there was very little difference between the two.

Looking at the influence of insulating the specimen when using the Type B apparatus they concluded that the temperature rise in the insulated specimen was lower than that in the non-insulated. They attributed this to the fact that some of the heat was consumed to heat the insulating material itself. The calorimeter developed by the Sumitomo Cement Company as a result of this work is shown in Fig. 1.1. The cylindrical specimen container is made of steel, the internal dimensions of which are 600 mm diameter x 600 mm in height. The container is placed inside a thermal jacket through which water is circulated at high speed and this thermal jacket is then itself placed inside an insulated container. The specimen container is separated from the thermal jacket by a 5 mm air gap. The water is heated or cooled to keep the surface of the sample of the same temperature as the core; it is claimed that the jacket temperature can be kept to within ± 0.1°C of the core temperature. Temperature measurements are made with platinum resistance thermometers and the sensitivity of the electronic control is quoted at ± 0.01°C. The apparatus can also be employed to take a smaller specimen container 300 mm in height and diameter; in addition, replacing the water with silicon oil as the thermal medium enables temperatures in excess of 100°C to be measured.

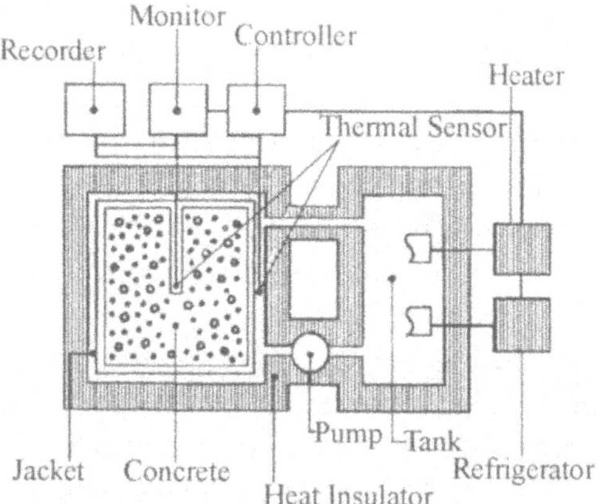

Fig. 1.1 Adiabatic calorimeter of the Sumitomo Cement Company.

A second commercially available calorimeter is marketed by ENEL spa [11] of Milan, Italy, and is similar in many respects to that of the Japanese described above. It uses a cylindrical specimen measuring 300 mm diameter x 360 mm high and the thermal jacket in this case is made of aluminium instead of steel. The only other significant differences are that in this apparatus the specimen container is separated from the thermal jacket by a layer of lightweight insulation, the jacket is made of aluminium and the temperature measurements are made with thermistors. The thermal jacket is heated by a mixing of water and glycol and the difference between the temperature of the concrete and the wall of the jacket is quoted at 0.1°C. The corresponding thermal loss is quoted at 0.002°C/h. To compensate for the high value of the apparent heat capacity of the calorimeter a thermal compensation technique has been applied during the hydration of the specimen; this consists of supplying to the sample by an electrical heater placed at the bottom of the sample container an external heat equal to that accumulated from the heat capacity of the calorimeter. From several tests performed on the same mix, the reproducibility in the measurement of the adiabatic temperature rise at seven days is quoted to be within 2.5%.

Similar but less sophisticated calorimeters have been developed elsewhere by a number of workers [5, 12-15] but none has been made available for production on a commercial basis.

A smaller but more versatile calorimeter has been developed in the UK by Coole [16] (see Fig. 1.2). A concrete sample weighing 1 kg is contained in a polypropylene cell within a larger insulated polypropylene container submerged in a well-stirred water bath.

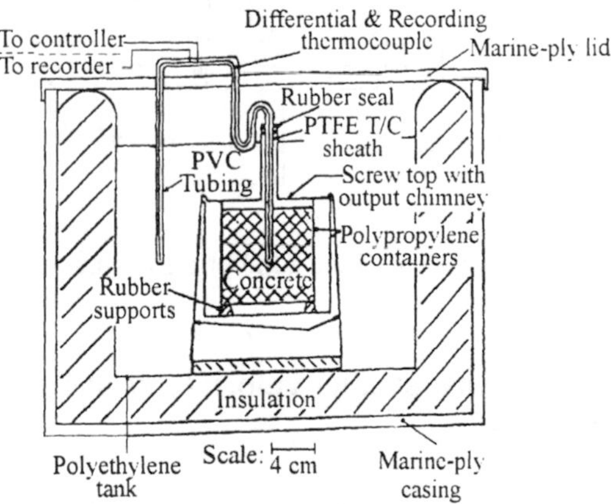

Fig. 1.2 Adiabatic calorimeter developed by Coole.

The significant difference between this and other calorimeters of a similar type is that there is no insulation between the specimen and the water bath. The temperature difference between the sample and the water is controlled by thermocouples connected to a thyristor control unit, this temperature difference can be varied to simulate heat losses from different size pours. When used as an adiabatic calorimeter the temperature difference is set at 0.01°C and the heat loss from the system has been measured at 0.01°C/h.

The calorimeters described previously have all been designed specifically to measure only the temperature rise in a concrete specimen. It is however often necessary to measure various properties of the concrete which has been subjected to such a temperature rise. The equipment developed by Wainwright [17] whose control system is shown in Fig. 1.3 was designed specifically to do this. As a calorimeter it is similar in many respects to those mentioned above with the 65 kg concrete specimen contained in an insulated mould which in turn is submerged in a temperature controlled water bath. The temperature of the water is controlled to follow that of the concrete but the volume of the bath is large enough to enable concrete specimens to be stored alongside the control cube.

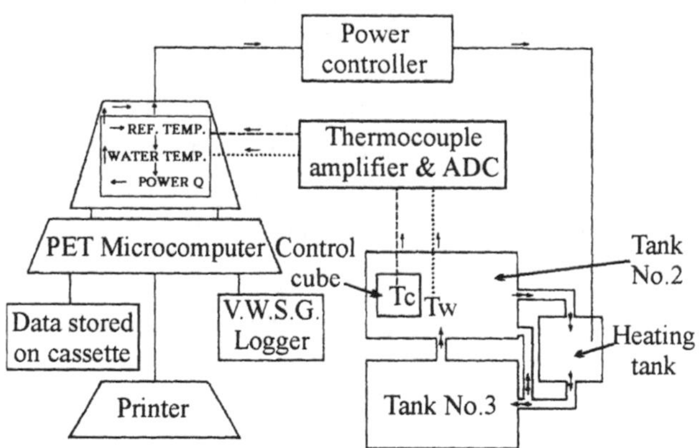

Fig. 1.3 Control and logging system of the adiabatic calorimeter developed by Wainwright.

A similar but somewhat smaller calorimeter has also been developed in Holland by Sarneel [18].

(b) Adiabatic calorimeters using air as the insulating medium

In terms of construction, calorimeters using air as the insulating medium tend to be somewhat simpler than those using water. In principle the calorimeter consists of an insulated container in which the air temperature is controlled to follow that of the concrete specimen placed inside it. The existing calorimeters in this category differ only with respect to the type and quantity of insulation, the method of heating the air and the sophistication and sensitivity of the control system.

Bamforth [19] used a standard laboratory oven as the control medium in which he placed a 300 mm cube of concrete contained in a plywood mould lined with 25 mm of polystyrene. The temperature of the oven was controlled to within 0.2°C of the concrete temperature using a balancing unit linked to electrical resistance thermometers located in the concrete and the oven. The heat loss from the system measured on fully hydrated concrete specimens heated to 80°C was quoted at 0.08°C/h.

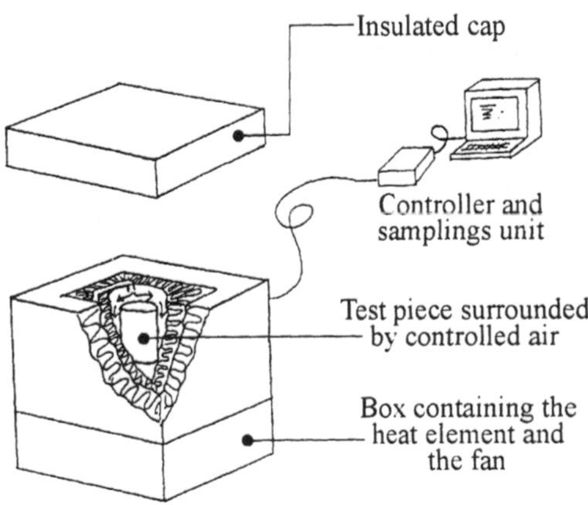

Fig. 1.4 Adiabatic calorimeter developed in Luleå University.

A similar but more sophisticated calorimeter has been developed in Sweden by Emborg [20] and is shown schematically in Fig. 1.4. The cylindrical test specimen weighing approximately 24 kg is placed inside a double insulated box with an air gap between the two. A box beneath the specimen contains the heating elements and circulating fan to control the temperature of the air surrounding the specimen. The temperature control for a known input of energy is calibrated using tests on aluminium specimens with known thermal properties. The reported maximum temperature loss at 70°C is 0.008°C/h.

According to Høyer [21] the Nordic draft test method for adiabatic calorimetry for the measurement of heat development in concrete proposes using air as the control medium. The draft gives no recommendations on specimen shape or size, design of calorimeter, control system, sensitivity etc. It simply states that the apparatus shall consist of a box in which the specimen is surrounded by air the temperature of which must always be the same as the temperature of the concrete. The passive stability of the box measured using an inert specimen (e.g. water, mature concrete) heated to approximately 60°C shall not be greater than + 0.5°C/day (i.e. 0.02°C/h).

Jaegerman [22] has reported the existence of an adiabatic calorimeter supplied by Toni Technik of Berlin. The volume of the test specimen is 5 litres and the "control precision" is quoted as better than 0.02°C/h. It is not clear what is meant by the term control precision and no results from tests run with the apparatus are available.

Several other workers [23-26] have developed calorimeters for their own use which are similar in most respects to those described above.

(c) Adiabatic calorimeters using heated container as the insulating medium

A clear distinction between this type of calorimeter and the type described previously in b) is sometimes difficult to make, particularly where the specimen container is heated and this in turn heats any air gap that might exist between it and the specimen.

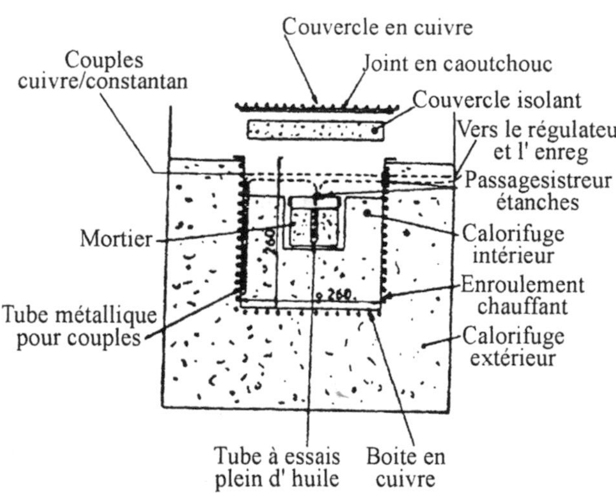

Fig. 1.5 Sketch of the CERILH calorimeter

The calorimeter used at the French Hydraulic Binders Research Institute (CERILH) and developed by Joisel [27] is shown in Fig. 1.5. The calorimeter consists of a cylindrical copper container which is enclosed in an electric heating coil. The sample used is a 2 kg cylinder of mortar or concrete which sits inside the copper container surrounded by an insulating layer of expanded polystyrene placed inside a second outer container. The temperature of the copper container and the test sample are monitored by thermocouples. The calorimeter is also fitted with a refrigerator for testing concrete to be used for dams. The heat loss from the apparatus (measured using an inert sample heated to approximately 60°C) is quoted at between 0.2 to 0.3°C/day (i.e. 0.008 to 0.013°C/h). It is also suggested that if the weight loss due to evaporation of mixing water during the test exceeds 1 g (i.e. 0.05 % by weight of the original sample) then the test should be considered unreliable.

A similar calorimeter has been developed in Italy by Costa [28] for Italcementi. The apparatus (Fig. 1.6) uses a cylindrical specimen 0.005m^3 in volume cast inside a mould made from tin plate and fitted with a water tight lid.

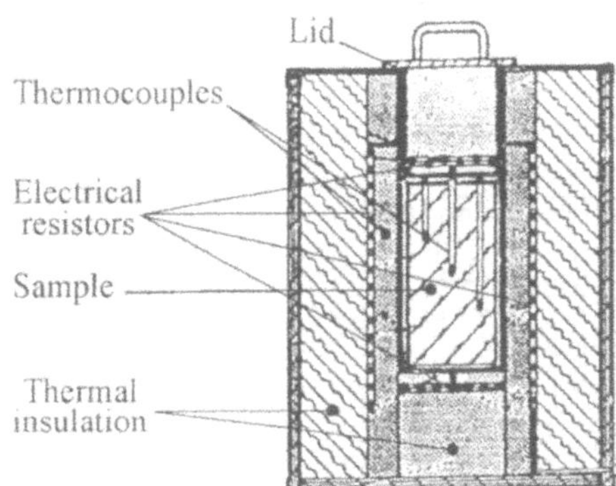

Fig. 1.6 Adiabatic calorimeter developed by Costa.

Five copper tubes are soldered to this lid and accommodate ten thermocouples measuring the temperature of the concrete at various locations. The sample container sits inside a second cylindrical container and is separated from it on all sides by a layer of insulation. This second cylinder together with two brass plates (situated at the top and bottom of the sample container) act as the isothermal surface which is attached to three electrical resistors connected in series. Ten more thermocouples are distributed inside the insulation to control its temperature and to match it to that of the concrete by means of a controller. The "error of adiabatism" was calculated using a concrete specimen heated to three different temperatures of 40, 50 and 60°C. Tests were carried out at each of these temperatures in which the concrete specimen was placed in the calorimeter and measurements were taken to determine how well the temperature was maintained. At 40 and 50°C there was no change but at 60°C the temperature of the block rose 0.1°C per 24 hours.

Another calorimeter [29] also developed in Italy uses a 300 mm aluminium cubic box as the isothermal surface with electrical resistance wires connected to each face. A 30 mm thick layer of insulation covers the inner surfaces of the box which is insulated from the environment by a 70 mm thick layer of insulation. The concrete specimen is contained within commercial polystyrene cubic moulds whose internal and external dimensions are 150 mm and 230 mm respectively. Two thermoresistances are used to detect the temperature of the concrete sample and the isothermal box. This calorimeter is available commercially produced and marketed by Controls.

A second calorimeter of this type is also available for purchase, the Stema calorimeter [30] developed and manufactured in Denmark. Fig. 1.7 shows a section of the equipment.

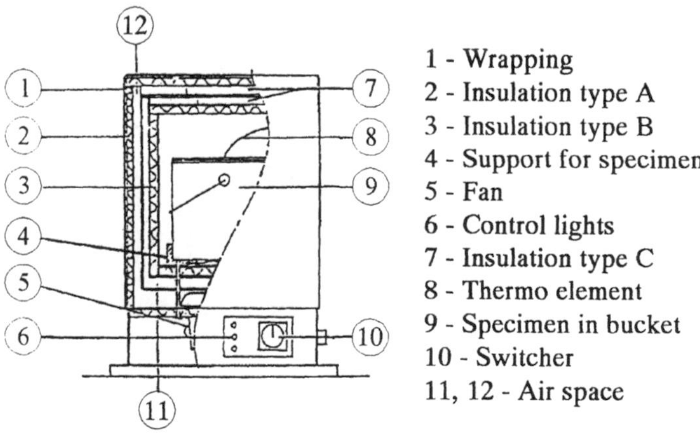

Fig. 1.7 Section of the Stema calorimeter.

1.3.2 Semi-adiabatic calorimeters

As mentioned in the introduction, this type of calorimeter is one in which the rate of heat loss from the concrete specimen is reduced only by some form of insulation, no external heat source is used to improve the efficiency of the apparatus.

These calorimeters can be broadly classified into two groups:

i) Those in which the insulation is provided by a thermos flask - the specimen size tends to be no more than about 2.5 kg.
ii) Those in which the specimen is surrounded by a layer of polystyrene or equivalent - the specimens tend to be cylinders or cubes and usually significantly larger than those used in (i).

According to the French standard, heat of hydration of cements can be measured using the Langavant test [6] which is essentially a semi-adiabatic calorimeter. The apparatus consists of a Dewar flask in which a 1.575 kg sample of 1:3 cement:sand mortar is placed within a water vapour tight cylindrical tin plate container. The temperature of the sample is measured using a thermometer connected to a recorder. No attempt is made to control the temperature of the environment in which the flask is placed to follow that of the sample. The temperature of the test specimen is recorded and compared with that of an inert specimen (i.e. one that is at least 3 months old) kept in a second reference calorimeter. Knowing the thermal capacity of the system and the recorded temperature differences between the reference and the active sample it is then possible to calculate the heat of hydration of the cement. There is no reason why such an apparatus should not be used for small concrete samples and why the adiabatic temperature curve should not be compacted from the semi-adiabatic curve.

The equipment developed by Grube [31] is in fact a modification of the Langavant test and is shown in Fig. 1.8.

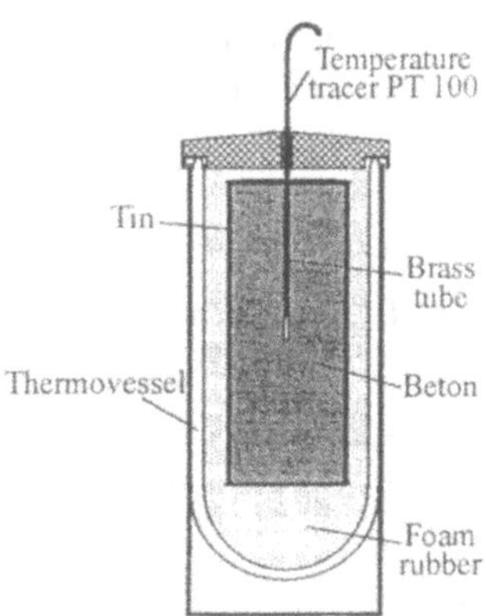

Fig. 1.8 Semi-adiabatic calorimeter developed by Grube.

The Dewar flask used is larger than in Langavant, the cylindrical concrete specimen measures 99 mm diameter x 230 mm high (i.e. approximately 4.25 kg) and is cast in a tin mould. The inside of the flask is protected by a lining of foamed rubber and the temperature of the concrete specimen is recorded using resistance thermometers. If required and as with Langavant the heat loss from the flask can be computed from a mature concrete specimen heated to a certain temperature and placed in the flask.

The calorimeters in which the specimen is surrounded by some form of insulating material differ only with respect to the size and shape of the test specimen and the thickness and type of insulation used. For example the test used at the Laboratoire Central des Ponts et Chaussées (LCPC) in Paris uses a relatively small sample (a 160 diameter x 320 mm cylinder) which is placed in an insulated box measuring approximately 400 x 400 x 550 mm high. The test is known as the "QAB" [32] semi-adiabatic test. At the other extreme the calorimeter developed by Breitenbücher [33] uses a cylindrical specimen measuring 360 mm diameter x 540 mm high which is placed in its mould inside a polystyrene mould which provides an insulation thickness of 56 mm top and bottom and 50 mm around the circumference. In addition an air gap of 2 mm is provided between the specimen mould and the polystyrene. The calorimeters developed by other workers [35-38] are all similar in principle to those described above.

The Nordic draft test method for an Insulated Block Calorimeter [34] states that a large scale well insulated concrete block shall be used for this test and it is assumed that the centre of the block will obtain approximately adiabatic conditions during the early stages of the hardening process. No technical specifications are given except that the cube should measure not less than 1 m and the coefficient of transmittance of the insulation shall be less than 5 kJ/(m^2.h.°C).

1.4 PREDICTION OF THE ADIABATIC TEMPERATURE RISE FROM ADIABATIC AND SEMI-ADIABATIC CALORIMETRY

The heat evolved during the hydration of a concrete sample in a calorimeter can be divided into the following three parts:

- heating of the specimen;
- heating of the calorimeter;
- heat loss.

The heating of the calorimeter is caused by the heat capacity of the components fitting the equipment. It should be kept as low as possible and should be accurately known. The heat losses play a different role for the two kinds of equipment. In adiabatic calorimeters they are usually small and almost constant along the duration of the test and give rise to a residual correction to determine the adiabatic temperature from the measured one. In semi-adiabatic calorimeters they are not constant throughout the runtime of the test as they increase with the increasing temperature of the sample and give the major contribution to the calculation of the adiabatic temperature rise. An example between measured and corrected curves is given in Fig. 1.9. The measured curve in an adiabatic calorimeter is usually only slightly lower than the corrected adiabatic curve whereas in a semi-adiabatic calorimeter the temperature of the sample is far below the adiabatic temperature; this in turn slows down the rate of hydration and, consequently, the corrections applied for heat losses will lead to a temperature rise which is still an underestimate of the adiabatic one.

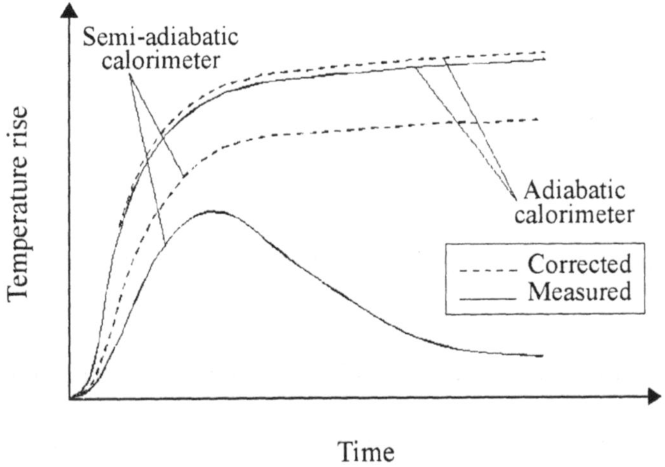

Fig. 1.9 Measured and corrected curves from adiabatic and semi-adiabatic tests.

To take account of the influence of the change in reactivity of the cement, some researches have adopted a further correction based on the assumption of a maturity function. This improves the estimate but however can give rise to some approximation in determining the true adiabatic temperature rise. Grube [31] considers the influence of temperature on hydration by means of a maturity function based on the temperature factor of Arrhenius:

$$k_T = e^{\frac{E_A}{R}\cdot\left(\frac{1}{293}-\frac{1}{273+T}\right)} \tag{1.1}$$

with: T temperature in °C;
E_A activation energy in J/mol;
R universal gas constant = 8.314 J/(mol·K);

$$\frac{E_A}{R}=\begin{cases}4000 & for \quad T \geq 20°C \\ 4000+175\cdot(20-T) & for \quad T<20°C\end{cases} \text{ for Portland cements;}$$

$$\frac{E_A}{R}=6000 \text{ for slag cements.}$$

The assumed maturity function is

$$M=\int_0^t k_T \cdot dt$$

In order to make comparisons between adiabatic and semi-adiabatic calorimeters a "Round Robin" test programme was conducted amongst some different organisations.

1.5 RILEM "ROUND-ROBIN" CO-OPERATIVE PROGRAMME

1.5.1 Adiabatic/semi-adiabatic calorimeter test programme

In order to obtain comparative data on the performance of a number of different types of calorimeters it was decided to carry out a "Round-Robin" test programme. The main aims of this programme were:

i) to compare the adiabatic temperature curves obtained from a number of different adiabatic calorimeters;
ii) to compare the predicted adiabatic temperature curves from semi-adiabatic calorimeters with those obtained from adiabatic calorimeters;
iii) to gain a greater understanding of the main factors affecting the results obtained from the different calorimeters.

1.5.2 Details of test programme

In all a total of 14 different organisations agreed to participate in the test programme. In order to minimise any errors all participants were sent appropriate quantities of the same materials and all were asked to use the same mix proportions.

Materials Used. - Samples of aggregate and cement were delivered in sealed air tight containers. The coarse aggregate was a 20 mm graded Thames Valley flint conforming to the requirement of BS 882 [39]. The fine aggregate was from the same source and was a mixture of five separate size fractions re-mixed before use to ensure conformity of grading. All aggregates were supplied in a saturated and surface dry condition.
The cement used was ordinary Portland cement (OPC) conforming to the requirements of BS 12 [40].

Mix Proportions. - Participants were requested to mix the materials in the following proportions by weight:

Cement	1.0
Fine	2.5
Coarse	3.5
Water	0.6

1.5.3 Results

The results from both the adiabatic and semi-adiabatic calorimeters are summarised in Tables 1.1 and 1.2 respectively.

Of the nine adiabatic results presented in Table 1.1, that of Luleå University is significantly lower than the rest; no reason can be given for this and for the purpose of this discussion this result will be discounted. The mean and the highest and lowest variability calculated at different ages are given in Tables 1.3.

Table 1.1 Summary of results from adiabatic tests

	Calorimeter			*Duration*	*Measured temperature [°C]*			*Corrected temperature [°C]*				
Laboratory	*Control Medium*	*Sample size [l]*	*Thermal loss [K/h]*	*of the test [h]*	*Start*	*Final*	*Total rise*	*Total Rise*	*Rise after 24 h*	*Rise after 48 h*	*Rise after 72 h*	*Comments*
A	Water	7.5	-	240	22	70	48	48	35.8	43.4	45.2	Average of 2 runs
B	Water	50	0.029	168	22.8	65	42.2	44.0	36.3	40.8	41.7	
C	Heated container	1.5	0.013	168	20.0	68.6	48.6	48.6	37.3	44.8	46.6	
D	Water	25	0.002	168	22.0	70.0	48.6	49	40	46	47.5	Correction for heat loss and heat capacity of the calorimeter
E	Water	4	-	160	25.9	69.8	43.9	43.9	36.1	40.9	42.1	
F	Water	0.5	0.01	264	22	62.3	40.3	42.5	31.9	38.6	40.4	Average of 2 runs; correction for heat loss
G	Water	170	-	319	20	68.5	48.5	48.5	38.5	44.7	46.3	
H	Air	10	0.008	50	22	54.8	32.8	32.8?	29	32.1	-	
I	air/water	30	0	168	20	72.5	52.5	52.5	38	45.1	47.6	Negligible heat loss

Table 1.2 Summary of results from semi-adiabatic tests.

	Calorimeter		*Measured temperature [°C]*			*Corrected temperature [°C]*			
Lab.	*Type of insulation*	*Sample size [l]*	*Start*	*Time to peak [h]*	*Rise at time to peak*	*Rise after 24 h*	*Rise after 48 h*	*Rise after 72 h*	*Comments*
B	50 mm polystyrene	50	22.6	32	34.1	36.9	43	44.6	Correction by the method proposed by Grube
J	Dewar Flask	2	20	17	18	38	44	46.6	Average of 2 runs; adjustments made using method proposed by Grube
C		6.4	22.8	21	23	31.4	36.6	-	Adjustment made using similar method to that proposed by Grube
K		13	20	19	26	35.5	41	42.5	Adjustments made according to a maturity function
H	1 m^3 foamed plastic	15	20.4	21	23.7	-	-	-	No adjustments performed
L	?	5	22.3	19	26.5	37.6	43.3	44.8	Adjustments made using a similar method to that proposed by Grube

Table 1.3 Variability of adiabatic and semi-adiabatic test results.

	Rise after 24 hours		*Rise after 48 hours*		*Rise after 72 hours*	
	adiabatic	*semi-adiabatic*	*adiabatic*	*semi-adiabatic*	*adiabatic*	*semi-adiabatic*
Mean values [K]	36.7	35.9	43	41.6	44.7	44.6
Highest variability	+8.9%	+4.8%	+6.9%	+5.8%	+6.3%	+4.4% *
Lowest variability	-13.2%	-12.5%	-10.3%	-12%	-9.6%	-4.8% *

* Calculated on 4 tests

Looking at the results from adiabatic calorimeters, some laboratories did not apply any correction to the measured curves whereas others adopted corrective methods which can differ each other and that in some tests represent an increase of 4-5% to the measured curves. It is not possible to say what the effect on the overall test variability would be if everybody had corrected their results although it is unlikely that it would have a significant effect. However, 50% of the adiabatic tests fall into a narrow range of only 2K of variation of temperature rise. There is no obvious correlation between the temperature rise and either the specimen size or the start temperature.

As concern the results from the semi-adiabatic calorimeters, the mean values of the results are only of the order of 2-3% below those derived from adiabatic calorimeters; of the five tests, the data supplied from LCPC are significantly lower than the remainders four whose range of variation is within 3 K. This would suggest that the adiabatic temperature rise could be accurately derived from a semi-adiabatic calorimeter. However, it must be remembered that the results are derived from one set of tests with one cement content and one cement type. A far more substantial test programme is required using different mix proportions and more importantly different cement types to substantiate this fact. No other meaningful conclusions it is possible to draw about the influence of specimen size, measured temperature or method of adjustments on the corrected temperature rise; it is however reasonable to think that a standard procedure to how calibrate a calorimeter and how calculate the adiabatic temperature rise from the measured curves should lead to a reduction of the variability among different laboratories.

1.6 DRAFT RECOMMENDATIONS

From discussions by all the experts of the RILEM Technical Committee 119 and from results and experiences of the Round Robin test it was agreed that the major influences over errors arising out of any form of adiabatic and semi-adiabatic calorimeter would be as follow:

a. The design of adiabatic calorimeter. Adiabatic calorimeters, by definition, must have a means by which the temperature of the enclosure surrounding the sample is carefully controlled. It was agreed that due to the physical and thermal properties of water that this should be the recommended medium with which to surround the sample for ease of circulation and close temperature control.

b. The thermal loss from adiabatic calorimeter. This should be of course be kept to a minimum in order to obtain as close as possible the adiabatic conditions required and to avoid large corrections to compensate for any non-adiabacity. Any large deviation from adiabatic conditions means that the cement will be subjected to a lower temperature than it should and, as cement hydration rate is accelerated by temperature to the degree of twice for every 10°C, the correct rate of reaction will not be obtained. It was agreed that the temperature loss of the sample should be less than 0.02 K/h.

c. The heat capacity of the calorimeter. This parameter produces the same effects of the thermal loss on the measured temperature rise. It must be kept much lower than the heat capacity of the sample. This can be achieved by testing large volumes of sample. The ratio between the apparent heat capacity of the calorimeter to the heat capacity of the sample should be limited to 0.1; if the condition is not satisfied a thermal compensation technique such that suggested by Morabito (1993) should be adopted.

d. The sensitivity of the temperature controller. This influences the amount by which the temperature of the control medium cycles around the set point and, of course, will even out to a certain extent. This variation must not exceed the 'adiabatic setting' otherwise there will be a net heat input to the system. The setting should be such to limit the thermal losses to 0.02 K/h. The layer of insulation between the sample and the control medium makes less critical the sensitivity of the temperature controller.

e. The concrete sample size. It was thought that any adiabatic or semi-adiabatic test method on concrete should take into account the thermal capacity effects of a maximum aggregate size of 32 mm and that the smallest dimension of the sample should be at least three times the maximum size of the aggregate.

At last, both types of calorimeter must be calibrated for the thermal losses and the effects of the heat capacity of the equipment. These factors are incorporated into the following relationship:

$$\vartheta_{ad} = \left(1 + \frac{C_{cal}}{C_c}\right) \cdot \left[\vartheta_S(t) + \int_0^t a(t) \cdot dt\right] \tag{1.2}$$

where t is the time, C_{cal} and C_S [J/K] are the apparent heat capacity of the calorimeter and the heat capacity of the sample respectively, $\theta_S(t)$ the measured temperature rise, $a(t)$ the coefficient of temperature loss [K/h]. If $\theta_S(t)$ is measured by an adiabatic calorimeter then $T_{ad}(t)$ represents the adiabatic temperature rise of the concrete; if the test is performed by a semi-adiabatic calorimeter equation (1.1) leads to a lower estimate of the adiabatic temperature rise. The estimate is improved by considering the influence of the temperature on the hydration by means of a maturity function. In general, the temperature factor of Arrhenius gives a good approximation for the temperature dependence of the hydration of the cement.

The ways to how calibrate the two types of calorimeter and how calculations can be carried out are given in a RILEM Technical Recommendation.

1.7 APPLICATIONS OF ADIABATIC AND SEMI-ADIABATIC CALORIMETRY

The choice of type and design of a calorimeter will depend almost entirely on the reasons why the test is being carried out. If an estimate of the temperature rise in a particular pour is all that is required then some form of semi-adiabatic calorimeter will be sufficient as it can be argued that the conditions of this test come closer to those existing in a real structure than any form of adiabatic test. However the accuracy with which the measurement is required will also dictate the type of equipment used.

If the true adiabatic temperature rise and heat of hydration are required then it could be argued that an adiabatic calorimeter will give a more accurate result. However, even though the temperature loss in adiabatic calorimeter is much lower than in a semi-adiabatic calorimeter, adjustments must be made to the measured curve to obtain the true adiabatic curve. To do this a measure of the heat loss from the calorimeter and the heat capacity of the calorimeter and concrete is required.

Additionally, because of the change in the rate of reaction of the cement with a change in temperature and the influence of the composition of the cement on this change, doubts exist to how accurately the true adiabatic curve can be predicted from a semi-adiabatic calorimeter. It is argued that such influence cannot really be taken into account by calculations and that the most accurate way of assessing this effect is to subject the concrete to a temperature rise which represent as closely as possible the true adiabatic curve.

APPENDIX A 'Round Robin' Test Programme

The laboratories which took part in the 'Round Robin' test programme are (in alphabetical order):

- AEC, Denmark;
- Blue Circle, England;
- ENEL Spa/CRIS, Italy;
- Hollandsche Beton Groep, The Netherlands;
- iBMB, Germany;
- LCPC, France;
- Luleå University of Technology, Sweden;
- NORCEM, Norway;
- Schretter & Cie, Austria;
- Sumimoto, Japan;
- Technische Universität München, Germany;
- VDZ, Germany.

REFERENCES

[1] British Standards Institution BS 4550 - "Methods of testing cement. Part 3. Physical Tests". 1978.

[2] American Society for Testing Materials. ASTM. C 186-86. "Standard test method for heat of hydration of hydraulic cement". 1986.

[3] DIN, 1164, Pt.8.

[4] Forrester, J. A. "A conduction calorimeter for the study of cement hydration". *Cement Technology*. **1**, No. 3, 1970, pp. 95-99.

[5] Karsch, K. H. and Schwiete, H. E "Adiabatisches Kalorimeter zur Bestimmung der Hydratationswärme eines Zementes (Adiabatic Calorimeter for determining the Heat of Hydration of a Cement)" *Zement-Kalk-Gips*, No. 5, May 1963, pp. 165-169.

[6] Association Française de Normalisation "Binders measurement of hydration heat of cements by means of semi-adiabatic calorimetry (Langavant Method)" AFNOR NF P 15-436, 1983.

[7] Suzuki, Y. *et al.* "Method for evaluating performance of testing apparatus for adiabatic temperature rise of concrete". *Concrete Library of JSCE*, No. 14, March 1990.

[8] Suzuki, Y. *et al.* "Evaluation of adiabatic temperature rise of concrete measured with the new testing apparatus" *Concrete Library of JSCE*, No 13, June 1989.

[9] Yokota, N. "Testing apparatus for measuring temperature rise". The Cement Association of Japan, Review 1985.

[10] Suzuki, Y. *et al.* "Applicability of adiabatic temperature rise for estimating time-dependent temperature changes in concrete structures". *Transactions of the Japan Concrete Institute*. **7**, 1985, pp 49-56.

[11] Morabito P. *et al.* "Measurement of adiabatic temperature rise in concrete". CONCRETE 2000. Proceedings of the International Conference, Vol. 1, 1993, pp. 749-759.

[12] Breitenbücher, R. Technical University Munich. Institute für Bauingenieurwesen II - Private communication.

[13] Rohling, S. *et al.* "Methods for determining heat evaluation of cement". RILEM recommendations for concreting in cold weather. Technical Research Centre of Finland. Research Notes 827, Appendix 5, 1988, pp 5/1-5/3.

[14] Pucher, S. "Die adiabate Kalorimetrie des Zementes" *Zement und Beton*, **3**, Jan. 1980, pp. 106-112.

[15] Catharin, V.P. "Die Hydratationswärme und ihre Bestimmung (The heat of hydration and its estimation). Sonderdruck aus *Tonind.- Ztg*, **90** (1966). No. 493, Vol. 12, pp. 554-559.

[16] Coole, M. J. "Heat release characteristics of concrete containing ground granulated blastfurnace slag in simulated large pours". *Magazine Concrete Research*. **40**, No. 144, Sept. 1988, pp. 152-158.

[17] Wainwright, P. J. and Tolloczko, J. J. A. "A temperature - Matched curing system controlled by microcomputer". *Magazine Concrete Research*. **35**, No. 124, Sept. 1983, pp. 164-169.

[18] Sarneel, R. Hollandsche Beton Groep. 2280 AB Rijswijk, Netherlands. Private Communication.

[19] Bamforth, P. B. "In-situ measurement of the effect of partial Portland cement replacement using either fly ash or ground granulated blastfumace slag on the early age behaviour of mass concrete". Taylor Woodrow Research Project, No 014J/77/1939, Nov. 1977, pp. 104.

[20] Emborg, M. University of Luleå Sweden. Private Communication.

[21] Høyer. O. AEC Consulting Engineers, 2950 Vedbaek, Denmark - Private Communication.

[22] Jaegermann, C. National Building Research Institute. Haifa, Israel. Private communication.

[23] Dezember. F. Vorläufiges Merkblatt für die Messung der Temperaturerhöhung des Betons mit dem adiabatischen Kalorimeter. Sonderdruck aus *beton*. Herstellung Verwendung, **20**, No. 12, 1970, pp. 545-549.

[24] Siemens, A. J. W. IBBC-TNO. 2600 AA Delft, Holland - Private communication.

[25] Di Pace, G. D. Institute Nacional De Tecnologia Industrial (INTI), Buenos Aires, Argentina - Private Communication.

[26] Basalla, A. "Ein adiabatisches Kalorimeter zur Bestimmung der Wärmeentwicklung im Beton (An adiabatic calorimeter for determining the evolution of heat in concrete)." *Zement-Kalk-Gips*, No. 3, 1962. pp. 136-140.

[27] Alegre, R. Cement Calorimetry at the French Hydraulic Binders Research Institute (CERILH). Report No. 333/HC/62 - Pt I Adiabatic Calorimeters.

[28] Costa, U. "A simplified model of adiabatic calorimeter". *Il Cemento.* **76**, No.2, April 1979, pp. 75-92.

[29] Morabito, P. - ENEL 20162, Milan, Italy. Private communication regarding adiabatic calorimeter manufactured by CONTROLS.

[30] Emborg, M. Technical University of Lulea, Sweden - Private communication regarding Stema adiabatic calorimeter.

[31] Grube, H. Forschungsinstitut der Zementindustrie Düsseldorf (VDZ). Germany - Private communication.

[32] Acker, P. "Thermal effects in concretes during manufacture and applications to engineering structures." *Annales de l'Institut Technique du Bâtiment et des Travaux Publics*, No. 442, February 1986. pp. 61-80 (English summary).

[33] Breitenbücher, R. Technical University Munich. Institut für Bauingenieurwesen II - Private communication.

[34] Høyer, O. AEC Consulting Engineers, 2950 Vedbaek, Denmark - Private Communication.

[35] Penwarden, A. D "Measurement of the heat of hydration of concrete using an insulated cube." Building Research Establishment. England. Internal Report N95/86. September 1986.

[36] Helland, S. and Maage, M. "Computer-based curing control." Proceedings FIP-XIth International Congress on Prestressed Concrete. Hamburg. June 1990, pp. Q7-Q11.

[37] Blue Circle (USA). "Semi-adiabatic Test-Newcem Concrete." Preliminary Report Project 259.

[38] Emborg, M. Technical University of Lulea, Sweden. Private communication.

[39] British Standards Institution, BS 882. Specification for aggregates from natural sources for concrete.

[40] British Standards Institution BS 12 Portland cement (ordinary and rapid-hardening).

2

Methods for Experimental Determination of Thermal Stresses and Crack Sensitivity in the Laboratory

M. Mangold

2.1 GENERAL ASPECTS

The calculation of thermal stresses and estimation of the cracking risk demands a lot of input data, such as the thermal and mechanical properties of the concrete. However, the most important concrete properties at early ages as the E-modulus and relaxation are changing continuously in a very wide range. Their developments cannot be predicted by generally valid formulas. This renders stress-computations for concrete at early ages vulnerable to unrealistic results. Because of this two ways have been pursued in the last few years:

2.1.1 Restraining Laboratory Tests

Restraint stresses and the cracking tendency of a concrete due to temperature changes as well as shrinkage are tested in direct restraining tests with special equipments (cracking frame or Temperature-stress testing machine). By this the influence of different concrete constituents, mixes and temperature on the development of the restraint stresses and on the cracking tendency can be directly measured.

In such restraining laboratory tests the overall behaviour of the concrete is investigated. They are a suitable tool to optimise the concrete mix as to a high cracking resistance. The experimental data can be used as a valuable information for checking theoretical numerical models. Thus the

Prevention of Thermal Cracking in Concrete at Early Ages. Edited by R. Springenschmid. RILEM Report 15. Published in 1998 by E & FN Spon, 11 New Fetter Lane, London EC4P 4EE, UK. ISBN 0 419 22310 X

models may be calibrated for concrete qualities, cement types, admixtures etc.

2.1.2 Determination of Concrete Properties

Another way is the exact determination and modelling of the development of single concrete properties such as E-Modulus relaxation, chemical shrinkage/swelling, coefficient of thermal expansion, hydration heat. Most of these parameters change during the first day of hardening over a wide range. Therefore, a time dependent formation has to be determined (also see chapters 2, 5 and 6). The results can be used as input data for temperature and stress computations. However it has to be kept in mind, that there are combined effects on the material parameters such as the influence of the loading age and temperature on the relaxation of concrete [24] or the stress induced thermal strain [2a, b].

As the total error of a stress computation adds up with the errors of the single parameters, the determination of the relevant parameters should be as precisely as possible. This may require extensive test series. Some features, such as the temperature dependency of chemical shrinkage/swelling [25], can be determined only approximately.

For the experimental determination of thermal stresses several adequate equipments have been developed. The following conditions have been considered:

- The tests should be carried out with concrete, i. e. a maximum aggregate size of at least 16 mm. For this purpose the specimen requires a minimum cross section of about 100 x 100 mm^2.
- During hydration the same temperature development must be modelled in the test specimen as in a concrete member in situ with larger dimensions.
- The restraint stresses must be measured continuously.
- Contraction as well as expansion of concrete must be restrained from the very beginning of hardening.
- The deformations of the specimen must be prevented to a very high degree, so that most of the deformations are transformed into stresses.
- In order to predict thermal restraint stresses as they occur in large concrete members it must be possible to prevent waterlosses from the specimen. However, also in sealed specimens volume changes due to the formation of hydration products (chemical shrinkage/swelling) contribute to the restrained deformations and thus to restraint stresses.

2.2 METHODS FOR DIRECT MEASUREMENT OF RESTRAINT STRESSES AND CRACKING TENDENCY

The first experimental investigations on the cracking tendency have been published about 50 years ago. A survey on the test methods for the experimental determination of load independent stresses is given in [1] and [2]. Many of these tests determine the cracking tendency due to drying shrinkage and/or chemical shrinkage of cement pastes or mortars. With a small cross-section of the specimen and no heat insulation the hydration heat dissipates almost immediately. Thus thermal stresses are not taken into account with these test-setups. The first test-setup which enabled the measurement of thermal stresses was the rigid cracking-frame [11]. The beam specimen in the cracking-frame heats up by its own hydration heat. For that, length of about 1.6 m and a cross-section of 150 x 150 mm^2 is required. Moreover the formwork must be heat insulated.

2.2.1 Ring tests

Originally, ring test were used to investigate the cracking tendency of cement pastes due to shrinkage.

One of the first equipments was a ring specimen with a rigid steelcore, which was developed in France. This was also specified in the French standard for hydraulic binders [3].

With this test device only the time of cracking can be noted. It is not possible to measure stresses. Only a contraction is restrained by the core. It is not possible to model hydration heat development.

In order to determine the stress development and the tensile strength of the concrete, the rigid core was substituted either by a thick-walled hollow cylinder (Bennet and Loat [4]) or a pressure capsule (see Fig. 2.1) (Mayfield [5], Ilantzis [6], Baron [7]).

Furthermore the dimensions of the specimen were enlarged so that mortars and concretes with a maximum aggregate size of about 18 mm could be tested (e.g. [5, 16]).

However, restraint stresses result in this case only from the restraining of the shrinkage induced deformations. Thermal effects are negligible because of the small specimen cross-section and the flow of heat off the surfaces. Furthermore a superposition of the restraint stresses and the

Eigenstresses is unavoidable, as the ring is exposed to evaporation on two outer surfaces only.

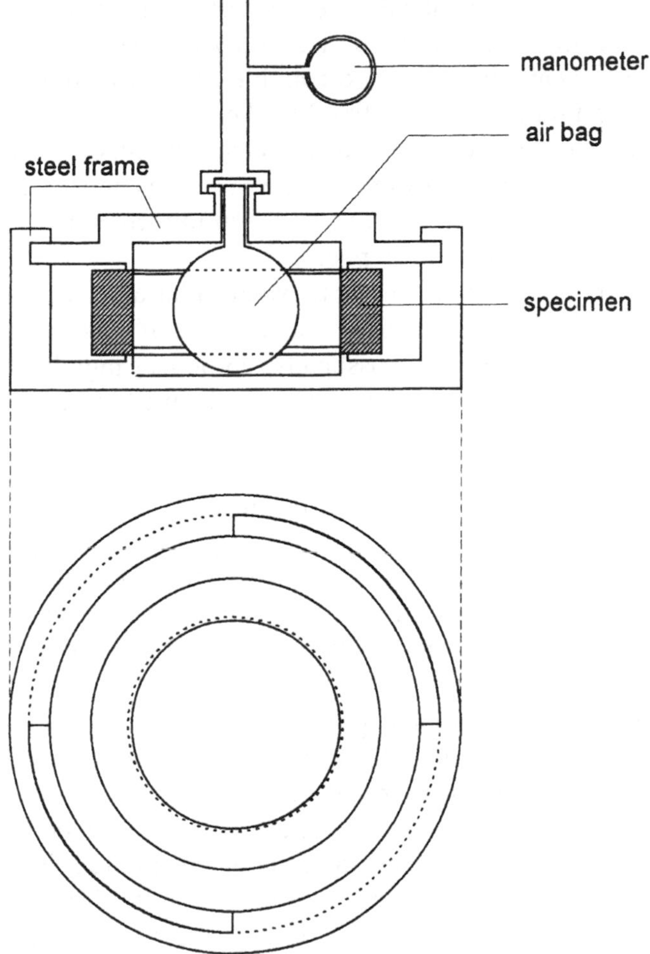

Fig. 2.1 Ring-test by Ilantzis [6]

Weigler and Nicolay [8] conducted tests on ring specimens for the first time with an adjustable core equipped with a control loop. This test apparatus works with pneumatically adjustable rubber bags situated in the core. The contraction of the specimens is controlled by generation of an internal pressure so that the ring expands again to its original diameter. Complete stress-strain curves can be obtained.

2.2.2 Beam tests

The simple principle of the ring specimen comes to its limit, when hydration heat plays the important role and concretes with a maximum aggregate size of 32 mm must be tested. This only can be attained with beam specimens. The beam ends must be capable of withstanding tension and compression and must be rigidly fixed to the crosshead of the frame. The frame prevents the deformations and takes over the function of the core in the ring specimen. The formwork of the specimen must be part of the equipment so that measurements can be taken immediately after the fresh concrete has been placed and compacted.

Temperature controlled either by thermal isolation (semiadiabatic) or Temperature of formwork. There are two variants of the cracking frame: The cracking frame with solid crossheads reproduces high but not complete restraint, whereas the cracking frame with an adjustable crosshead can give full restraint.

2.2.2.1 Cracking frame with solid crossheads

The cracking frame with solid crossheads (rigid cracking frame, Fig. 2.2) was developed by Springenschmid [9] Breitenbücher [10] and Nischer [11]. This frame can be used in tests on concrete specimens with a cross-section of 150 x 150 mm², and a maximum aggregate size of 32 mm. The frame is made of steel. For the longitudinal bars a special steel is used having a coefficient of thermal expansion of $\alpha = 1.0\ 10^{-6}\ K^{-1}$. This is only the twelfth part from a of the ordinary steel of the crossheads. The ends of the specimens are fixed in the crossheads.

Between the two crossheads the length of the specimen must be kept constant even if the ambient temperature changes. In order to achieve this, the crossheads are screwed to the bars in a definite distance of the bar ends, see [10]. The reaction force in the massive steel bars (actio = reactio) causes a small elastic deformation of the bars which is recorded continuously by the strain gauges. By this also the stress-development in the specimen is obtained continuously. Due to the small elastic deformation of the steel bars the degree of restraint of the concrete specimen lowers from 100 % for the fresh mix to about 80 % for the hardened concrete. For obtaining a temperature rise in the specimen as in an about 50 cm thick member a thermal insulating formwork is used. The formwork can also be connected with an external heating/cooling system. Thus it is possible to measure thermal stresses also for an arbitrary course of concrete temperature [26].

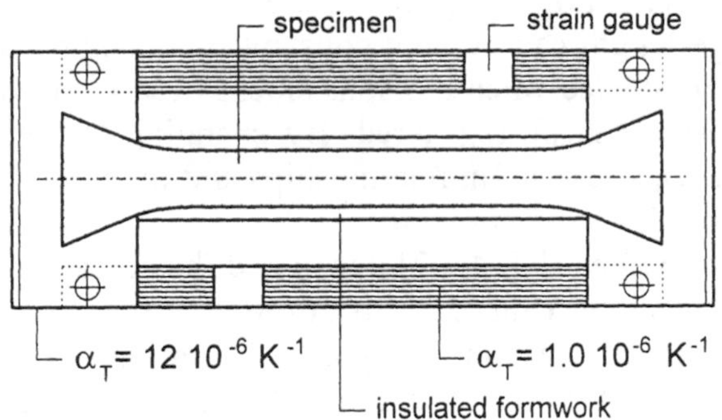

Fig. 2.2 Principle of a rigid cracking frame

2.2.2.2 Cracking frame with an adjustable crosshead

The cracking frame with an adjustable crosshead was developed in various places.

One apparatus for testing cement or mortar specimens was designed by D. Orr and G. Haigh [12]. Since the specimen has a cross-section of only 5 x 25 mm², it can only be used to measure shrinkage stresses.

In 1976 A.-M. Paillère and J.-J. Serrano [13] described a machine with which a concrete specimen 175 x 7 x 5 cm³ can be tested starting from the early age on (Fig. 2.3). The swallow tailed specimen is cast horizontally and tested in vertical position. The lower end is fixed, the upper end remains movable and is connected to an air pressure test equipment.

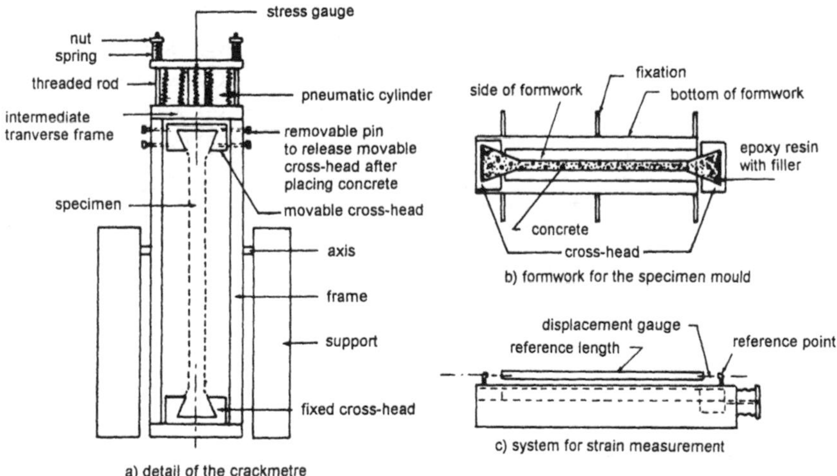

Fig. 2.3 Test device by Paillère and Serrano [13]

Changes in length of the specimen are measured with inductive gauges. The gauge-length can be kept constant by changing the pressure in the load bearing device when a displacement of ± 1 mm occurred. Thus shrinkage stresses can be registered. By changing the length (stress or strain controlled), stress-strain relationships and the modulus of elasticity can be determined.

Springenschmid *et al.* [14, 27] developed the so-called Temperature - stress testing machine (Fig. 2.4) to study restraint stresses caused by hydration heat. In this machine an approx. 1.5 m long beam with a cross-section of 150 x 150 mm² is cast and tested horizontally. A temperature controlled formwork is used to achieve an arbitrary course of hardening temperatures within the concrete specimen. The ends of the beams are fixed in the crossheads. One cross-head is adjustable, the other is rigidly connected to the massive longitudinal steel bars. For the length measurement, two steel bars are concreted in transversely to the main specimen axis at a distance of 500 mm (see Fig. 2.4). On both sides of the specimen the distance of these bars is measured with deformation transducers on carbon fibre bars with a sensitivity of 0.0001 mm. Another length measurement system has been developed using optical transducers [23].

As soon as the deformation of the concrete between the bars exceeds 0.001 mm, the adjustable crosshead is moved back with a step motor so that the total deformation is restored again to zero. By that, full longitudinal restraint is achieved. The restraining force which is produced as a result of the crosshead control is measured continuously by a load cell.

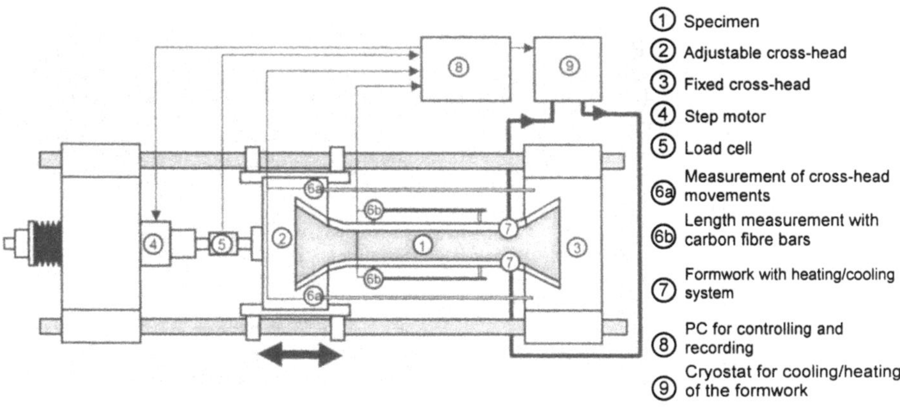

Fig. 2.4 Temperature-stress testing machine Munich

2.2.3 Tests with Cube/Cylinder Specimen

In some places instead of a special cracking frame a commercially available material testing machine is used. The specimens (cubes or cylinders) are either cast in a mould within the machine (EMPA [15]) or installed in the machine after a sufficient hardening period (Thomas [16], Dinh [17], Emborg [18]). An advantage of using a commercial material testing machine is that the electronic control equipment needed for the test may already be built-in in the machine. This may offer a simpler and cheaper way to determine thermal stresses in the laboratory. Usually commercial material testing machines have an adequate stiffness and symmetry of the loading unit. However, the use of such a testing machine may have also disadvantages compared to the cracking frame above. It is not possible to measure the stresses during early hardening. Moreover the tests can only be performed vertically.

2.2.4 Comparison of equipments (Table 2.1)

2.2.4.1 Degree of restraint

The restraint strain is expressed as the difference between the free, not restrained strain of the concrete member or the lab specimen (ε_O) and the strain (ε_S) which the situation on site or the test configuration allows. The degree of restraint δ is obtained as a quotient

$$\delta = (\varepsilon_O - \varepsilon_S)/\varepsilon_O \cdot 100\ [\%] \quad (2.1)$$

and depends on the stiffness of the concrete (c) and that of the testing machine(s):

$$\delta = 100/(1 + [E_C A_C / E_S A_S])\ [\%] \quad (2.2)$$

It follows, therefore, that the degree of restraint diminishes with an increasing E-modulus of the specimen. While, at first, in the still plastic concrete with $E_c = 0$ the deformations are practically 100 % restrained, the degree of restraint for a 4 day old concrete with $E_c = 20\ 000$ N/mm^2 becomes e.g. about 82 % in the rigid cracking frame. In this instantaneous consideration the viscoelastic behaviour of the concrete has not been taken into account: The restraining force reduces considerably, especially in young concrete, due to stress relaxation. For this reason the deforma-

Table 2.1 Comparison of equipments.

Criterion	*Ring Specimen*		*Beam Specimen*		*Cube/Cylinder specimen*
	Core rigid	Core adjustable	Crosshead solid (rigid cracking frame)	Crosshead adjustable (Temperature stress testing machine	Commercial material testing machine
Test specimen	Cementpaste, Mortar, Concrete *)		Mortar and Concrete		Mortar and Conrete
Extent of testing	low	high	low	high	low
Limit of maximum aggregate size	*) limitation by specimen size		Enlargement of the equipment is possible		Enlargement of specimen possible
Degree of restraint	Function of $E_b+\Psi$	adjustable up to 100 %	Function of $E_b+\Psi$	adjustable up to 100 %	adjustable up to 100 %
Type of restraint stresses	only tension		compression and tension		compression and tension
Influence of fixing	no		minimal	no	no
Stress distribution	non uniform		uniform		uniform
Measurement of stresses	not possible	by internal pressure	by deformation of steel bars	by load cell	by load cell
Measurement of E-Modulus and coefficient of thermal expansion	not possible	possible	not possible	possible	possible
Literature	[3], [4]	[5], [6], [7]	[9], [10], [11]	[12], [13], [14], [27]	[15], [18], [18a]

tion of the test device itself is reduced again so that the degree of restraint rises.

For theoretical considerations a test device providing complete (100 %) restraint is useful. On construction site, the restraint depends on the specific situation. For many practical cases it can be assumed that the degree of restraint is about 80% [20]. Therefore the behaviour of an element in situ is sufficiently well reproduced in the rigid cracking frame.

2.2.4.2 Influence of fixing

A reliable and fixing of beam specimens without any slippage is most essential. Most configurations of the cracking frame use test specimens in the shape of bars with dovetail shoulders which can be fixed to the crossheads by additional use of adhesive or bolting. The dovetail grading of the specimen leads to compressive forces in the fixing zone during the tensile phase of a test. By means of cross ties the opening of the crossheads can be prevented. Thereby, any slippage can be reduced to a negligible value. In ring specimens such problems do not arise. However, it must be taken into account that only contractions of the specimen can be restrained. Therefore, only tensile stresses are produced.

2.2.4.3 Stress distribution over the cross-section

In the central part of a beam specimen in the cracking frame the stresses are uni-axial. A constant stress distribution can be assumed there. In ring specimens, however, stresses tend to concentrate on the inside of the core with increasing specimen thickness [21].

2.2.4.4 Deformation measurements

For conventional testing machines working with control loops (e.g. tests with adjustable core or crosshead) a high precision of deformation measurements is necessary. To obtain total restraint (100%), the loading device should - depending on length of measuring distance - react for a change of the deformation with about 0.00015 - 0.0005 mm. Otherwise the adjustable crosshead cannot be steered precisely enough and slightly cyclic loading or unloading of the specimen may result.

2.2.4.5 Moisture and temperature conditions

For determination of thermal stresses drying shrinkage of the specimen must be prevented. Therefore in tests with cracking frames [10, 14] the concrete usually will be sealed after casting with a foil, which is impermeable to water vapour. Other procedures like placing the specimen under water or sealing only after removal of the form are not suitable to restrict early drying shrinkage.

The hardening temperatures of beam specimens in the cracking frame are fundamentally easier to control than in the case of the ring specimens. The beam specimen is in contact with the frame only at the crossheads. Along the free length a formwork can be used which can have any desired heat-insulating properties and/or additional heat exchange properties. In the investigation of concretes for large elements (mass concrete) the concrete mix can be first tested in an adiabatic calorimeter. Then the temperature rise can be calculated for the dimensions of the element [14]. Later on this calculated temperature development is used for the external temperature control of the specimen (external heating and cooling).

In this way, the influence of casting sequences, form stripping, curing, cooling with embedded pipes etc., which affect temperatures and thermal stresses in the actual concrete member, can be examined by small beam specimens.

REFERENCES

[1] Springenschmid, R., Adam, G.(1980) Mechanical properties of set concrete at early ages. Test Methods, in *Matériaux et Constructions*, **13**, No. 77, pp. 391-399.

[2] Schrage, I., Breitenbücher, R. (1989) Experimentelle Ermittlung lastunabhängiger Spannungen in jungem Beton, in *Betonwerk + Fertigteil-Technik*, **11**, pp. 48-55.

[2a] Thelanderson (1987) Modelling of combined thermal and mechanical action in concrete, in *Journal of ASCE*, **113**, No. 6, pp. 893-906.

[2b] Bazant and Chern (1985): Concrete creep at variable humidity, in *Materials and Structures*, RILEM, **18**, No. 103, pp. 1-19.

[3] AFNOR (1960) FDP 15-434, 8., in Liants hydrauliques - Technique des essais - Essai de fissurabilité.

[4] Bennet, E. W., Loat, D. R. (1970) Shrinkage and creep of concrete as affected by the fineness of Portland cement, in *Mag. Concr. Research*, **22**, No. 71, pp. 69-78.

[5] Mayfield, B. (1970) A new tensile strength test for concrete - or a new control test, in *Mag. Conr. Research*, **22**, No. 71, pp. 107-108.

[6] Ilantzis, M. A. (1958) La résistance en traction et la fissuration des pâtes pures de ciment, in *Ann. ITBTP*, No. 131, pp. 1253-1254.

[7] Baron, J., Mme. Brachet-Rolland, Pidoux J. C.(1969) Choix des ciments pour diminuer le risque de fissuration des revêtements en béton, in First Europ. Symp. on Concrete Pavements, Paris, Sess. A 2, pp. 1-18.

[8] Weigler, H., Nicolay, J. (1975) Konstruktionsleichtbeton - Temperatur und Rißneigung während der Erhärtung, in *Betonwerk + Fertigteil-Technik*, **41**, pp. 226-232, pp. 295-298.

[9] Springenschmid, R., Nischer, P. (1973) Untersuchungen über die Ursache von Querrissen im jungen Beton, in *Beton- und Stahlbetonbau*, **68**, pp. 221-226.

[10] Breitenbücher, R. (1989) Zwangsspannungen und Rißbildung infolge Hydratationswärme. Dissertation, TU München.

[11] Springenschmid, R., Nischer, P. (1973) Untersuchungen über die Ursache von Querrissen im jungen Beton, in *Beton- und Stahlbetonbau*, **68**, H. 9, pp. 221-226.

[12] Orr, D. M. F. and Haigh, G. F. (1971) An apparatus for measuring the shrinkage characteristics of plastic mortars, in *Mag. of Concr. Res.*, **23**, No. 74, March.

[13] Paillère, A.-M. et Serrano, J.-J. (1976) Appareil d'étude de la fissuration du béton, in *Bull. Liaison Labo. P. et Ch.*, **83**, mai - juin, pp. 29-38.

[14] Springenschmid, R., Gierlinger, E., Kiernozycki, W. (1985) Thermal stresses in mass concrete: A new testing method and the influence of different cements, in 5th Int. Conf. Large Dams (ICOLD), Lausanne Q. 57-R. 4, pp. 57-72.

[15] EMPA-Prüfbericht Nr. 78 960 (1971)

[16] Thomas (1936) Cracking in reinforced concrete, zitiert nach: L'Hermite, R.: Le retrait des ciments, mortiers et bétons. Circ. ITBTP, Ser. F - No. 37 (1947).

[17] Dinh, T. L. (1975) Etude des traitements mixtes par l'essai de retrait thermique empêché, *in Bull Liaison Labo. P. et Ch.*, **78**, pp. 103-116.

[18] Emborg, M. (1985) Temperature stresses in massive concrete structures. Licentiate Thesis, Luleå (Sweden).

[18a] Emborg M., Johansson H. O.(1988) Relaxation tests on young concrete. Development of test methods and methods of measurements for thermally induced loading (in Swedish) Div. of Stud. Eng. Luleå Univ. of Technology. Techn. Rep. DIT., 97 pp.

[19] Kasai, Y., Yokohama, K., Matsui, J. (1975) Study on properties of shrinkage of early age - Concrete under restraint conditions, in Rev. 29th Gen. Meeting Cem. Ass. Japan, pp. 194-196.

[20] Wischers, G. (1965) Betontechnische und konstruktive Maßnahmen gegen Temperaturrisse in massigen Bauteilen, in *Betontechn. Berichte 1964*. Düsseldorf, Beton-Verlag.

[21] Autorenkollektiv (1974) *Das Grundwissen des Ingenieurs*. 9. Aufl., Leizig: VEB Fachbuchverlag.

[22] Rostásy, F.S., Laube, M. (1988) Verformungsverhalten und Eigenspannungsrißbildung von jungem Beton; Institut für Baustoffe, Massivbau und Brandschutz, Forschungsbericht, TU Braunschweig.

[23] Thielen, G., Grube, H. (1990) Maßnahmen zur Vermeidung von Rissen im Beton. *Beton- und Stahlbetonbau*, H 6, pp. 161-167.

[24] Shen, H.-H. (1992) Lineare und nichtlineare Theorie des Kriechens und der Relaxation von Beton unter Druckbeanbeanspruchung. Deutscher Ausschuß für Stahlbeton, Heft 432, Beuth Verlag, Berlin.

[25] Schöppel K. (1993) Entwicklungen der Zwangspannungen im Beton während der Hydratation. Dissertation, TU München.

[26] Mangold, M. (1993) Einfluß der Nachbehandlung auf die Zwang- und Eigenspannungen von Bauteilen aus Beton. Beiträge zum 28. Forschungskolloquium des DAfStb, München

[27] Schöppel, K., Breitenbücher, R., Springenschmid, R. (1990) Investigations of mass concrete in the temperature-stress testing machine. *Concrete Precasting Plant and Technology.*

3

Influence of Constituents, Mix Proportions and Temperature on Cracking Sensitivity of Concrete

R. Springenschmid and R. Breitenbücher

3.1 INTRODUCTION

Thermal cracking on site is determined by thermal and other early age deformations of the concrete on the one hand and by the restrain of these deformations on the other hand. Therefore the following two parameters have to be considered for the judgement of cracking on site:

- the cracking sensitivity of a concrete mix, which depends on the constituents, the mix composition and the fresh concrete temperature, and
- the boundary conditions (dimensions of the member, degree of restraint, temperatures, etc.).

Much progress have been made in the past decade as to the understanding of the cracking sensitivity of a concrete mix. Since engineers have learned to consider thermal and other early age stresses instead of solely temperatures much consideration is given to the development of Young's modulus and temperatures in the decisive early period up to a concrete age of 24 to 36 hours. In this period most of the hardening and most of the temperature development takes place.

3.2 TECHNOLOGICAL MEASURES UP TO NOW

From the technological point of view only measures for the limitation of the maximum temperature during hydration were usually taken into

Prevention of Thermal Cracking in Concrete at Early Ages. Edited by R. Springenschmid. RILEM Report 15. Published in 1998 by E & FN Spon, 11 New Fetter Lane, London EC4P 4EE, UK. ISBN 0 419 22310 X

account. For this purpose besides a low temperature of the fresh concrete and a reduced cement content especially the use of low-heat cements was favoured. For such cements normally the heat of solution of an isothermally hardened cement paste after 7 days must not exceed a certain value. For the judgement of the temperature rise within the first one to three days the single value after 7 days is not significant.

With most measures only the temperature rise is considered. However, cracking does not occur in concrete members by the temperature change itself. Cracking results from stresses which exceed the strength of the material. Owing to this reason it is necessary to consider besides the temperature change all factors which influence the stress and strength developments. In particular the Young's modulus, the coefficient of thermal expansion and the relaxation including their developments within the early age have to be taken into account [1].

3.3 RESTRAINT STRESSES AND CRACKING IN CONCRETE AT EARLY AGES

The main problem of thermal cracking in concrete at an early age is that during the high temperature rise mainly within the first day Young's modulus is very low and therefore only small part of the inhibited thermal expansion is transformed into compressive stresses (fig. 3.1). Furthermore in this stage a high relaxation reduces the thermal prestress. On the other hand during subsequent cooling, the increased Young's modulus and a reduced relaxation contribute to relative high tensile stresses [2, 3].

Although restrained thermal contractions - resulting from the hydration heat or other temperature changes - can be denoted as the main reason for cracking at early ages in members with a thickness of about 20 cm and more, deformations caused by autogenous shrinkage cannot be neglected in general. Especially in high strength concretes containing silica-fume high autogenous shrinkage can occur (fig. 3.2).

However, there are also cement brands which cause a non - thermally induced expansion. Such expansion can be connected e.g. with chemical expansion during the formation of ettringite. By such deformations additional compressive stresses are raised, which compensate the later tensile stresses to some extent and reduce the cracking sensitivity of the concrete.

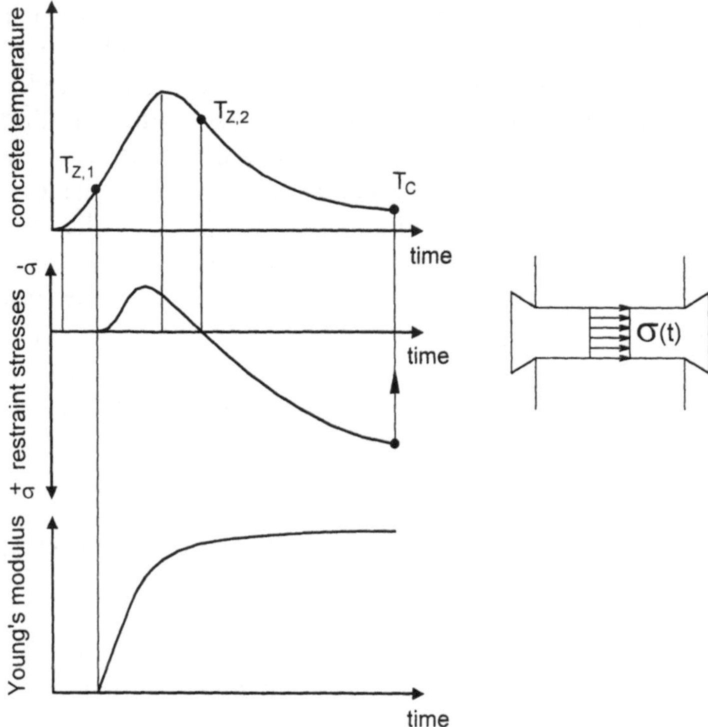

Fig. 3.1 Temperature, stress and Young's Modulus development in a restrained concrete member (thickness about 0.2 to 1.0 m) at early ages ($T_{Z,1}$: first, $T_{Z,2}$: second zero-stress temperature; T_C: cracking temperature).

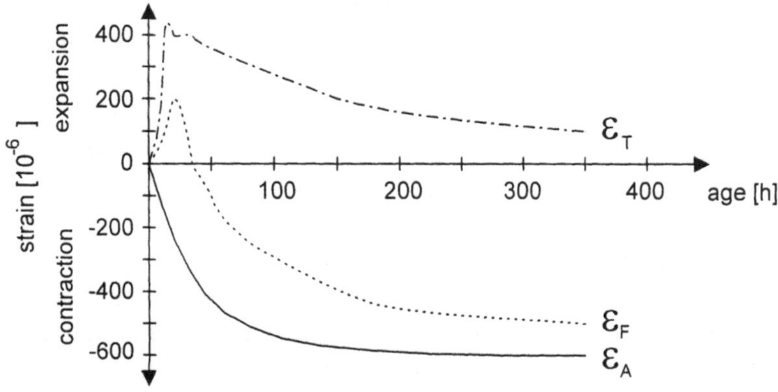

Fig. 3.2 Unrestrained deformations of a high strength concrete (w/c = 0.22) 1.5 m below the surface of a massive structure due to temperature changes (e_T) and autogenous shrinkage (e_A). e_F = superposed deformations, according to E. Tazawa *et al.* [4].

3.4 METHODS TO QUANTIFY FACTORS OF INFLUENCE

In order to recommend useful technological measures the influence of all relevant parameters on thermal stresses and thermal cracking must be studied and quantified. For such tests cracking frames or thermal stress testing machines as described in chapter 2 of this State of the Art Report are appropriate measures, because the temperature and stress development in restrained members can be studied in the laboratory simultaneously from the very beginning.

For the judgement of the cracking sensitivity of a concrete the cracking temperature as determined in the cracking frame test [5] is helpful. The **cracking temperature** is a global indicator of all influences (including concrete tensile strength) of a concrete mix to its cracking sensitivity under defined restrainment. Low cracking temperatures are an indicator of low cracking sensitivity.

3.5 INFLUENCE OF FRESH CONCRETE TEMPERATURE

A very important role plays - as expected - the **temperature** of the **fresh concrete**. In members with a high pouring temperature, e.g. 30°C, also the temperature increases faster, because of the acceleration of the hydration. If the lowest ambient temperature is low as compared with the second zero-stress-temperature, the thermal contraction during cooling down to the ambient temperature is much higher in comparison to concrete with low pouring temperature.

Additionally it has to be considered, that the tensile strength of a concrete hardening at a high temperature level is lower due to a weaker connection of shorter hydration products (CSH) [6]. Both effects lead to a significant higher cracking temperature in concretes with high fresh concrete temperature (fig. 3.3).

As a general rule it can be stated: When the fresh concrete temperature is 10 K higher, the cracking temperature increases by 13 to 15 K. For the cracking sensitivity the absolute cracking temperature is decisive. It has to be correlated with the temperatures to be expected on site.

3.6 INFLUENCE OF CONCRETE CONSTITUENTS

The use of **aggregate** with a low coefficient of thermal expansion (tab. 3.1) lowers the thermal deformation of the concrete. With this, the level of restraint stress can be reduced and a lower cracking sensitivity can be expected.

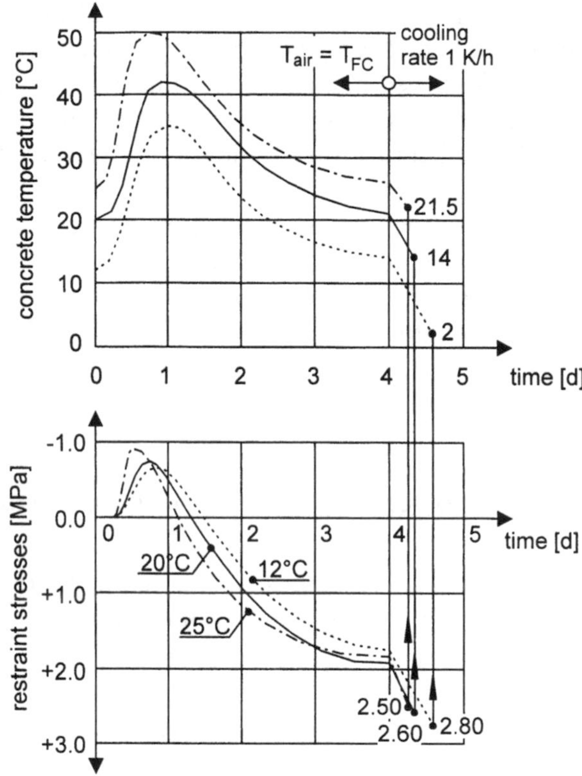

Fig. 3.3 Influence of fresh concrete temperatures on the temperature and stress development in concrete under restraint.

Table 3.1 Coefficient of thermal expansion of different aggregates, according to [7]

Aggregate type	*Coefficient of thermal expansion [10^{-6} . K^{-1}]*
Granite	7.4
Diorite, gabbros	6.5
Quarzitic porphyry	7.4
Basalt	6.5
Quartzite, graywacke	11.8
Quartzitic sandstone	11.8
Other sandstone	11.0
Limestone	5.0 to 11.5
Slag	5.5

Crushed aggregates are usually advantageous, because the concrete tensile strength is higher due to their rough surface. It is also convenient to use a high maximum size of the aggregates, because then less cement paste is necessary for a sufficient workability. A low cement content results in a low temperature rise caused by hydration heat. However, with a high maximum size of aggregate, the tensile strength can be somewhat lower.

For a low cracking sensitivity **cements** would be favourable, which develop their Young's modulus relatively fast within about the first 12 hours so that relative high compressive (pre-) stresses can be obtained and develop only little hydration heat during this period. However, very little is known as how to transform this theoretical consideration into practice.

Ordinary Portland cements (OPC) from various plants lead to very different cracking sensitivity due to their different chemical composition. To some extent the different behaviour already is reflected in a different developement of temperature. Additionally the various cements show different non-thermally induced deformations and restraint stresses. In isothermal tests some "favourable" cements showed a slight swelling - due to expansion by ettringite - while with other cements of high cracking sensitivity tensile stresses, resulting from a restrained contraction due to autogenous shrinkage, were observed (fig. 3.4).

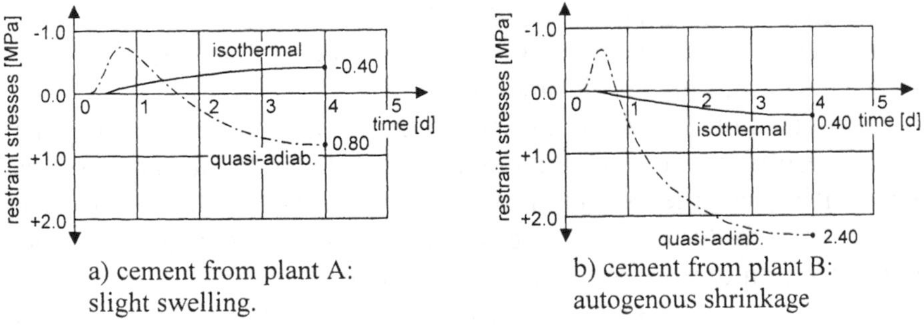

a) cement from plant A: slight swelling.

b) cement from plant B: autogenous shrinkage

Fig. 3.4 Restraint stresses of concrete with OPC from two different plants at isothermal and quasi-adiabatic conditions.

Whereas for a non-thermal expansion the thermally induced tensile stresses can be compensated partly, autogenous shrinkage increases the cracking sensitivity. Favourable OPC cements were found which had a low alkali content (K_2O, Na_2O), high sulphate content in relation to the C_3A-content and which were not too fine. The cracking susceptibility of a cement can be assessed in a cracking frame standard test using a standardised concrete mix composition, as defined in [5]. A maximum cracking temperature of 10°C for structural concrete and of 7°C for mass concrete cements has been successfully specified.

Low heat cements have often been used for mass concrete in the past. Most of them contain granulated slag and often less than 50 % Portland cement clinker. In this case in spite of a moderate temperature rise, cracking may occur, because the Young's modulus develops too slowly in the early hardening phase. In recent years for many larger constructions and mass concrete ordinary Portland cement was partly (20 to 50 %) replaced by puzzolanes. The question, which cement or cement puzzolane mixture would result in a high cracking resistance has often been investigated in cracking frame standard tests, see [3].

Usually for the selection of the cement and the concrete mix, other requirements such as early compressive strength have to be observed. E.g. for tunnel linings of the German Railway where the formwork have to be taken away after 12 hours only, a concrete mix with a 12 hour compressive strength of thermally insulated cubes of 3 MPa and a low cracking sensitivity was specified. Fig. 3.5 shows that concrete mixes with various Portland cement brands, partly with a replacement of cement by fly ash and fresh concrete temperature of 12, 20 and 25°C show a wide range of cracking temperatures. The required compressive strength between 3 and 6 MPa was obtained with cracking temperatures between 2 and 18 °C. Since high early strength results in a high cracking sensitivity, it was specified that the 12 hour compressive strength must not exceed 6 MPa. Therefore, a higher amount of Portland cement was replaced with puzzolanes during summer.

The question to which extent a high slag content of a low heat cement results in a low cracking temperature depends - as discussed earlier - mainly on the development of the Young's modulus and of the tensile strength during the first days and therefore is strongly influenced by the reaction of the clinker and sulphate at early ages.

In mass concrete structures with a concrete lift height of 3 m and more, the maximum temperature is reached only after more than 4 days and it takes several weeks when a high tensile strength is necessary to avoid transverse cracks. Mass concrete cements react slowly and gain an adequate tensile strength only after weeks. Therefore their cracking temperature at an age of 4 days could underestimate the cracking resistance of low heat cements for many concretes.

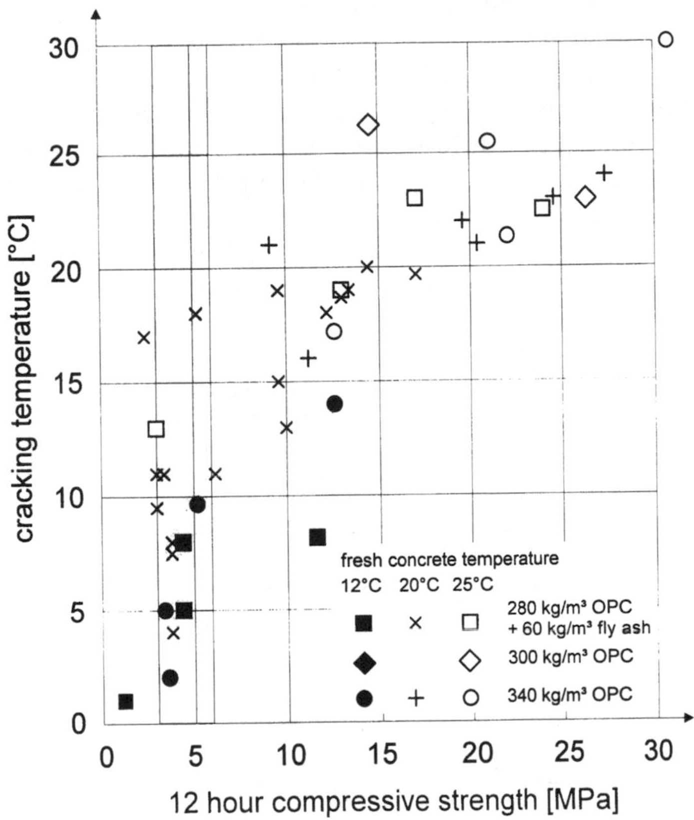

Fig. 3.5 Influence of the fresh concrete temperature, cement content, fly ash and compressive strength on the cracking temperature.

Comparative tests in the cracking frame representing 50 cm thick concrete walls and in the temperature-stress-testing machine with a temperature/time curve used equal to that in 3 m high concrete lifts have shown however that concrete mixes with mass concrete cements of high cracking resistance give also low cracking temperatures at early age (obviously due to the development of elastic properties at the decisive early age). Slowly hardening mass concrete is rather sensitive to environmental temperature changes. This contributes to the fact that extreme low heat cements are seldom used.

It is current practice to reduce the **cement content** as far as possible and to ensure the sufficient workability by the addition of puzzolanes, e.g. fly-ash. So the heat development can be reduced and a sufficient strength is obtained by the puzzolanic reaction.

In **air-entrained** concretes the ultimate tensile strain is about 20% higher than for normal concrete. After hardening the Young's modulus is reduced, so that lower restraint stresses are raised. Both effects result in a reduced cracking sensitivity of air-entrained concrete.

The cracking sensitivity of concrete can be reduced nearly to zero by the use of **expansive additives**. Such additives mostly are based on calcium sulphoaluminates or free lime. During hydration they lead to an expansion, e.g. due to the formation of ettringite. When this expansion is restrained, compressive stresses occur which compensate the thermal stresses during cooling to a great extent (fig. 3.6).

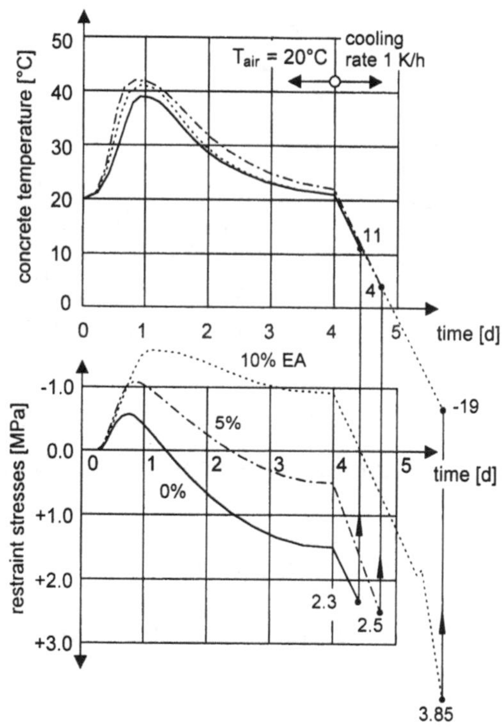

Fig. 3.6 Temperature and stress development for concretes containing expansive additives in the cracking frame.

However, such expansive additives are very sensitive. The degree of expansion is influenced by many parameters: temperature, moisture, dosage, cement, mix composition etc. When the dosage of the additive is not optimised individually, either very little expansion or too much expansion which results in deterioration can occur. Therefore such additives are not licensed in many countries.

3.7 QUANTIFICATION AND SELECTION OF MEASURES

The influence of the various technological parameters on thermal cracking can be approximately quantified by a comparison of the range of the cracking temperature. Following ranges can be quoted (listed with decreasing effects):

- Reduction of the fresh concrete temperature from 25 to 12°C: ΔT_C = 15 - 18 K
- Use of a favourable cement type and brand: ΔT_C = up to 20 K
- Maximum aggregate size of 32 mm instead of 8 mm <u>and</u> an adequate reduction in cement paste content: ΔT_C = 5 - 10 K
- Use of aggregates, which lead to a low coefficient of thermal expansion in the concrete: ΔTC= up to 10 K
- Addition of air-entraining agents (Air-content ≈ 3 - 6 % vol.) ΔT_C =3 - 5 K
- Use of crushed aggregate instead of gravel: ΔT_C = 3 - 5 K
- Reduction of the cement content (OPC) e.g. from 340 kg/m³ to 280 kg/m³ and replacement by 60 kg/m³ fly ash ΔT_C = 3 - 5 K

By tests with cracking frames the effect of each measure can be quantified. The reduction of the cracking temperature achieved by simultaneous application of different of some of the above mentionned measures is usually lower than the sum of the effects of each measure separately and must be determinated in particular. Considering the boundary conditions the adequate measure(s), by which cracking can be prevented to a great extend, can be selected.

Also the costs for investigations and measures as well as the possibilities on the particular site have to be related to the effect of the measures and the necessity of crack prevention. While in a small foundation the use of a normal hardening or a suitable low-heat-cement can be sufficient - a few cracks can be sealed by injection, in more critical members, e.g. linings of large tunnels or watertight-constructions, a special selection of constituents can be useful. For particular projects, e.g. large dams, intensive investigations are necessary, e.g. to test the sensitivity of the concrete. In such cases also special measures like the use of crushed ice for reducing the fresh concrete temperature is often done. Such a measure in normal walls would be exaggerated.

REFERENCES

[1] Springenschmid R.; Breitenbücher, R.: Are Low Heat Cements the Most Favourable Cements for the Prevention of Cracks Due to Heat of Hydration? *Concrete Precasting Plant Technology*, **52** (11), 1986, pp. 704 - 711

[2] Springenschmid R.; Nischer, P.: Untersuchungen über die Ursache von Querrissen im jungen Beton. *Beton und Stahlbetonbau*, **68** (9), 1973, pp. 221 - 226

[3] Breitenbücher, R.: Zwangsspannungen und Rißbildung infolge Hydratationswärme. Doctoral thesis, Technical University of Munich, 1989.

[4] Tazawa, E.; Masuoka, Y.; Miyazawa, S.; Okamoto, S.: Effect of Autogenous Shrinkage on Self Stress in Hardening Concrete. Thermal Cracking in Concrete at Early Ages, Proceedings of the International RILEM Symposium. Edited by R. Springenschmid, E &FN Spon, 1995, pp. 221 - 228

[5] RILEM Technical Recommendation Testing of the Cracking Tendency of Concrete at Early Ages in the Cracking Frame Test, *Materials and Structures*, **30**, October 1997, pp. 461-464. Reprinted in Chapter 10.

[6] Verbeck, G. J.; Helmuth, R. H.: Structure and physical properties of cement paste. Proceedings of 5th International Symposium on Chemistry of Cement. Tokyo, 1968, Part III, pp. 1 - 32

[7] Dahms, J.: Normalzuschlag. *Zementtaschenbuch*, **48** (1984), Bauverlag GmbH, Wiesbaden/Berlin, 1983, pp. 133-157

4

Prediction of Temperature Development in Hardening Concrete

K. van Breugel

4.1 INTRODUCTION

From decades of practical experience and from theoretical considerations, it is well known that early age temperatures and temperature induced stresses may have a great influence on the *quality* of the concrete and concrete structures. In order to enhance the quality and performance of concrete structures, control of these temperatures and of the associated thermal stresses is of paramount importance. For quantification of the effectiveness of technical and technological measures for improving the quality, *numerical simulation techniques* can be used. Simulation programs must be able to predict the development of temperatures, strength, stresses and the risk of early age cracking [54]. Although thermal stresses and the risk of cracking are not correlated linearly with the temperatures in the concrete [4], there is no doubt that temperatures and temperature differentials are among the key-factors in a thermal analysis of concrete structures [47].

May be Tetmayer [51] was one of the first who recognised the importance of temperature effects in early-age concrete. He performed temperature measurements as early as 1883. Since Tetmayer's measurements many authors have discussed thermal problems. Major items in these discussions, of which many date back to the twenties and thirties, referred to either optimisation of pouring sequences in view of reducing maximum hydration temperatures in *mass concrete* structures, like dams [11], and the avoidance of thermal cracking [26,10]. Thermal problems and recommendations for formwork removal in case of *thin* and *slender* elements were discussed in the forties by Forbrich [16].

Prevention of Thermal Cracking in Concrete at Early Ages. Edited by R. Springenschmid. RILEM Report 15. Published in 1998 by E & FN Spon, 11 New Fetter Lane, London EC4P 4EE, UK. ISBN 0 419 22310 X

In the last decade the risk of thermal cracking is discussed particularly in connection with *durability* and *functionalability*. Incompatibility of thermal strains of the cement matrix and aggregate and the associated risk of microcracking and loss of durability has pushed forward the need for control of maximum temperatures in the concrete's early life [15,55]. This in particular with regard to high strength concrete, which concretes are known to be very susceptible to early age cracking. A sub-item in this respect refers to the *avoidance of frost damage* in early age concrete [31]. More recently, the consequences of thermal cracking and associated loss of durability have been discussed in view of the *shortage of raw materials* [50].

4.2 MODELS FOR TEMPERATURE PREDICTION

Numerical models for the prediction of temperature development in cement-based systems like cement paste, mortar and concrete, can be subdivided into *micro level* and *macro level* models [6]. Micro level models are developed in the area of materials science. They allow for physico-chemical processes and mechanisms which occur on the micro- or even nano-level scale. The potentialities of those models have been discussed by Garboczi *et al.* [19]. Macro level models are developed for the prediction of temperature fields in real concrete structures. These models are used in the engineering practice as an aid in decision making regarding time of formwork stripping, application of insulation or just cooling of the concrete.

In this report emphasis will be on macro-level engineering models. Due attention will be given to the way in which these models allow for the continuous change of the thermal properties and to the modelling of the progress of the hydration process, i.e. to the consistency of the models with approved materials science concepts.

4.3 MODELS FOR TEMPERATURE PREDICTION - HISTORICAL DEVELOPMENTS

The first attempts to predict hydration temperatures in hardening concrete have been ascribed to Yoshida and date back to 1921 [48]. A *graphical method* for the prediction of maximum temperatures has been elaborated by Werner (in [17]). His method has been applied successfully for the prediction of maximum temperatures in the piers of the Vaster Bridge in Stockholm. Maximum temperatures could be predicted within a range of ±5°C.

Several *analytical methods* for the determination of temperature fields were proposed and discussed in the thirties [9, 34]. Almost all those methods were based on the differential equation of Fourier:

$$\frac{\partial T}{\partial t} = a_c * \left[\frac{(\partial^2 T)}{(\partial x^2)} + \frac{(\partial^2 T)}{(\partial y^2)} + \frac{(\partial^2 T)}{(\partial z^2)}\right] + Q(t,x,y,z) \tag{4.1}$$

where a_c [m^2/s] is the coefficient of thermal diffusivity of concrete, T [°C] the temperature, x,y,z are the coordinates of a particular point in the structure and Q(t,x,y,z) [J/g] is the source term, i.e. the heat liberation function.

In the early predictive models the coefficient of thermal diffusivity a_c was mostly assumed constant. The heat production curve Q(.) was often described with a predefined exponential curve of the shape [11,23,29,32]:

$$Q(t) = Q_\infty * (1 - \exp(-r.t)) \tag{4.2}$$

where Q(t) [J] is the heat of hydration liberated at time t, Q_∞[J] the liberated heat when hydration has practically ceased and r an empirical constant. In order to account for the typical S-shape of the actual hydration curve Helland [21] has proposed the function:

$$Q(t) = Q_\infty * \exp(-(\tau / t)^\beta) \tag{4.3}$$

with Q(t) and Q_∞ as in eq. (4.2), and t and b empirical constants.

As early as 1937 it was Carlson [9], who pointed out that the heat liberation curve Q(t) is affected by the actual reaction temperature of the concrete. This temperature is not known at the outset of the calculation. This unknown temperature will affect the heat evolution curve, which can not, therefore, be prescribed in a form as presented with eq. (4.2). Rastrup [41] was probably the first who actually took the influence of the reaction temperature on the rate of reaction into account explicitly. A most important feature in Rastrup's approach was that not only the temperature of the concrete was calculated, but also the *development of the degree of hydration* in the concrete.

The recognition, that the degree of hydration is a predominant parameter in numerical procedures for the evaluation of early-age thermal problems, including its importance for the development of the materials properties and the risk of early age thermal cracking, reveals itself as a distinct trend in modern numerical predictive models. In essence a reliable prediction of temperatures *without* a simultaneous calculation of the progress of hydration process is impossible.

4.4 HYDRATION OF CEMENT-BASED SYSTEMS

4.4.1 Degree of hydration

The state of the hydration process at time t is generally indicated with the degree of hydration $\alpha(t)$. A comprehensive and in all respect satisfactory definition of the degree of hydration is hard to give [6]. Parameters applied to indicate the degree of hydration are, among other things, the amount the liberated heat of hydration and the amount of chemically bound water. It must be born in mind, however, that the four major cement compounds, viz. C_3S, C_2S, C_3A and C_4AF, hydrate at different rates [38], while binding different amounts of water per unit mass and liberating different amounts of heat per unit mass. This means that the degree of hydration of cement, if quantified in terms of liberated heat or bound water, must be considered as a *weighted* average of the degrees of hydration of the constituents. Still there appears to be a fairly good correlation between the amount of cement that has reacted and the amount of liberated heat (Fig. 4.1). For many practical engineering purposes it is justified, therefore, to equalise the degree of hydration with the degree of heat development:

$$\alpha(t) = \frac{Q(t)}{Q_{max}} \tag{4.4}$$

in which Q(t) [J] and Q_{max} [J] are the amounts of heat liberated at time t and at complete hydration ($a = 1$), respectively. Describing the hydration process in the *heat domain*, as it is done with expression (4.4), substantially eases the step to temperature calculations.

In case of non-Portland and blended cements, the correlation between the heat of hydration and the degree of hydration is even more complicated than in the case of Portland cement [2]. In a silica-fume containing mix, for example, the degree of reaction of the cement and the silica fume will differ significantly [22]. But even in many of those cases the amount of released heat appears to be good parameter for characterising the actual state of the reaction process, be it that the correlation may not be a linear one. More research in this field is recommended.

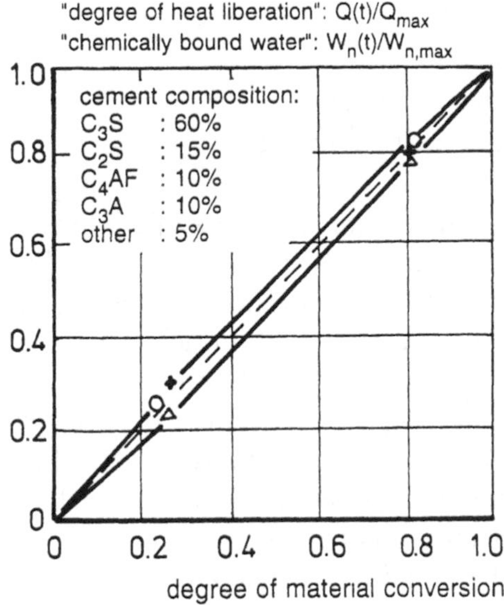

Fig. 4.1 Correlation between material conversion (+), heat of hydration (Δ), amount of chemically bound water (o) [44].

4.4.2 Potential amount of heat of hydration

The amount of heat Q_{max} liberated at complete hydration of the cement, depends on the chemical composition of the cement. For a given chemical composition the total heat of hydration can be determined by adding the amounts of heat liberated by the individual compounds. The compound composition of the cement can be determined according to the classic Bogue method or, more precisely, according to Taylor [49]. According to Wang *et al.* [52] noticeable variations in the clinker quality have to be considered, even when produced by one manufacturer. This may have significant consequences for the potential heat of hydration and hence for the determination of the degree of hydration according to eq. (4.4). These variations may be attributed to, among other things, variations in the alkali content of the cement.

4.4.3 Factors influencing the hydration process

The hydration process is predominantly determined by the following parameters:

- the chemical composition of the cement (of blended cement)
- the fineness and particle size distribution of the cement
- the water/cement ratio
- the (initial) reaction temperature
- the presence and type of admixtures

These parameters determine the shape of the hydration curves α(t). The rate of hydration is faster the finer the cement, the higher the temperature and the higher the water/ cement ratio (w/c ratio) (Fig. 4.2). It is noticed, that the maximum degree of hydration achieved in an ordinary concrete mix is determined predominantly by the fineness of the cement and the w/c ratio. In practice the degree of hydration will generally not exceed 70....80% (Fig. 4.3), even if the w/c ratio exceeds the critical value of 0.4, the latter value being considered the lowest value at which complete hydration would theoretically be possible.

The effect of the initial mix temperature on adiabatic hydration curves is shown in Fig. 4.4. Hydration curves for different types of cement, i.e. for Type I, II, III and IV cement, are shown in Fig. 4.5. It is noticed that the shape of these curves is not only a function of the chemical composition of the cement, but also of its particle size distribution.

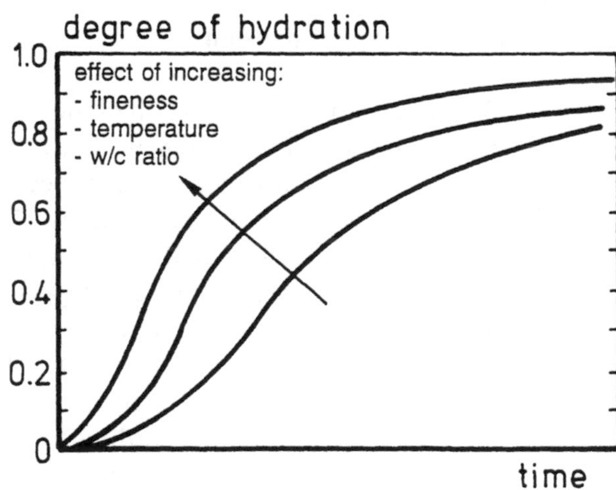

Fig. 4.2 Effect of variations in cement paste parameters on the rate of hydration.

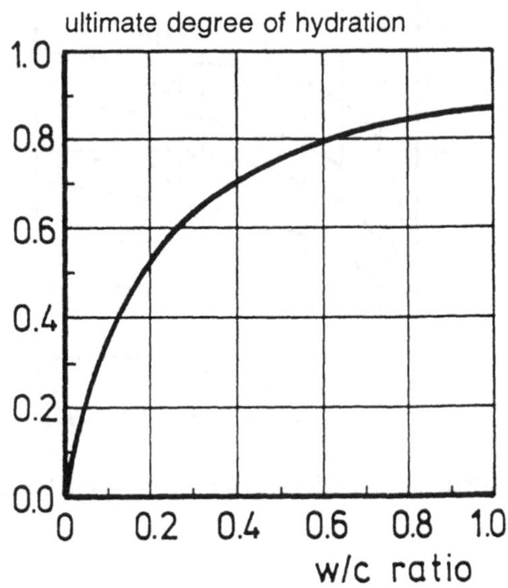

Fig. 4.3 Ultimate degree of hydration as a function of the w/c ratio [35].

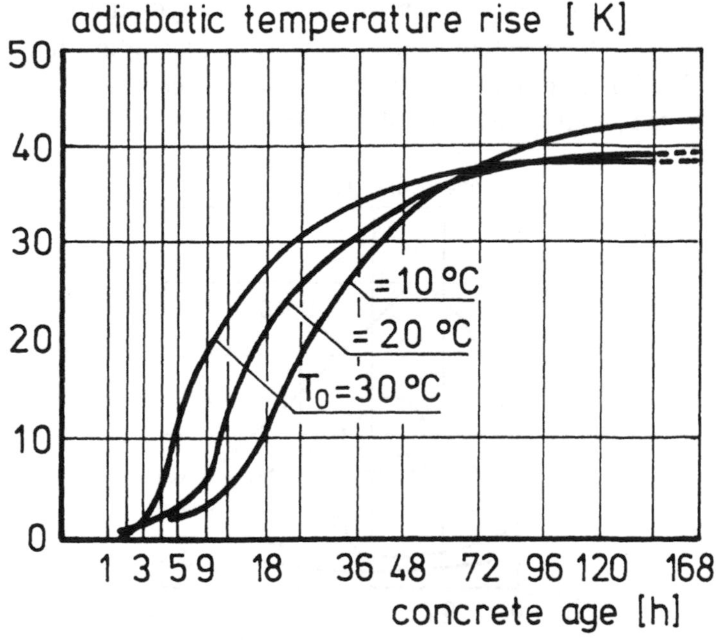

Fig. 4.4 Adiabatic temperature rise as a function of mix temperature [30].

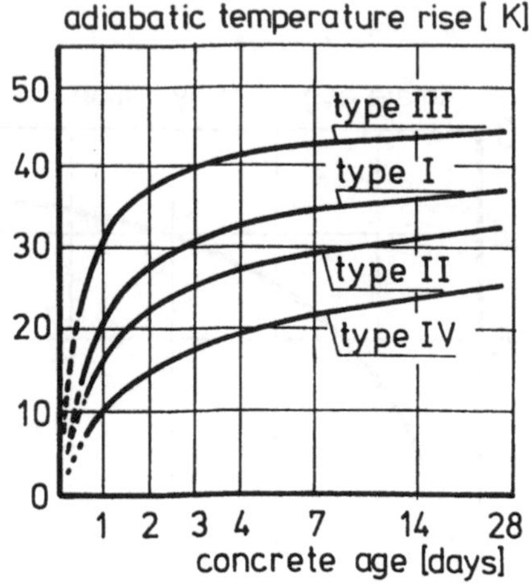

Fig. 4.5 Adiabatic hydration curves. Different cement types. 223 kg/m^3 cement [1].

4.4.4 Temperature effects and temperature functions

The rate of most chemical and physical processes increase with increasing temperature. The temperature dependency of the hydration process can be described with a temperature factor f(α,T), which factor is a function of the degree of hydration α and the actual reaction temperature T [6, 44]. Examples of the temperature factor are shown in Fig. 4.6.

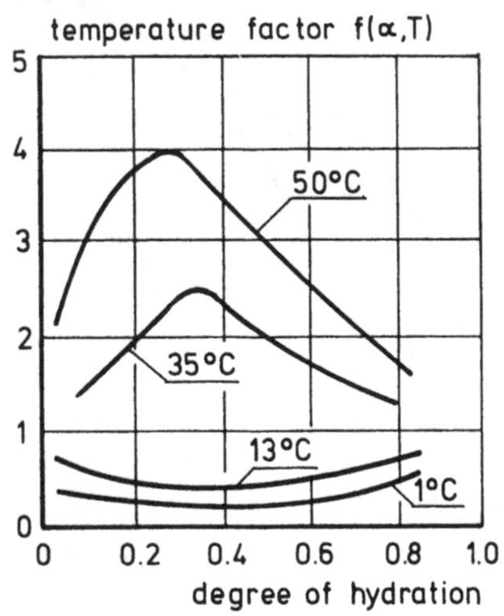

Fig. 4.6 Temperature factor f(α,T) as a function of the degree of hydration. (after [44]). Reference temperature T_{ref} = 20°C.

The factor $f(\alpha,T)$ indicates the rate of hydration at temperature T compared to the rate of hydration at a certain reference temperature T_{ref}. In Fig. 4.6 a reference temperature $T_{ref} = 20°C$ is used. A mathematical formulation of the temperature factor $f(\alpha,T)$ will be given in paragraph 4.5.2.

4.4.5 Effect of elevated hydration temperatures on microstructure

Curing at elevated temperatures does not only affect the rate of hydration but also the ultimate strength. This strength is generally lower if the curing temperature is higher [27]. This phenomenon must be attributed to differences in the microstructure when formed at different temperatures. High reaction temperatures result in a denser reaction product and, consequently, a paste with a higher capillary porosity. Since the strength is closely related to the capillary porosity, the strength reduction at elevated temperatures is plausible.

4.5 MATHEMATICAL MODELS FOR QUANTIFICATION TEMPERATURE DEVELOPMENT

4.5.1 Input curves

The majority of currently used mathematical models for the determination of temperature effects are based on the D.E. of Fourier (eq. (4.1)). The fact that the source term Q(t) in this equation is a function of the unknown reaction temperature requires to solve the D.E. using a step-wise procedure. The state parameters at the end of each step are the input for the calculations in the next step.

The source term is determined step wise by adjusting the so called *input curve.* This input curve represents the course of the hydration process under well-defined temperature conditions, for example an adiabatic, isothermal or semi-adiabatic temperature regime. Adiabatic and semi-adiabatic hydration and temperature curves are presented in Fig. 4.7. For adjustment of the input curve the *temperature sensitivity*, i.e. the temperature factor $f(\alpha,T)$, of the (physico-)chemical reactions and processes involved in the hydration process must be known. Provided that accurate information about the temperature sensitivity is available, there is, in essence, no preference for either the adiabatic, isothermal or semi-adiabatic input curve! From the pure materials science point of view an isothermal curve would be preferable. For practical purposes the adiabatic and semi-adiabatic curves appear to be more convenient.

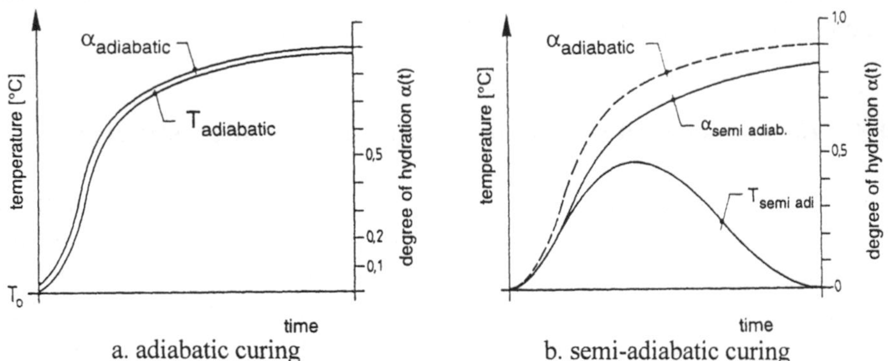

Fig. 4.7 Isothermal, adiabatic and semi-adiabatic hydration curves - schematic.

Considerations which make us deciding in favour of one of the possible input curves may refer to, firstly, the experimental difficulties with the determination of the input curves and, secondly, the size of the concrete elements to be analysed.

4.5.1.1 Experimental difficulties with determination of input curves.

Adiabatic input curves can be determined quite accurately with adiabatic calorimeters. After about 72 hours hydration, however, the reliability of the measurements suffer from experimental difficulties. An advantage of adiabatic input curves is that the curves can be determined for the concrete mix to be used on site. *Isothermal* input curves generally refer to cement paste and not to concrete. Since paste hydration proceeds differently from hydration in concrete, isothermal paste hydration is not unconditionally representative of isothermal hydration of the cement in a concrete mix. *Semi-adiabatic* input curves are generally easy to determined. Concrete specimens of the same mix as will be used on site can be tested.

4.5.1.2 Size of the concrete elements

For thermal analysis of mass concrete structures it seems reasonable to use adiabatic input curves. In smaller structures , where a substantial loss of heat to the environment is to be expected in the early stage of hydration, a semi-adiabatic input curve may be favourable. Isothermal hydration is hardly imaginable in real concrete structures. Only in very small sections and for a constant ambient temperature isothermal hydration may occur. Isothermal hydration is also possible in the close vicinity of cast-in cooling pipes. At some distance from the cooling pipes, however, the conditions are already far from isothermal. This means that in the greater part of the cross section substantial adjustment of the input curve will be required, which may lead to less accurate results.

4.5.2 Adjustment of the input curves for actual curing temperatures

Assume the hydration process starts adiabatically, i.e. according to an *adiabatic input curve* A_0 (Fig. 4.8). As soon as heat dissipates to the environment, the reaction temperature will become less compared to the adiabatic temperature rise. Consequently, the rate of hydration decreases and the curve which indicates the actual degree of hydration, i.e. the "process curve", starts to deviate from the adiabatic input curve.

An arbitrary stage in the semi-adiabatic process, viz. at time t_j, is schematically shown in Fig. 4.8. At time t_j the reaction temperature, or process temperature, is $T_{p,j}$, and the amount of liberated heat $Q_{p,j}$. In an increment of time $\Delta\tau_{j+1}$ hydration according to the adiabatic curve A_0 would result in the liberation of an amount of heat $\Delta Q_{a,j+1}$. Since, however, the temperature at which the reaction occurs is lower than in the adiabatic process, i.e. $T_{p,j} < T_{a,j}$, the hydration process will be retarded. For the rates of hydration $k_a(T_{a,j+1})$ and $k_p(T_{p,j+1})$ at the temperatures $T_{a,j}$ and $T_{p,j}$ during the increments of time Δt_{j+1} (= Δt_{j+1}) and Δt_{j+1}, respectively, and for the quotient of these rates, the following expression holds (see Fig. 4.8):

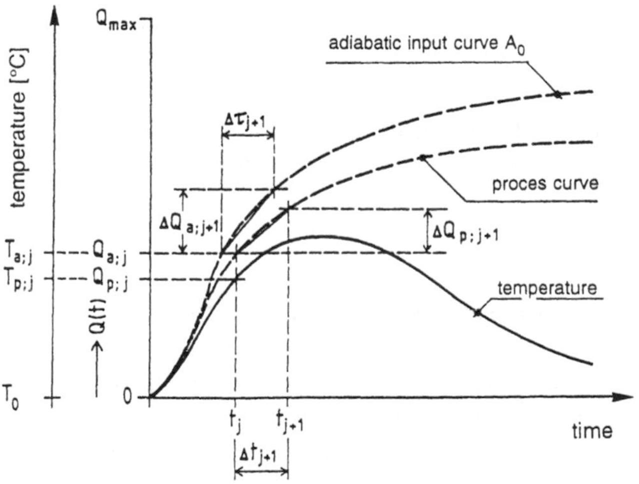

Fig. 4.8 Procedure for adjustment of the rate of hydration in case of non-adiabatic hydration.

$$\frac{k_a(T_{a,j+1})}{k_p(T_{p,j+1})} = \frac{\dfrac{\Delta Q_{a,j+1}}{\Delta t_{j+1}}}{\dfrac{\Delta Q_{p,j+1}}{\Delta t_{j+1}}} = \frac{\Delta Q_{a,j+1}}{\Delta Q_{p,j+1}} \tag{4.5}$$

In many numerical models temperature effects are quantified on the basis of the Arrhenius equation:

$$k(T) = A \cdot e^{-\frac{E_A(T)}{R.T}} \tag{4.6}$$

in which A is an experimental constant, k(T) a rate constant [J/s], E_A the (apparent) activation energy [J/mol], R the gas constant [8.31 J/mol.K] and T the temperature in K. For $k(T) = k_a(T_{a,j+1})$ and $k_p(T_{p,j+1})$, respectively, we get for the temperature function $f(\alpha,T_p)$ ((4.6) in (4.5)):

$$f(\alpha,T_p) = \frac{k_a(T_{a,j+1})}{k_p(T_{p,j+1})} = e^{+\frac{E_A(T)}{R}\cdot\frac{T_a-T_p}{T_a \cdot T_p}} \tag{4.7}$$

The temperature sensitivity is, in essence, determined by the apparent activation energy E_A, the latter being a function of the temperature and the degree of hydration [44]. For many practical purposes, however, the following expressions would hold quite good, at least for Portland cement [18].

$E_A = 33.5$ kJ/mol for $T \geq 20°C$
$E_A = 33.5 + 1.47\ (20\text{-}T)$ kJ/mol for $T < 20°C$

For further information about the activation energy and for other methods for quantification of temperature effects reference is made to literature [6, 40].

4.5.3 Output of numerical temperature predictions

The output of a calculation procedure as described in the preceding paragraph consists of the evolution of both the temperature *and* the degree of hydration. *It is emphasised that the degree of hydration, or degree of conversion of the anhydrous cement into gel, is not a by product, but in essence the backbone of the temperature calculation.* In fact, a temperature prediction of a hardening system without calculation the evolution of the degree of hydration, i.e. the rate of heat evolution, is impossible. The degree of hydration is to be considered as the most suitable state parameter for describing the evolution of materials properties in a hardening cement-based material. In that manner the degree of hydration can bring about the consistency in the mathematical treatment of early-age concrete behaviour.

4.5.4 Prediction of the degree of hydration in hardening concrete structures

As indicated in the forgoing, the degree of hydration is part of the output of a numerical temperature prediction. In the engineering practice, however, we are not only interested in predictions, but may be even more in the actual state of hardening, i.e. of hydration, in a concrete structure under real site conditions. For determination of the actual degree of hydration on site the calculation procedure described in paragraph can be applied in a modified way. The only modification concerns the way in which the process temperature T_p is determined. In paragraph 4.5.2 the process temperature has to be *calculated*, for which purpose finite difference methods and finite element packaged can be used as will be explained further in paragraph 4.7.1. Instead of the predicted process temperature we can, however, also take the actual concrete temperature as measured on site as input for the calculation. The input of the calculation now exists of the adiabatic hydration curve and the measure temperature, whereas the evolution of the degree of hydration is the output. The degree of hydration in the hardening structure being known, the evolution of materials properties can be determined as a function of this state parameter. Implementation of this "degree-of-hydration-concept" for monitoring the evolution of the hydration process and the associated material properties has been described by van Beek [3].

4.6 THERMAL PROPERTIES OF CONCRETE

4.6.1 Coefficient of thermal conductivity

The coefficient of thermal conductivity λ_c of concrete depends on the *moisture content*, *content* and *type of aggregate*, *porosity*, *density* and *temperature*. For ordinary concretes in a normal temperature and moisture state the coefficient of thermal conductivity varies between 1.2 and 3.0

[W/m.K] [14]. Values beyond 3.0 W/m.K, even up to 3.5 W/m.K, are found as well (see Table 4.1).

Table 4.1 Typical values of the coefficient of thermal conductivity of concrete made with different types of aggregates [1].

Type of aggregate	*Thermal conductivity of concrete [W/m.K]*
Quarzite	3.5
Dolomite	3.2
Limestone	2.6 - 3.3
Granite	2.6 - 2.7
Rhyolite	2.2
Basalt	1.9 - 2.2

For most concretes the conductivity increases with increasing moisture content [1, 24]. For two different concretes the effect of the moisture content on the conductivity is shown in Fig. 4.9.

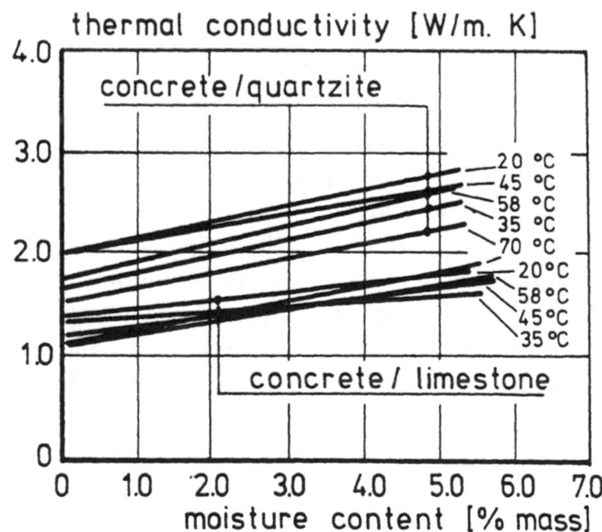

Fig. 4.9 Influence of moisture content on thermal conductivity of two different concretes (after Hundt [24]).

Since the free moisture content changes with progress of the hydration process and the conductivity is a function of the moisture content, the conductivity will change with the degree of hydration. A substantial increase of the conductivity in the early stage of hardening has been reported by Maréchal [33] and Sassedateljev [45]. Hundt *et al.* [24] and

Brown *et al.* [8], however, have found a decrease by 20% to 30% (Fig. 4.10). An analytical evaluation of the effect of transformation of cement and water into gel, carried out on the basis of a structural model for porous materials proposed by [53], did not reveal a significant drop of the conductivity [5]. An increase of the conductivity with age was found, which is in agreement with the findings of Maréchal, Sassedateljev and of Morabito [36].

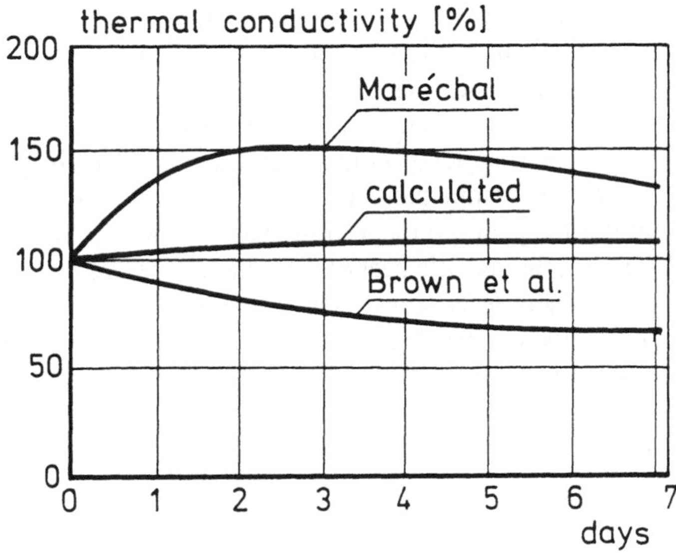

Fig. 4.10 Development of coefficient of thermal conductivity at early ages. Different authors.

4.6.2 Specific heat of concrete

Typical values of the specific heat c_c of ordinary concrete vary between 0.850 and 1.150 kJ/kg.K [12, 32, 37]. For fresh mixes and water saturated concretes the higher values apply, the lower values for dry concretes and light weight concrete. The specific heat slightly decreases, by about 5...8%, with increasing degree of hydration [5]. It increases with increasing temperature. An increase of about 10% for an increase in temperature from 10 to 66°C has been reported in [1].

4.6.3 Coefficient of thermal diffusivity a_c

The dependency of the coefficient of diffusivity a_c [m^2/h] on the moisture content, temperature and degree of hydration follows from the dependency in the coefficient of conductivity λ_c and the specific heat c_c from these parameters. As both the conductivity and the specific heat increase with increasing moisture content, the net influence on the coefficient of thermal diffusivity is only marginal. The thermal diffusivity increases with an increase of the dry density of the concrete [1, 36].

The influence of temperature on the thermal diffusivity depends upon the corresponding influence on both the thermal conductivity and the specific heat. Since the conductivity decreases and the specific heat increases with increasing temperature, the thermal diffusivity decreases more with increasing temperature than the thermal conductivity (up to about 15% in the temperature range from 10°C to 66°C). With increasing degree of hydration the specific heat of the concrete decreases. As indicated in the foregoing, there is no consensus yet about the dependency of the thermal conductivity on the degree of hydration. In case the diffusivity becomes a decisive factor in the evaluation of temperatures and thermal stresses, it is recommended to perform sensitivity studies with different assumptions for the relationship between diffusivity and degree of hydration.

The main influence on the thermal diffusivity of normal weight concrete comes from the type of aggregate (Table 4.2). For quantification of the effect of the aggregate on the thermal diffusivity of the concrete reference is made to Dettling [13].

Table 4.2 Diffusivity for different types of aggregate [1].

Coarse aggregate	*Diffusivity of concrete*	
	ft²/h	*m²/h*
Quartzite	0.058	0.0054
Limestone	0.051	0.0047
Dolomite	0.050	0.0046
Granite	0.043	0.0040
Rhyolite	0.035	0.0033
Basalt	0.032	0.0030

4.7 DETERMINATION OF TEMPERATURE DISTRIBUTION IN HARDENING CONCRETE STRUCTURES

4.7.1 Boundary conditions

At the surface of a continuum the flow of heat depends on the heat transfer properties of the boundary layer between surface and surrounding area. Factors to be considered are the type and thickness of the formwork, the time of formwork removal, the curing method, wind velocity, the ambient temperature and solar radiation. It are these parameters which determine the coefficient of heat transfer α_{eff}:

$$\frac{1}{\alpha_{eff}} = \frac{1}{\alpha_w + \alpha_r} + \sum \frac{d_i}{\lambda_i} \tag{4.8}$$

in which α_{eff} it the effective coefficient of heat transfer [W/m².K], α_w the coefficient of heat transfer accounting for flow or convection [W/m².K], α_r the coefficient of heat transfer which allows for the effect of radiation [W/m².K], d_i the thickness of formwork or insulation in layer i [m] and λ_i the thermal conductivity of formwork or insulating material in layer i [W/m.K]. Values for the average surface heat transfer coefficient can be taken from literature (for example [28]).

Solar radiation, particularly in case a steel formwork is used, may cause a quite significant increase in the maximum temperature reached during hydration. An example of the effect of solar radiation on the concrete temperatures is shown in Fig. 4.11. The curves a and b represent the concrete temperature at the surface of a wall made of an ordinary Portland cement-based concrete mix. Comparison of the curves a and b reveals a difference in peak temperatures at 18 h of about 6°C. On cooling this will give rise to higher (tensile) stresses and enhancement of the risk of cracking. Nocturnal emission at night may further enhance this risk of cracking.

Even more traitorous and detrimental than solar radiation may be the effect of wind. An indication of the effect of wind velocity on temperature development at the surface and in the centre of a 0.7 m thick concrete wall can be deduced from Fig. 4.12. Curves 1a and 1b are obtained for a heat transfer coefficient α_{eff} = 3.12 W/m².K (no wind), curves 2a and 2b for α_{eff} = 26.3 W/m².K (wind velocity 5 m/s). It is evident that ignorance of wind effects may completely invalidate the interpretation of otherwise sophisticated calculations.

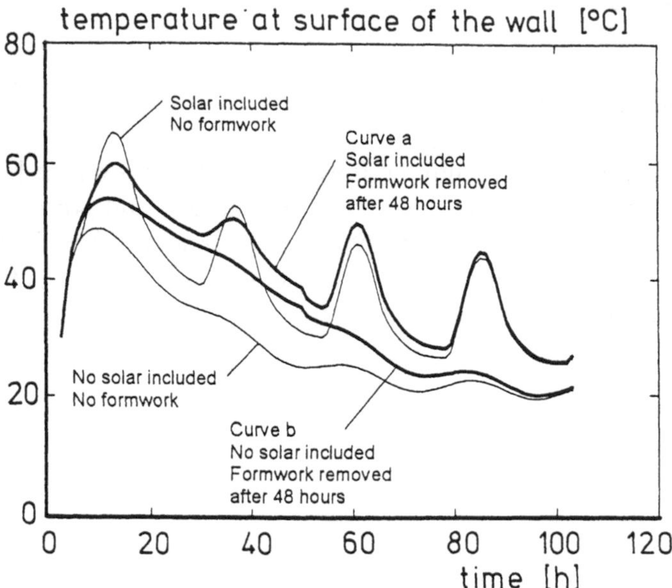

Fig. 4.11 Effect of solar radiation on temperatures in hardening concrete structures. Maximum flux 830 W/m². Ambient temperature: variable.

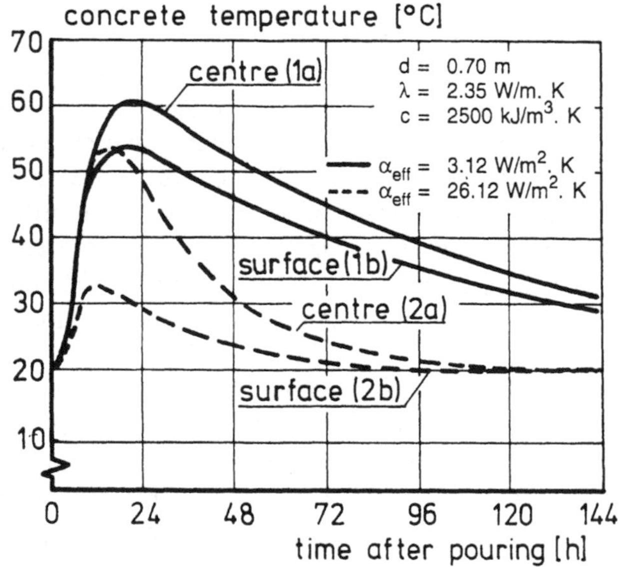

Fig. 4.12 Effect of heat transfer coefficient on temperature development. Wall thickness 0.7 m. FEM-calculations [20].

Strong winds may lead to substantial evaporation at the surface after demoulding. Early evaporation may jeopardise the hydration process in the surface layer of the concrete. Improper curing may result in a loss of durability. The importance of early age evaporation and moisture movements caused by internal temperature gradients is dealt with in more detail by Roelfstra *et al.* [43]

4.7.2 Numerical methods for solving the D.E. of Fourier

For solving the D.E. of Fourier finite difference methods and finite element are used. The former methods, extensively discussed in e.g. [39], are used in many special purpose programs. For one and two dimensional problems these methods work quite satisfactory, be it that for two dimensional temperature problems the computation time may become quite considerable. For complex geometries finite element methods, comprehensively discussed by Hamfler [20], are preferable. In case of pipe cooling finite element methods seem the proper choice [42].

As pointed out by Wang *et al.* [52] complex geometries can often be schematised to one or two dimensional problems. With those approximative calculations preliminary results may give interesting and meaningful results already. Interesting studies in this area are reported by de Schutter *et al.* [46]. Also in case of pipe cooling simplifications of the in fact three-dimensional problems to two dimensional ones have turned out to give satisfactory approximative results.

4.8 CONCLUDING REMARKS

The number of papers reporting successful temperature predictions is steadily increasing. It must be admitted, however, that accurate temperature predictions were possible in the thirties already. This does not mean that no progress has been made in this area in past decades. Progress in temperature predictions first of all concerns the *increased consistency* of modern numerical models. Increased knowledge about the temperature dependency of rate processes, in combination with the nowadays available powerful computer hardware, enable the development of computer-based models which are firmly rooted in approved materials science concepts. In this respect it is noticed that in many currently available numerical programs the degree of hydration is explicitly determined. It is strongly emphasised, that the degree of hydration is not a "by product" of a temperature calculation, but the *backbone* of it, and the basis for the quantification of thermal, physical and also time dependent materials properties.

On the basis of a detailed description of microstructural development, including porosity, with progress of the hydration process it has become possible to model the combined processes of temperature development and moisture movements and their mutual interaction in step-wise calculation procedures. Interesting contributions to this theme come from Jonasson [25] and Roelfstra *et al.* [43]. The former author also points to the effect of autogenous shrinkage as a most important factor that contributes to the risk of cracking. A reliable analysis of the contribution of autogenous shrinkage to the risk of cracking requires information about the progress of the hydration process, i.e. the degree of hydration, to which autogenous shrinkage is closely related.

In conclusion it can be stated, that temperature predictions in hardening concrete can be performed with relative high accuracy. Certainly when they are used as a basis for numerical analyses of the risk of thermal cracking. In those cases it are the modulus of elasticity, plastic deformations and early-age creep and relaxation which constitute the major sources of uncertainty. The author is convinced that accurate predictions of early-age concrete behaviour, including predictions of the risk of early-age cracking, will remain wishful thinking unless these predictions are based on sound material models, in which the progress of the hydration process and the associated microstructural development are modelled explicitly on the microlevel scale.

Acknowledgement
I wish to thank MSc. E.A.B. Koenders for conducting numerical analysis concerning the effects of solar radiation and nocturnal emission. The financial support of the Dutch Technology Foundation is gratefully acknowledged.

REFERENCES

[1] American Concrete Institute. (1987) Manual of Concrete Practice. Part I. ACI-CI 213R-79, ACI 207.1R-70, ACI 207.2R-73, ACI 207.4 R-80.

[2] Atlassi, E. (1992) Desorption isotherms of silica fume mortar. 9th Int. Conf. on the Chemistry of Cement, Vol. IV, 634-639.

[3] Beek, A. van, Lokhorst, S.J., Breugel, K. van. (1996) On-site determination of the degree of hydration and associated properties of hardening concrete. Proc. Int. Congress on Non-Destructive Testing, Colorado, USA.

[4] Bernander, M. (1982) Temperature stresses in early age concrete due to hydration. Int. Conf. on Concrete at Early Ages, Paris, Vol. II, pp. 218-221.

[5] Breugel, K. van (1980) Artificial cooling of hardening concrete. Delft University of Technology, Research Report No. 5-80-9, Delft.

[6] Breugel, K. van (1991) Simulation of hydration and formation of structure in hardening cement-based materials. PhD thesis, Delft. 295 p.

[7] Breugel, K. van (1993) Bridging the gap between material science and engineering practice, in *Concrete 2000, Economical and durable construction through excellence* (Eds. R.K. Dhir and M.R. Jones), E.& FN Spon, Vol. II, pp. 1677-1692.

[8] Brown, T.D., Javaid, M.Y. (1970) The thermal conductivity of fresh concrete. *Matériaux & Construction*, **3**, No. 18, pp. 411-416.

[9] Carlson, R.W. (1937). A simple method for the computation of temperatures in concrete structures, *ACI-J.*, **34**, 89-102. Disc. Newmark, pp. 104-1/5.

[10] Carlson, R.W. (1938) Temperatures and stresses in mass concrete, *ACI-J.*, **34**, pp. 497-515.

[11] Clover, R.E. (1937) Calculation of Temperature Distribution in a Succession of Lifts Due to Release of Chemical Heat, *ACI-Journal*, **34**, pp. 105-116.

[12] Cummerer, J.S. (1962) *Wärme- und Kälteschutz in der Industrie.* Springer-Verlag, Berlin-Göttingen-Heidelberg.

[13] Dettling, H. (1961) *Wärmedehnung des Zementsteines, der Gesteine und der Betone.* Stuttgart.

[14] Eibl, J., Waubke, N.V., Klingsch, W., Schneider, U., Rieche,G. (1979) Studie zur Erfassung spezieller Betoneigenschaften im Reaktorbehälterbau, Deutscher Ausschuß für Stahlbeton, Heft 237, Berlin.

[15] El-Jazairi, B. and Illston, J.M. (1980) The hydration of cement paste using the semi-isothermal method of derivative thermogravimetry, in *Cement and Concrete Research*, **10**, pp. 361-366.

[16] Forbrich, L.R. (1940) The Effect of Various Reagents on the Heat Liberation Characteristics of Portland Cement, *ACI-Journal*, **37**, pp. 161-184.

[17] Forsen, L. (1935) Zur Chemie des Portlandzementes, in *Zement*, **24**, pp. 191-197.

[18] Freiesleben-Hansen, P. and Pedersen, J. (1977) Maleinstrument til kontrol af betons hardening, in *Nordisk Betong*, **21**, pp. 21-25.

[19] Garboczi, E.J., Bentz, D.P. (1992) Computer-based models of the micro structure and properties of cement-based materials, in Proc. 9th. Int. Conf. on the Chemistry of Cement, New Delhi, Vol. VI, pp. 3-15.

[20] Hamfler, H. (1988) Berechnung von Temperatur-, Feuchte- und Verschiebungsfeldern in erhärtenden Betonbauteilen nach der Methode der finiten Elemente, in Deutscher Ausschuß für Stahlbeton, No. 395, Berlin.

[21] Helland, S. (1986) Curing control by micro computer, in *Nordisk Betong*, **1-2**, pp. 63-70

[22] Helland, S. *et al.* (1990) High strength concrete. State of the Art Report, FIP SR90/1, CEB Bull. No. 197.

[23] Hellmich, K. (1984) Einfluß der Hydratationswärme des Betons auf die Lagerverschiebungen bei Spannbetonbrücken, in *Beton- und Stahlbetonbau*, **2**, pp. 42-46.

[24] Hundt, J. and Wagner, A. (1978) Einfluß des Feuchtigkeitsgehalts und des Reifegrads auf die Wärmeleitfähigkeit von Beton, in Deutscher Ausschuß für Stahlbeton, Heft 297, Berlin.

[25] Jonasson, J.E., Groth, P., Hedlund, H. (1994) Modelling of temperature and moisture field in concrete to study early age movements as a basis for stress analysis. Proc. RILEM Symp. Thermal cracking in concrete at early ages. Munich, 1994, pp. 45-52.

[26] Keller, H. (1937) Der Einfluß silikatischer Beimischungen auf die Bildung von Kalkhydrat auf Abbindewärme und Festigkeit von Portlandzement. *Beton und Eisen*, **36**, No. 231-235, No. 249-255.

[27] Kjellsen, K.O., Detwiler, R.J., Gjørv, O.E. (1991) Development of microstructures in plain cement pastes hydrated at different temperatures. *Cement and Concrete Research*, **21**, pp. 179-189.

[28] Kehlbeck, F. (1975) *Einfluß der Sonnenstrahlung bei Brückenbauwerken*. Werner-Verlag, Düsseldorf.

[29] Kratzer, K. (1981) Verfahren zur Berechnung von instationärer Temperatur im Beton. *Zement und Beton*, **26**, No. 3, pp. 118-123.

[30] Laube, M. (1990) Werkstoffmodell zur Berechnung von Temperaturspannungen in massigen Betonbauteilen im jungen Alter. PhD, TU-Braunschweig.

[31] Maage, M. and Helland, S. (1988) Cold weather concrete curing planned and con-trolled by microcomputer. *Concrete International*, pp. 34-39.

[32] Mandry, W. (1961) *Über das Kühlen von Beton*. Springer Verlag, Berlin-Göttingen-Heidelberg.

[33] Maréchal, J.C. (1972) Détermination simultanée de la diffusivité et de la conductivité thermique du béton pendant son hydratation. Contribution to the Task Group W 40 of the Conseil International du Bâtiment (CIB), Holzkirchen.

[34] McHenry, D. (1937) Measured and Computed Temperatures of Concrete at Norris Dam, in *ACI-Journal*, **34**, pp. 117-125.

[35] Mills, R.H. (1966) Factors influencing cessation of hydration in water cured cement pastes. ACI-SP 60. Washington, pp. 406-424.

[36] Morabito, P. (1989) Measurement of the thermal properties of different concretes. *High Temperatures - High Pressures*, **21**, pp. 51-59.

[37] Neville, A.M. (1975) *Properties of Concrete*. Pitman Publishing Ltd. London.

[38] Parrott, L.J. and Killoh, D.C. (1984) Prediction of cement hydration. *The chemistry and chemically-related properties of cement* (Ed. D. Glasser, Univ. Aberdeen), pp. 41-53.

[39] Press, W.H., *et al.* (1992) Numerical recipes in C, Cambridge Univ. Press.

[40] Rastrup, E. (1956) The temperature function for heat of hydration in concrete. Int. Symp. Winter Concreting, Copenhagen, Session B-II.

[41] Rastrup, E. (1954) Heat of hydration in concrete. *Mag. of Con. Res.*, 79-90.

[42] Reinhardt, H.W., Blaauwendraad, J. and Jongendijk, J. (1982) Temperature development in concrete structures taking account of state dependent properties. Proc. Int. Conference on Concrete at Early Ages, Paris, Vol. I, 211-218.

[43] Roelfstra, P.E. and Salet, T.A.M. (1994) Modelling of heat and moisture transportation in hardening concrete. Proc. RILEM Symp. Thermal cracking in concrete at early ages. Munich, 1994, pp. 37-44.

[44] Röhling, S., Nietner, M. (1990) A model to describe the kinetics of structure formation and strength development. Report submitted to the RILEM-TC 119.

[45] Sassedateljew, J.B., Mischin, G.W. (1970) Einfluß der Strukturbildung im Beton auf die Wärmeleitung. *Baustoffindustrie*, pp. 116-119.

[46] Schutter, G. de, Taerwe, L. (1994) Influence of the geometry of hardening concrete elements on the early age thermal crack formation. Proc. RILEM Symp. Thermal cracking in concrete at early ages. Munich, 1994, pp. 53-60.

[47] Springenschmid, R. and Breitenbücher, R. (1992) Influence of cement on thermal cracking of concrete at early ages. 9th Int. Conf. on the Chemistry of Cement, New Delhi, Vol. V, 122-128.

[48] Swenson, E.G. (1956) Weather in relation to winter concreting. Proc. RILEM Symp. Winter Concreting, Copenhagen, pp. 3-38.

[49] Taylor, H.F.W. (1989) Modification of the Bogue calculation. *Advances in cement research*, **2**. pp. 73-77.

[50] Tazawa, E. and Iida, K. (1983) Mechanism of thermal stress generation due to hydration heat of concrete. *Trans. Jap. Con. Inst.*, **5**, pp. 119-126.

[51] Tetmayer (1883). *Deutsche Töpfer- und Ziegler Zeitung*, p. 234

[52] Wang, C. and Dilger, C. (1994) Prediction of temperature distribution in hardening concrete. Proc. RILEM Symp. Thermal cracking in concrete at early ages. Munich, 1994, pp. 21.28.

[53] Weigler, H. and Nicolay, J. (1973) Temperaturdehn- und Wärmeleitzahlen von gefügedichtem Leichtbeton. *Beton-Herstellung und Verwendung*, **11**, pp. 486-491.

[54] Yamazaki, M., Harada, H., Tochigi, T. (1994) Low heat Portland cement used for silo foundation mat. Proc. RILEM Symp. Thermal cracking in concrete at early ages. Munich, 1994, pp. 29-36.

[55] Zimbelmann, R. (1987) Frost-Tausalz Widerstand von Beton im Lichte neuerer Erkenntnisse. *Cement & Concrete Research*, **17**, pp. 651-660.

5

Development of Mechanical Behaviour at Early Ages

M. Emborg

5.1 INTRODUCTION

The accuracy of thermal stress analysis depends mainly on how the thermal properties and specially the mechanical properties are described for early ages. It has thus been demonstrated in several reports and articles that criteria, that are purely based on temperature computations, offer only a poor basis for crack prediction in hardening concrete. Therefore, apart from the *thermal properties of young concrete* defining the temperature development (see Chapters 1 and 4), the *influence of restraint and the transient mechanical properties* of the young concrete *should be considered in analyses providing a base for accurate crack predictions*.

In this chapter the following mechanical properties of the young concrete are treated: (a) strength development (b) elastic and time-dependent behaviour at normal and high stress levels (c) thermal expansion and thermal contraction and (d) fracture mechanics behaviour.

The *strength growth* in young concrete has been very extensively studied and modelled in literature. The compressive strength gain and its dependence on amount and type of cement and admixtures, temperature and other curing conditions, specimen size etc. are modelled as dependent on age or on the degree of hydration and various expressions are available, see e.g. Kasai *et al.* [1], ACI-207 [2], Weigler and Karl [3], Freisleben-Hansen and Pedersen [4], Byfors [5], Horden and Reinhardt [6], Jonasson [7], Maatjes and van Breugel [8], Helland [9], and van Beugel [10].

The influence of temperature on strength growth, important for the estimations of cracking risks, is discussed in other chapters in this report and in investigations including those by Kasai *et al.* [11], Sadgrove [12],

Prevention of Thermal Cracking in Concrete at Early Ages. Edited by R. Springenschmid. RILEM Report 15. Published in 1998 by E & FN Spon, 11 New Fetter Lane, London EC4P 4EE, UK. ISBN 0 419 22310 X

Haugland *et al.* [13], Freiesleben-Hansen and Pedersen [4], Byfors [5], Jonasson [7], van Breugel [10], Kjellsen and Detwiler [14] and Jonasson [15]. Among different formulae for the maturity function, the formula based on the Arrhenius concept has shown to give the best agreement with tests, Byfors [5]. Applying the maturity concept on strength development we may note that at the same maturity, independently of the temperature history, the same strength must have been reached. However, at higher temperatures, above 30 a 40°C, this is not always the case due to the losses of final strength values (see Fig 5.1) - a phenomenon that has been observed in various tests and the causes of which still remain rather unclear. Among others, van Breugel [10] discusses causes to the effect of strength losses at higher temperatures and gives a comprehensive survey of references dealing with the phenomenon.

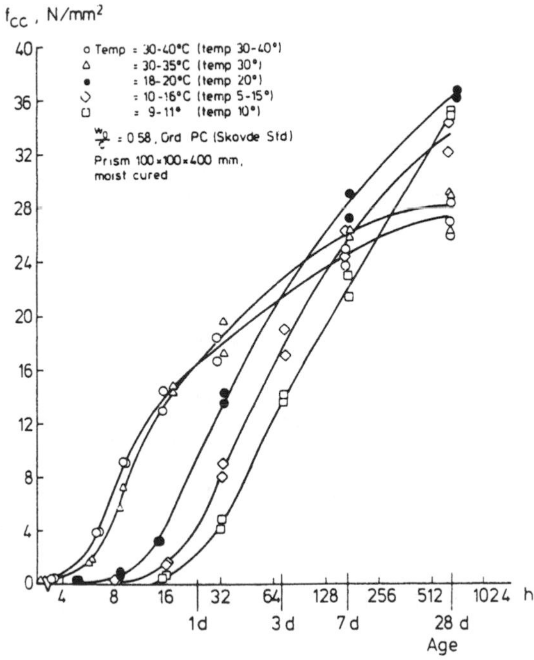

Fig. 5.1 Compressive strength gain at different curing temperatures (from Byfors, [5]).

Regarding early-age *tensile strength* of concrete several authors report test results and theoretical models e.g. Hellman [16], Kasai *et al.* [1], Weigler and Karl [3], Bellander [17] and Byfors [5]. The results presented by Kasai indicate that the tensile strength tends to grow faster than the compressive strength. Similar findings were reported by Byfors.

Concerning the *elastic and time-dependent behaviour* (i.e. the viscoelastic behaviour) *of hardened concrete*, numerous reports and books are found in literature, for instance Neville [18], Neville, Dilger and Brooks [19], Bazant [20], Bazant and Wittman [21], Rostásy and Budelmann [22], and Bazant *et al.* [23].

However, when confining the scope to elastic and time-dependent behaviour at early ages (< 5 days after pouring) and in particular, at very early ages (<1-2 days after pouring), available experimental data and theoretical approaches become more scarce. The fact is that the noticeable age-dependence of the elastic and creep behaviour, see Fig 5.2, is normally modelled and verified by tests for ages at application of load not less than 2-3 days.

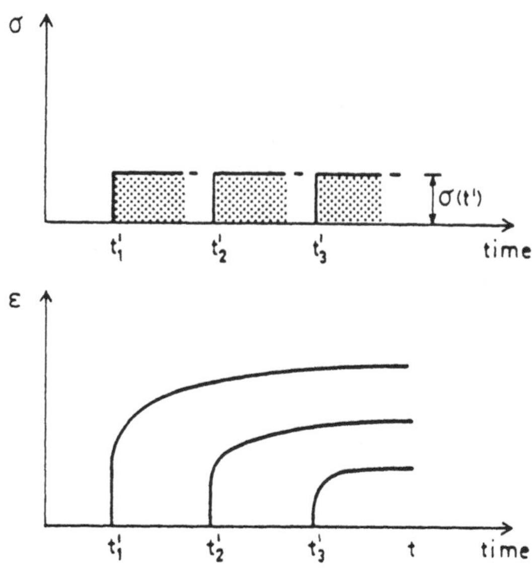

Fig. 5.2 Age-dependence of the elastic and creep behaviour shown schematically (Emborg, [27])

For earlier ages of load application, the scatter in reported creep test data is considerable, see Fig 5.3 which shows the relation between creep behaviour and age at application of load for tests of various investigators. It appears that modelling of creep in very young concrete is difficult and only few attempts are found in literature. For instance, Byfors [5] has proposed a correction formula for early age creep based on results from tests with ages at application of load in the range 8-650 hours. However, Byfors' tests were carried out with concretes of only two different water-cement ratios and few conclusions can be drawn. Another attempt to consider early-age properties were made by van Breugel [24]. He proposed a model where effects of progressing hydration on early-age stress relaxation was incorporated in an existing relaxation formula.

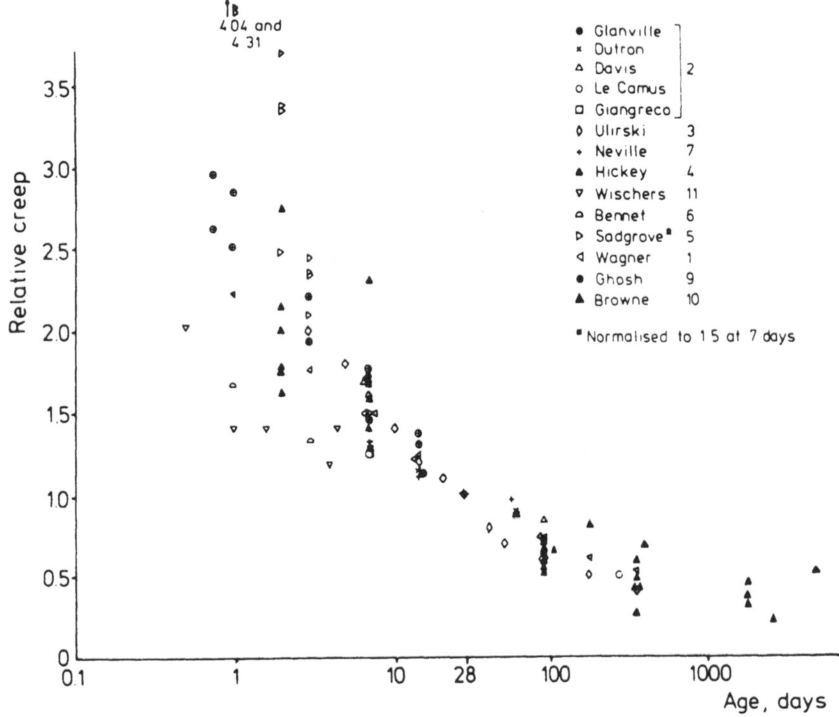

Fig 5.3 Creep at different loading ages relative to creep at 28 days. Result from a compilation made by Parrott [28] (quoted from Byfors, [5]).

In Bazant and Carol [25] some contributions on creep at early ages are reported, among them Tanabe and Ishikawa [26] who present a creep model based on pore water migration.

The temperature dependent *volume change* of concrete (thermal dilatation) is of major importance in the thermal stress analysis. It is shown in Fig 5.4 that in particular the stress decrease in the cooling phase is affected by the magnitude of the thermal dilatation. A concrete with a low coefficient of thermal expansion is thus less prone to early-age thermal cracking. The value of the coefficient of thermal expansion is dependent on many factors as treated in Browne [29], Emanuel and Hulsey [30], Harrison [31], Nolting [32]. The thermal expansion coefficient for hardened concrete may vary in the range of 5-15 microstrain/°C but is often set at 10 microstrain/°C. However, it will be shown in section 5.6 that it is appropriate for young concrete to define the thermal deformations not only by a coefficient of thermal expansion, α_e, but also by a coefficient of thermal contraction, α_c.

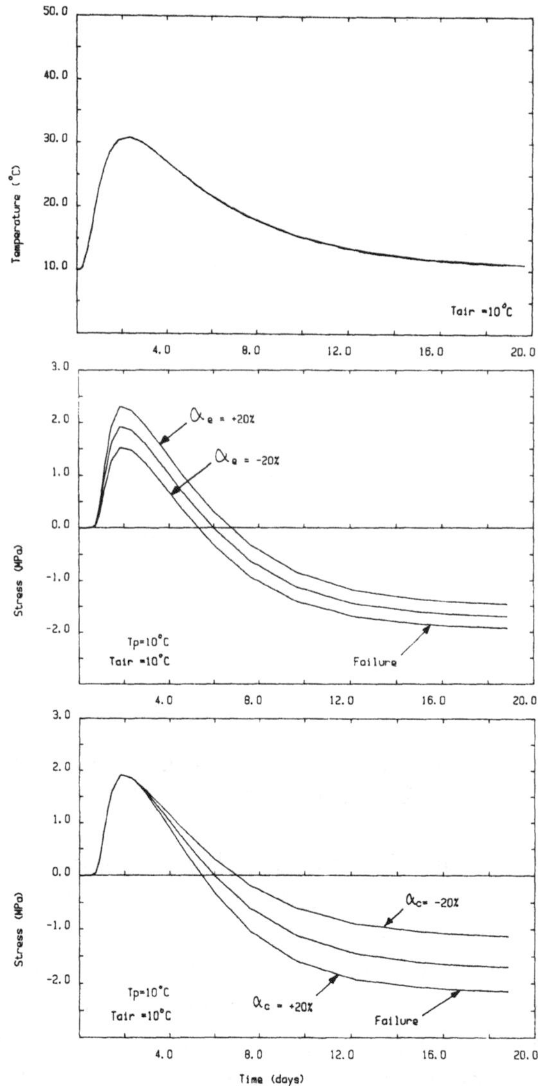

Fig. 5.4 a) Development of mean temperatures in a concrete wall. **b)** Development of thermal stresses at inflexible supports (100% restraint) for various values of thermal expansion (α_e) and contraction coefficients (α_c), see section 5.6. Both temperatures and stresses are computed (Emborg, [27]).

Further, in the thermal stress analysis we must consider *non-linear phenomenon* at high stress levels in tension. These phenomena occur mainly due to the growth of microcracks and may be modelled with the *fracture mechanics approach.* Non-linear effects also occur in compression, both at low and high stress levels.

5.2 COMPRESSIVE STRENGTH

5.2.1 General

The compressive strength is the mechanical property which has been most studied for early ages. It is easy to determine and provide information on other properties e.g. tensile strength, stiffness, durability.

There exist at least five applied concepts at engineering level (macro level) for the description of the development of strength with time (van Breugel, [10]):

- porosity concept
- gel-space ratio concept
- degree of hydration concept
- maturity laws
- chemistry-oriented strength laws

In this section, some examples of formulas based on the maturity laws (equivalent time laws), degree of hydration concept, gel-space ratio concept and porosity concepts are described.

When studying expressions for the strength gain we should note the differences in strength values obtained at various test conditions (size and shape of specimen, rate of loading, humidity conditions at tests and at curing, temperature at test, etc.). For example, cylinders (150 x 300 mm) are used in some countries to establish the compressive strength and in other countries 100 mm, 150 mm and/or 200 mm cubes are used.

Concerning the size effect, the ratio between the strength of 100 and 200 cubes is in the range of 1.05 - 1.20, Neville [18] and Graf *et al.* [33]. For comparisons between cube strength and cylinder strength (diameter/height =1/2), a factor of 0.8 to the cube strength is often applied (CEB/FIP Model Code 90, [36]). According to the Norwegian Standard (NS 3473, [34] - Kanstad, [35]) the cylinder strength may be obtained from the cube strength f_{cc} for normal strength concrete:

$$f'_c = 0.80\ f_{cc} \text{ for } f_{cc} \leq 55 \text{ MPa}$$

$$f'_c = f_{cc} - 11 \text{ MPa for } f_{cc} > 55 \text{ MPa} \quad (5.1)$$

Depending on curing conditions (temperature effects and shrinkage), the strength of structural concrete is often lower than the strength estimated in laboratory. In the Norwegian Standard, the structural strength is expressed from cylinder strength as (Kanstad, [35]):

$$f^*_c = 0.775\ f'_{cc} + 3.1 \text{ MPa for } f_{cc} \leq 55 \text{ MPa}$$

$$f^*_c = 0.840\ f'_{cc} \text{ for } f_{cc} > 55 \text{ MPa} \quad (5.2)$$

In next section, some formulas for the compressive strength at early ages are given.

5.2.2 Mathematical models

As been mentioned earlier, various expressions exist for the compressive strength gain and its dependence on temperature, amount and type of cement and admixtures etc. It is here only possible to describe some of the models for the strength gain found in literature, see Table 5.1.

Alternatively, van Breugel [41, 10] gives a relation of strength to the gel-space ratio, x

$$f = f_o\ x^3 \text{ where } f_o \text{ is the intrinsic strength.} \quad (5.3)$$

This concept was launched by Powers [42]. For mortars the intrinsic strength f_o is between 80 and 342 MPa, Fagerlund [43]. Van Breugel [41] proposes a value of fo = 240 MPa and relates the gel-space ratio to the degree of heat production H_w ($0 < H_w < 1$):

$$f_{cc} = 240\ x^3 \quad ; \ x = \frac{H_w}{0.82 + 0.46\ H_w} \quad (5.4)$$

Fitting of results from expression of Byfors in Table 5.1 to recorded strength gain of modern Swedish concretes (Fagerlund, [44]), shows that the expression is not sufficient accurate for the whole time of interest i.e. for $1 \leq t_e \leq 28$ days.

Table 5.1 Examples of equations for the compressive strength gain

Refe-rence	$\beta_{cc}(t) = \frac{f_{cc}(t)}{f_{cc}(28d)}$ = *relative compressive strength (cylinder, cube)*	*Coefficients etc.*	*Comments*
CEB/FIP MC90 [36]	$\exp[s(1 - 5.3/t^{0.5})]$	s=0.20 rapid hard. cement =0.25 normal hard. cement = 0.38 slowly hard. cement	
Hamfler [37]	$b\left(\frac{\alpha(t)/\alpha(28d) - a}{1 - a}\right)$	$\alpha(t)$, $\alpha(28d)$ degree of hydration a=0.15-0.17 for w_o/c=0.4-0.58 b=1.0; α=0.75, b<1;a<0.25	
Wesche [38]	$a\,e^{b}$ with $b = c_1 \frac{w_0}{c}(t)^{d}$	$d \approx -0.55$ for a and c_1, see Table 5.2	
Byfors [5]	$\frac{a_1\, t^{b_1}}{1 + \frac{a_1}{a_2} t^{(b_1 - b_2)}}$	a_1, a_2, b_1, b_2, see Table 5.3	Coefficients by least-square fitting (Jonasson, [7])
Carino [39]	$\frac{k(t - t_0)}{1 + k(t - t_0)}$ with $k = k_m(T - T_0)$	k_m=0.81, T_o = 0°C t_o=B/T; B=6.9°C, T=temperature	Hyperbolic function
Laube [40]	$\frac{f'_c(\alpha)}{f'_c(\alpha = 1)} = 0.85\left[\frac{rel \cdot f_{ct}(\alpha)}{c_D \cdot f_{ct}(\alpha = 1)}\right]^{3/2}$	rel f_{ct}, f_{ct} see Table 5.4 c_D=0.236 $(N/mm^2)^{1/3}$	Dependent on chosen expression for $\alpha(t)$.

Table 5.2 Coefficient for use in Equation of Wesche [38] for different grades of concrete and cement types (cylinder strength).

Cement type	*a*			*c*
w_o/c	*0.35*	*0.55*	*0.75*	
Z55 , Z45F	1.10	1.15	1.20	-1.50
Z45L, Z35F	1.20	1.30	1.40	-2.80
Z35L	1.30	1.50	1.70	-4.40
Z25	1.50	1.90	2.30	-7.10

Table 5.3 Coefficients for use in expression by Byfors [5], Table 5.1, for different grades of concrete and cement types. (Cube strengths, from Jonasson, [7]).

Grade of concrete [MPa]	*Cement*	a_1	a_2	b_1	b_2
K 20	OPC	0.00220	0.684	1.48	0.0629
K 25	"	0.00220	0.516	1.54	0.105
K 30	"	0.00220	0.449	1.61	0.125
K 35	"	0.00141	0.379	1.80	0.15
K 40	"	0.000932	0.354	2.0	0.16
K 45	"	0.000620	0.331	2.2	0.17
K 50	"	0.000437	0.331	2.4	0.17
K 55	"	0.000500	0.353	2.4	0.16
K 60	"	0.000426	0.402	2.5	0.14
K 20	RHPC	0.00178	0.434	1.65	0.13
K 25	"	0.00213	0.433	1.65	0.13
K 30	"	0.00115	0.342	1.95	0.165
K 35	"	0.00124	0.331	2.00	0.17
K 40	"	0.00136	0.331	2.0	0.17
K 45	"	0.000812	0.353	2.25	0.16
K 50	"	0.00104	0.365	2.25	0.155
K 55	"	0.00148	0.429	2.2	0.13
K 60	"	0.000467	0.458	2.7	0.12
K 20	OPC + 1.5% $CaCl_2$	0.0017512	0.67740	1.48	0.0629
K 25	"	0.0017351	0.50772	1.54	0.105
K 30	"	0.0017165	0.44043	1.61	0.125
K 35	"	0.0010684	0.37034	1.8	0.15
K 40	"	0.0006847	0.34538	2.0	0.16
K 45	"	0.0004417	0.32244	2.2	0.17
K 50	"	0.0003019	0.32244	2.4	0.17
K 55	"	0.0003454	0.34440	2.4	0.16
K 60	"	0.0002898	0.39342	2.5	0.14

This may be solved by calibrating the parameters of the equation to discrete parts of the gain of equivalent age. Instead, it has been found to be more convenient to use discrete values of the strength and to interpolate intermediate values for linear relations in logarithmic time scale, Fig 5.5. Evaluations of trend curves according to the figure are performed with use of knowledge on strength gain for different cement types, concrete qualities and ages. For example, influence of concrete quality (water cement ratio) on rate of strength gain may be obtained from well known diagrams such as Figs 5.6 and 5.7. Thus, only a small amount of test result is needed in the evaluation of a "Compressive Strength Gain

Spectra" for each cement type, see Fig 5.8, which may be a base for the trend curves such as the ones in Fig 5.5.

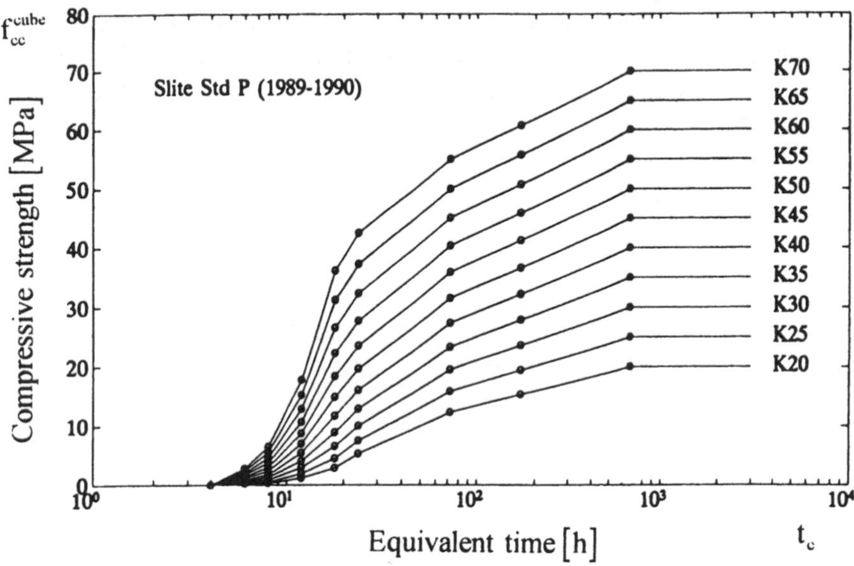

Fig. 5.5 Compressive strength gain for different grades of Swedish concretes expressed as linear relations in logarithmic time scale (based on data from Jonasson, [15]).

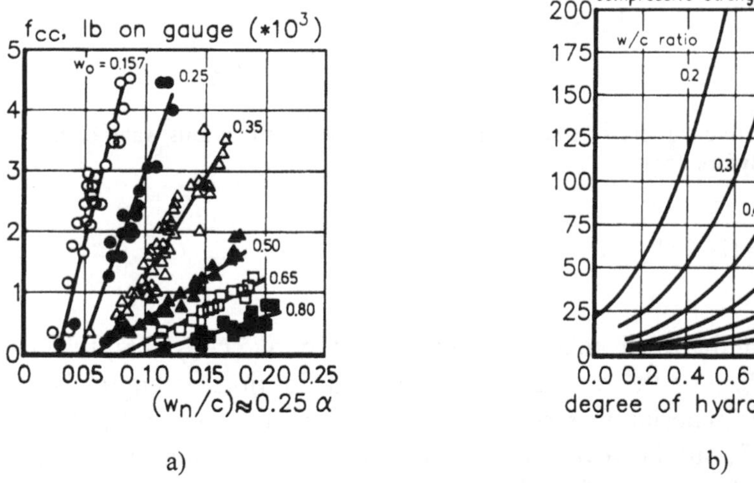

Fig 5.6 Relations between compressive strength in concrete and degree of hydration at a very early age (a) based on data by Taplin [45] and (b) according to Locher [46] (from van Breugel, [10]).

Simple relations between strength and porosity and between strength and hydration state have been derived by Fagerlund [43]. Analysis of experimental data by Taplin [45] and Powers & Brownyard [42] show, according to Fagerlund, that linear relations are valid after a certain critical hydration or a certain decrease in porosity, P, has occurred. With the basic formula of a porous material, the strength f is proportional to the load carrying area A_{solid}:

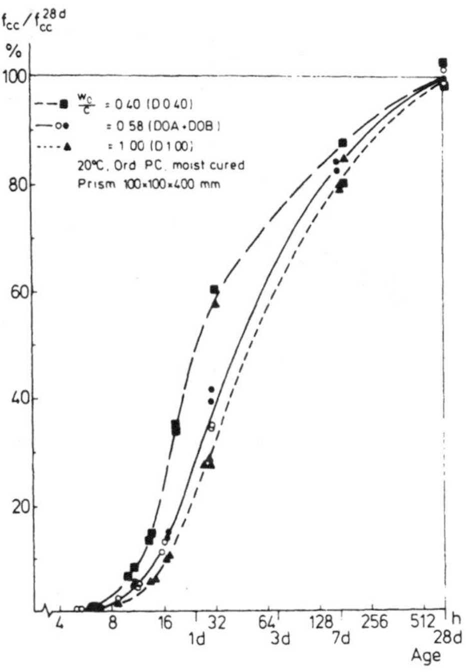

Fig 5.7 Relative compressive strength gain in concrete with various water cement ratio $w_{o/c}$ (from Byfors, [5]).

$$f = f_0 \frac{A_{solid}}{A_{total}} = (1 - P / P_{cr}) \tag{5.5}$$

We see that the linear relation is determined by two constants; the critical porosity P_{cr} (or the critical hydration) and fictitious strength fo of the non-porous material. Fagerlund gives equations for the porosity P as well as P_{cr} and f_o, but, however, states that further investigations are needed for more reliable relations for different concrete mixes, cement types etc.

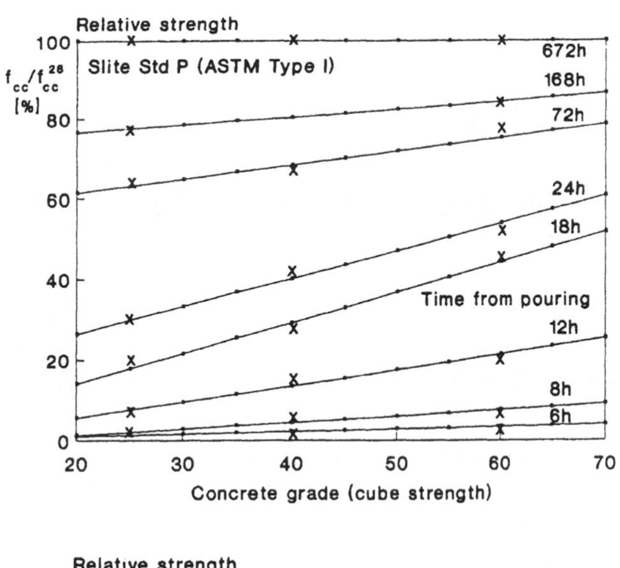

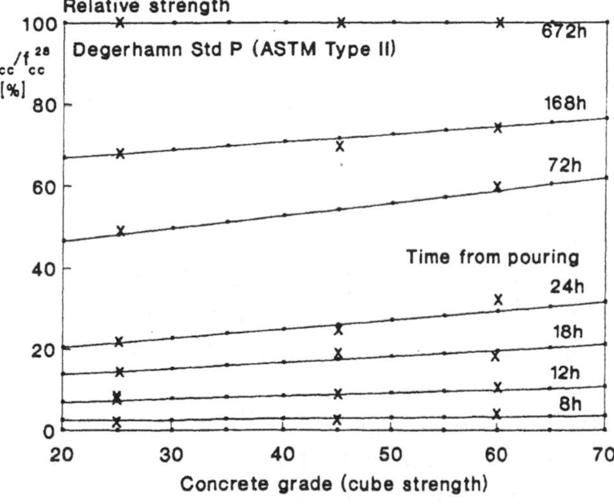

Fig 5.8 Relative compressive strength gain as a function of concrete quality - "Compressive Strength Gain Spectra" for two Swedish cement types. Values obtained at tests (x) and interpolated/extrapolated in the diagram (•). (Based on Fagerlund, [44] and Jonasson, [15]).

5.2.3 Temperature effects

Different equations for the equivalent age, t_e, have been proposed in literature, see also chapter 4. The concept of activity energy according to the Arrhenius function has shown to give the best agreement with test results in several investigations, see e.g. Freiesleben-Hansen, Pedersen [4] and Byfors [5]:

$$t_e = \int_0^t \beta_t \, dt \quad ; \quad \beta_t = \exp\left[\frac{U_h}{R}\left(\frac{1}{293} - \frac{1}{T_K}\right)\right] \tag{5.6}$$

where the activating energy of hydration U_h may be expressed as (Freiesleben-Hansen, Pedersen, [4])

$$\begin{aligned} U_h(T_K) &= 33.5 \text{ kJ/mol} && ; T_K \geq 293 \text{ K} \\ U_h(T_K) &= 33.5 + 1.47\,(293 - T_K) \text{ kJ/mol} && ; T_K < 293 \text{ K} \end{aligned} \tag{5.7}$$

or alternatively according to Jonasson [7]

$$\frac{U_h}{R} = Q_{ref}\left(\frac{30}{T_K - 263}\right)^{\kappa_3} \tag{5.8}$$

where κ_3 = 0.59 and Q_{ref} = 4600 K for (Swedish) Standard Portland Cement.

It seems that the expressions for the equivalent age and compressive strength work satisfactory for up to 50% of the 28-day strength (about 2 day of equivalent age). After the 50% level in strength have been reached the strength is influenced so that higher temperatures lead to lower strengths and vice versa. A higher strength at lower curing temperatures implies no problem, but the loss of final strength at higher temperatures could be of such magnitude that it should be considered in thermal cracking analysis and, in fact, generally in design.

A number of investigations indicate that the reduction of final strength due to high curing temperatures is present already at temperatures of 30-35°C, see e.g. Fig 5.9. In thermal stress analysis such high temperatures (above 35°C) almost always are present and for accurate knowledge of strength at early ages it is thus necessary to adjust the strength development with regard to the curing temperatures. It is here observed that for high strength concrete (strengths over 60-70 MPa), the temperature effect on final strength is very small or not present at all, as recently reported by some researchers.

Van Breugel [47] states that the origin of this phenomenon could be explained on a microstructural basis. High temperatures result in a denser packing of the hydration products which give an increase of the capillary porosity. This reduces the final strength. Besides a lower strength and a corresponding increase of the possibility of cracking, the higher porosity according to above implies a more permeable concrete thus reducing the durability of the concrete. Therefore, as concluded by van Breugel, the positive effect of a lower casting temperature and hence lower maximum temperature in a structural element is threefold: a small temperature decrease, a high strength and a low permeability. This temperature effect on final strength has been simulated by the program HYMOSTRUC developed by van Breugel.

The effect of temperature on the final strength is also discussed in Jonasson [15], Kjellsen and Detwiler [14], Emborg [27] and Laube [40]. In Appendix A one suggestion of modelling this effect is described.

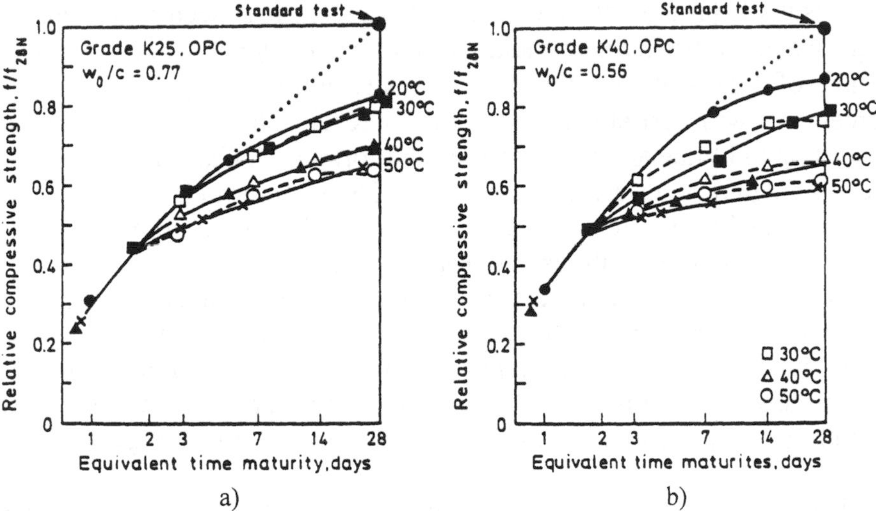

Fig. 5.9 Measured strength for concrete of grade K25 (a) and for grade K40 (b) at different curing temperatures, 150 mm cubes, tested wet and warm. Type I Cement (Slite). (From Jonasson, [48]).

5.2.4 Influence of rate of loading

The effect of the load rate on the compressive strength may be significant for low rates of loading due to creep phenomenon during the test. It may be assumed that this effect is similar for compressive strength and tensile strength and the effect is further described in section 5.3.4.

5.2.5 Concluding remarks - comparisons of models

In the section, only some examples of the various formulas for strength development found in literature are given. Almost every national standard include an own expression for the strength growth and a large amount of other formulas are used in research which all of them could not be listed here.

It is, of course, of interest to know how well the expressions in section 5.2 describe strength growth at a very early ages (t >1-1.5 day), at early ages from (day 1 to about 7 days) and at later ages. Fig 5.10 shows an example of an implementation of some formulas of Table 5.1 for two Swedish grades of concrete: K 25 and K 45 (cube strengths, tested and cured wet).

Results from laboratory tests are also given. It is seen that there is a large scatter between the formulas, specially at early ages. For example, at an age of 2 days, the scatter in relative strength is nearly 0.25! The best agreements with test at early ages are for K 25 obtained with expression according to Laube and for K 45 with the formula of Carino. At later ages the best agreements are obtained for K25 with formulas of CEB/FIP, Byfors, and Carino. For later ages of K 45 rather good conformity with tests are obtained with almost all proposed formulas.

However, it is very important to note that this comparison is carried for only two concrete qualities with a certain cement type. Other concrete mixes surely imply quite different agreements between formulas and tests. So, in the scope of this report, we cannot propose any general recommendations on which formula that should be used for strength gain at early ages.

But one *most essential* conclusion may be drawn: *We must always be very careful when using compressive strength gain formulas.* We cannot trust in every sophisticated expression for influence of cement type, water-cement ratio etc. Therefore, during design, we must be aware of the sensibility of the used formula and hence the consequences on tensile strength and thermal cracking. After the design phase it is therefore most advisable to, if it is possible, for each type of concrete mix of interest, perform a set of computations and comparisons to a limited test series to the ones in Fig 5.8 thus evaluating "Strength Gain Spectra". If necessary, coefficients in formulas used may be obtained by e.g. least-square optimisation.

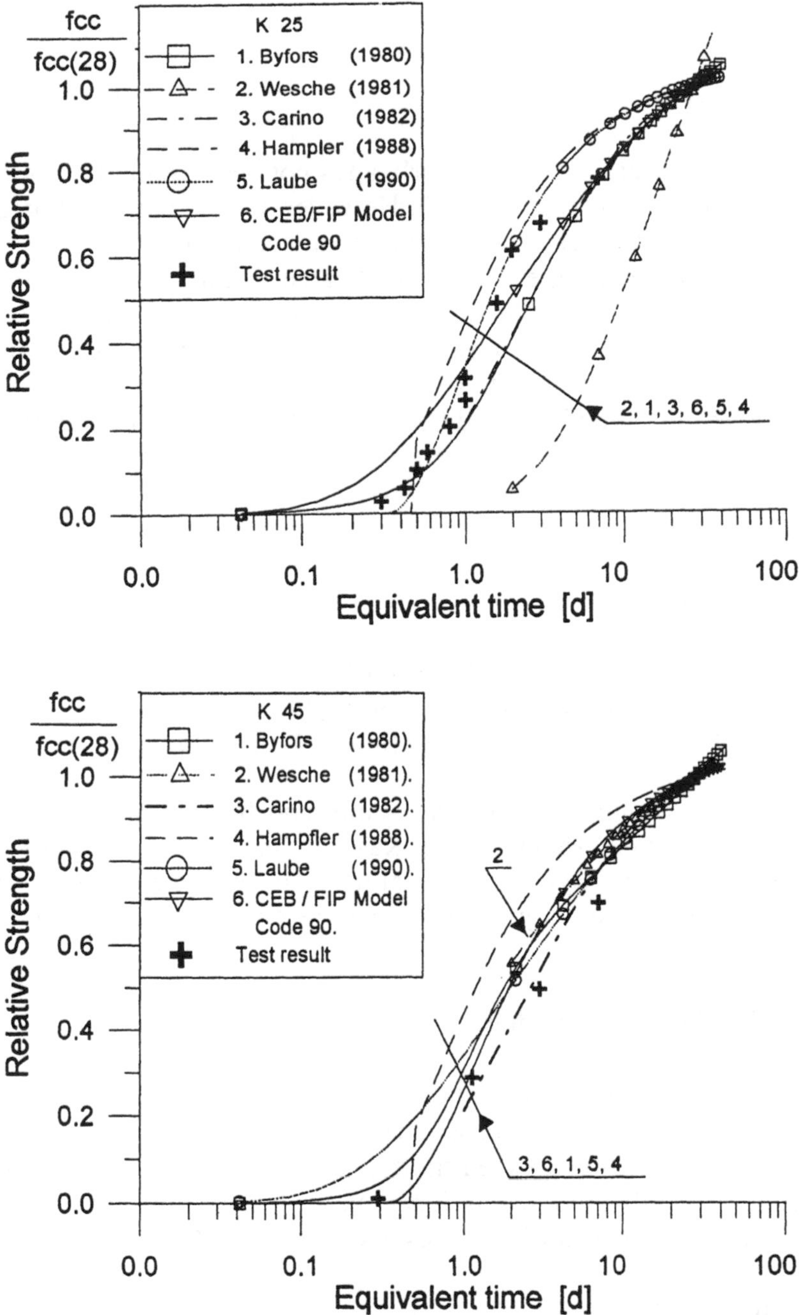

Fig 5.10 Example of relative compressive strength growth according to Table 5.1 as a comparison with cube strength of two grades of concrete. Cement: Standard Portland, cement content K 25: 248 kg/m³ (w_o/c =0,78), K 45: 425 kg/m³ (w_o/c =0,40), T=20°C.

5.3 TENSILE STRENGTH

5.3.1 General

The tensile strength has for early ages not been examined to the same extent as the compressive strength. However, it is an important parameter when assessing the risk of cracking in young concrete.

As uniaxial tensile tests are complicated to carry out, other methods for determining the tensile strength are used such as splitting test and flexural test. The uniaxial tensile strength may be related to the splitting strength and flexural strength as follows (according to Swedish Concrete Handbook [49] - Eq (5.9a), and CEB/FIP MC 90, [36]) - Eq (5.9b)

$$f_{ct} = \frac{2}{3} f_{ct,spl} \tag{5.9a}$$

$$f_{ct} = f_{ct,fl}\, 0.08\, d^{0.7} / (1 + 0.08\, d^{0.7}) \tag{5.9b}$$

where $f_{ct,spl}$ is the splitting tensile strength
$f_{ct,fl}$ is the flexural tensile strength
d is the depth of beam [mm] ($d > 50$ mm)

Several investigations report test results and theoretical models for the tensile strength e.g. Hellman [16], Kasai *et al.* [1], Weigler and Karl [3], Bellander [17] and Byfors [5]. From many of the tests it is seen that the tensile strength tends to grow faster than the compressive strength.

5.3.2 Mathematical models

Some relations between uniaxial tensile strength and compressive strength, found in literature, are listed in Table 5.4.

Further, a relation between splitting strength and compressive strength of cubes, recently proposed by Jonasson [15], can be mentioned by which quite good agreements with test results were obtained

$$f_{spl} = 0.18\, (f_{cc})^{0.75} \tag{5.10}$$

We may note that neither of the formulas in Table 6.4 nor Eq (5.10) consider the influence of aggregate on the tensile compressive strengths relation. Observations have namely shown that the aggregate (type, form etc.) has a marked influence on the strength in tension due to e.g. bond. This calls for future refinements of tensile strength equations.

5.3.3 Temperature effects

No information has been found in literature about temperature effects on tensile strength (final strength, strength at tests etc.) and the same influence on tensile strength as on compressive strength may be assumed.

5.3.4 Influence of rate of loading

Wittman *et al.* [50] have studied the effects of rate of loading, age of loading, w/c-ratio and of other parameters on the behaviour of concrete beams at three-point bending mode tests. Although the flexural strength behaviour of bending tests cannot directly be compared to the behaviour of uniaxial tension tests, it is nevertheless assumed that the peak tensile strength is the same for the two types of tests. Thus, the effect of load rate on uniaxial tensile strength may be obtained from the bending tests.

In Wittman *et al.* [50] the influence of rate of loading on tensile strength was studied for six different loading rates and for two w/c ratios of concrete as well as two types of beam geometry, see Fig 5.11. For lower rates of loading (less than about 0.001 mm/min), tensile strength is reduced markedly, see the figure. The dependence on loading rate was approximated with an empirical expression:

$$f_{ct} = d_1 \, \dot{\delta}^{d_2} \tag{5.11}$$

in which $d_1 = 5.64$ $d_2 = 0.048$ for w/c = 0.40 and $d_1 = 4.01$ $d_2 = 0.023$ for w/c = 0.65. $\dot{\delta}$ (mm/min) is the rate of deflection of the beam of the test.

If the strength at $\dot{\delta} = 0.1$ mm/min is considered as a "normal" strength value a reference is obtained to which the strengths at other loading rates may be related, (Emborg, [27]). For the low loading rates, often present during the cooling of concrete at a temperature cycle, it has been found that the reduction in tensile strength is about 0.7. This reduction leads to fairly good agreements between computed tensile strengths and failure stresses in laboratory tests where thermal stresses and cracking are examined.

Table 5.4 Examples of equations for tensile strength at early ages

Reference	*Relation*	*Coefficients etc.*	*Comments*
Byfors [5]	$f_{ct}(t) = 0.115 \cdot f_{cc}^*(t) - 0.022$ [MPa] $f_{cc}^*(t) \leq 20$ MPa $f_{ct}(t) = 0.105 \cdot (f_{cc}^*(t) - 20)^{0.839} + 2.28)$ [MPa] $f_{cc}^*(t) > 20$ MPa		f_{cc}^* - strength of prisms 100 . 100 . 400 mm (≈ cylinder strength)
CEB/FIP [51]	$f_{ct} = 0.30 \cdot f_{ck}^{2/3}$ $f_{ck} = f_{cm} - 8.0$ MPa		f_{cm} - mean value cylinder strength f_{ck} - characteristic cylinder strength
CEB/FIP [36]	$f_{ct,\,min} = 0.20 \cdot f_{ck}^{2/3}$ $f_{ct,\,max} = 0.40 \cdot f_{ck}^{2/3}$		upper and lower bond values
Weigler and Karl [3]	$f_{ct} = 0.100 \cdot f_c'$		f'_c = cylinder strength
DIN 1045 [52]	$f_{ct} = 0.24\,(f_{cc})^{2/3}$	$f_{cc} = 1.20\, f'_c$	f_{cc} - cube strength
ACI [53]	$f_t = 6\,[f'_c]^{0.5}$		f'_c - cylinder strength in psi (6.895 KPa)
Oloukun *et al.* [54]	$f_t = 0.584\,[f'_c]^{0.79}$; $f'_c < 1000$ psi $f_t = 0.928\,[f'_c]^{0.6}$; $f'_c > 1000$ psi		both normal and high strength concrete, f'_c in psi
Laube [40]	$\text{rel}\, f_{ct}\; f_{ct}(\alpha) = \dfrac{f_{ct}(\alpha)}{f_{ct}(\alpha=1)} = (a_z + 1)\alpha_e(t) - a_z$	$a_z = 0.21$	*) see below the table

*) Some difficulties may occur when determining f_{ct} (α=1) as for many concrete mixes (low water-cement ratio etc.) the degree of hydration equal to unity is not reached. The value f_{ct} (α=1) is in these cases evaluated by extrapolation in the f_{ct} - α relation. Further, the relative tensile strength is strongly dependent on which expression for the degree of hydration versus time that is used. In Fig 5.12 an expression for α according to Jonasson [7] was entered in the expression of Laube.

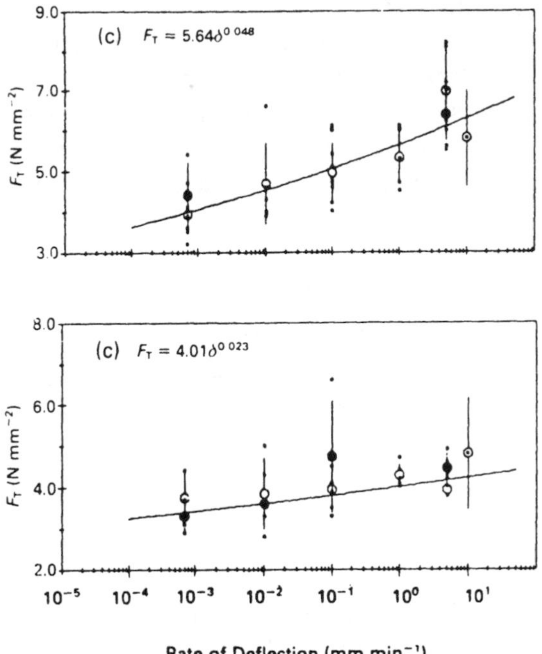

Fig 5.11 Tensile strength, determined directly from load-deflection diagrams by means of fits with FEM-calculations, as a function of rate of loading. Three-point bending mode test **(a)** w/c = 0.40 **(b)** w/c = 0.65. Full circles represent data of large beams (span = 1150 mm) and hollow circle those of small beams (span = 800 mm). (From Wittman *et al.*, [50]).

5.3.5 Concluding remarks - comparisons of models

Similarly to compressive strength, we may compare the formulas for the uniaxial tensile strength growth, Fig 5.12, where a large scatter between the expressions is observed. For example, for both concrete qualities, the computed tensile strength at 10 MPa compressive strength varies between 0.25-1.25 MPa. Hence, it is concluded that it is advisable to be *very careful when using formulas for tensile strength gain* found in literature and codes. The expression chosen should always be calibrated to actual concrete mixture.

Further, it is noted that other concrete qualities than in Fig 5.12 (other type of binder, aggregate etc.) may lead to different relationships between the formulas which may be examined in more comprehensive comparisons, however beyond the scope of this report.

It is also observed that when comparing with tests (that not have been done in Fig 5.12) the experiences are that conditions at uniaxial tests (loading rate, eccentricity and stiffness of loading machine, curing conditions etc.) have a strong effect on the strength recorded and may vary widely between test-setups. Hence, tests at different laboratories may give various results for the same concrete mix and experimental values found in literature may therefore sometimes be taken with some suspicion.

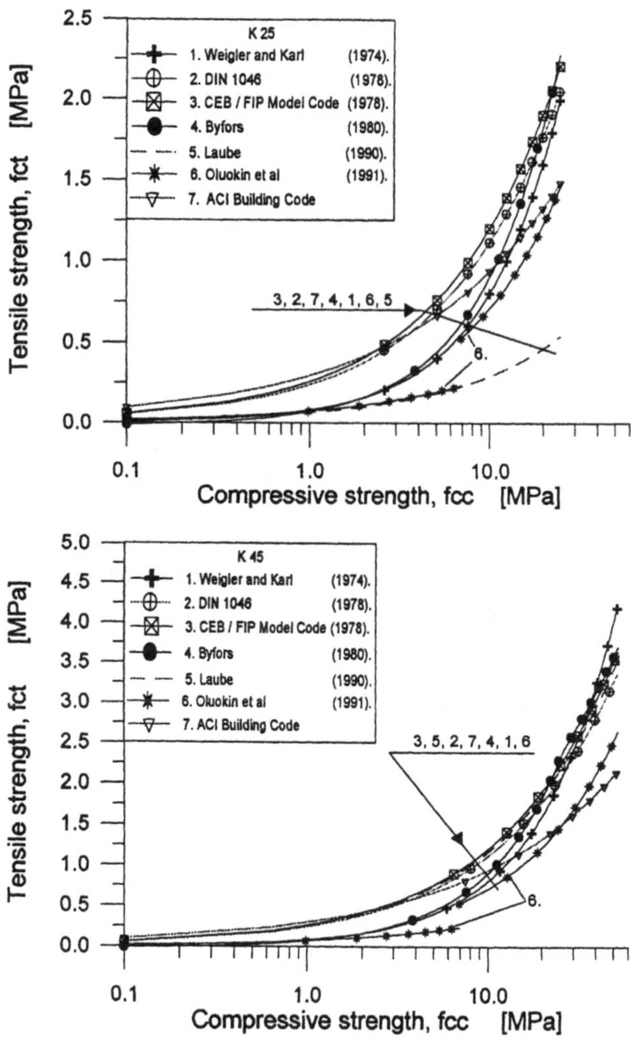

Fig 5.12 Uniaxial tensile strength versus compressive strength according to Table 5.4. Concrete mixes, see Fig 5.10.

5.4 VISCOELASTIC BEHAVIOUR OF EARLY-AGE CONCRETE

5.4.1 General

Three fundamental types of deformations may occur when concrete is subject to loading: elastic, plastic and viscous deformation, see Fig 5.13. Combinations of these types of deformations often are present such as visco-elastic and elasto-plastic deformation. The deformations are usually referred to an uncracked material. However, cracking occurs at high tensile stress levels leading to large deformations additive to the ones above which are related to the fracture mechanics behaviour described in next section.

The viscoelastic behaviour of the young concrete is the most important parameter in the thermal stress analysis. The validity and reliability of such an analysis depends mainly on how well the viscoelastic model used describe the real behaviour of the young concrete. Hence, a rather detailed description of the viscoelastic behaviour will be given here, both for the early ages and for hardened concrete.

The viscous flow component, see Fig 5.13, is irrecoverable and may be defined as the time-dependent deformation occurring for normal working stress levels. The viscous flow component as well as the elastic and delayed elastic deformations may be modelled with the principle of superposition. However, it seems that there is some change in the microstructure of the cement paste while under load affecting the behaviour of the concrete. Thus, the superposition principle has been shown to overestimate creep recovery at unloading and creep strain at further loading see Fig 5.14. It appears that the irrecoverable time-dependent deformation is not purely viscous but also includes a plastic flow component depending on this change of the cement paste (Neville, Dilger and Brooks, [19]).

In engineering practice creep is mainly used to denote time-dependent deformations under load, e.g. the viscous flow and the delayed elastic deformation in Fig 5.13. The creep phenomenon is very complicated and is not yet fully understood and thus a number of theories of creep mechanism have been proposed, see e.g. Neville *et al.* [19].

Factors influencing creep and elasticity may be subdivided into:

- intrinsic factors: design strength, type of cement, elastic modulus of aggregate, fractions of aggregate in the mix
- extensive factors: age at loading, load duration, type of loading (tension or compression), temperature, size of specimen

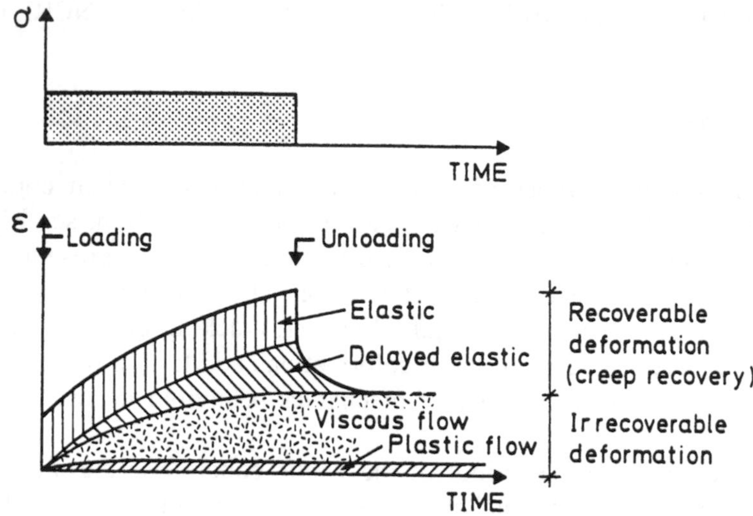

Fig. 5.13 Typical strain-time curves showing fundamental types of deformation at loading and unloading.

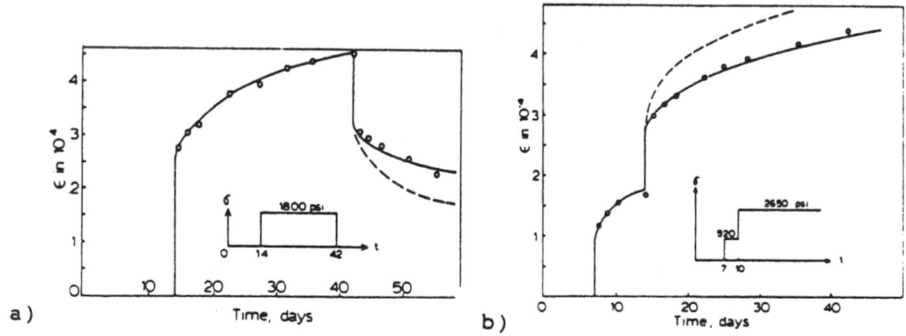

Fig. 5.14 Deviations from the superposition principle due to low-stress nonlinearity when a loaded concrete is further loaded: **a)** creep recovery **b)** stepwise loading (from Bazant and Kim, [64]).

For the analysis of thermal stresses, the influence of age at application of load on the viscoelastic behaviour, Figs 5.2 and 5.3, is one of the most important extensive factor.

The phenomenon of ageing of a concrete under load implies several complications for the modelling of concrete creep. The progress of the hydration in a loaded concrete means that cement gel is continuously formed in a stress-free state but will gradually carry the load, see Fig 5.15 (Bazant, [20]). This means that the stress in the gel that was originally loaded decreases (see the figure) and hence also the rate of creep. Compare for instance the creep strain of a hydrating specimen with the creep of a specimen that not hydrate after the load have been applied, see Fig 5.16, (van Breugel, [24]).

The physical nature of creep at early ages may be studied in Neville *et al.* [19] but also in Tanabe and Ishikawa [26] who point out the role of pure water migration. At early ages, a strong effect of pure water pressure on the creep behaviour is present according to Tanabe and Ishikawa.

Regarding modelling of viscoelastic response at early ages, several approaches have been suggested. Byfors [5], Bernander and Gustafsson [55], Wierig [56], Emborg [27], Laube [40], Acker and Eymard [57], Tanabe and Ishikawa [26], Rostásy *et al.* [58], Dameros *et al.* [59], Ohshita *et al.* [60], Serra *et al.* [61] include some examples of creep models. Some are treated in section 5.4.3.

At the RILEM Conference on thermal cracking in concrete at early ages, several authors presented test results and models for viscoelastic behaviour see e.g. Paulini and Gratl, [62], Lokhorst and van Breugel [63], Umehara, Uehara, Isaka and Sugiyama [65], Westman [66], Morimoto and Koyanagi [67], Guénot, Torrenti and Laplante [68], Gutsch and Rostásy [69].

The temperature has a considerable effect on creep in thermal stress analysis. A higher temperature tends to accelerate creep (i.e. increase the creep rate) but will also indirectly reduce the creep as a result of accelerated hydration - an effect which is large at early ages. Several authors report on effects of temperature on creep e.g. Wierig [70] who has studied effects of heat treatment of concrete.

The curve shown in Fig 5.13 is valid for moderate stress levels. In case of high stress levels creep and elastic deformations increase due to nonlinear effects, Fig 5.17. A linear relation between creep in compression and applied stress have been observed for stresses up to 0.3 - 0.75 of strength of the concrete, Neville *et al.* [19]. However, at early ages, nonlinear effects on creep are probably stronger, especially in case of tensile stresses. Regarding nonlinear effects on creep in compression, the observed phenomenon with creep recovery due to plastic flow, mentioned above, is often denoted the low-stress nonlinearity. The phenomenon has been observed in various deviations from the principle of superposition, see Bazant and Kim [64] and Neville *et al.* [19]. The phenomena with nonlinear behaviour at high tensile stresses and at creep recovery are present in a concrete subjected to early age thermal loading and should be considered in an accurate thermal stress analysis.

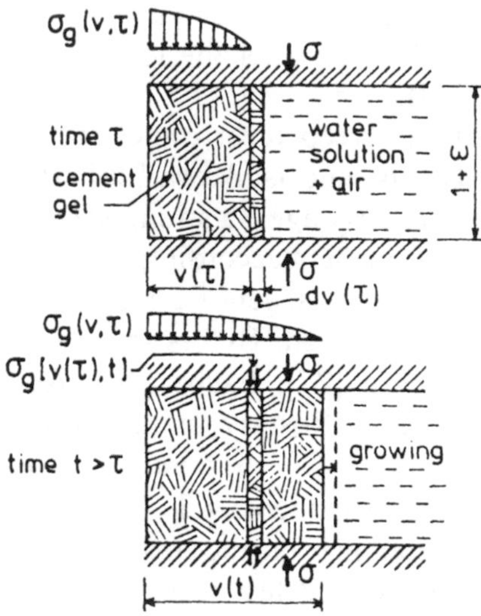

Fig 5.15 Parallel coupling model for describing creep in a hydrating concrete - a solidifying porous material according to Bazant [20] (from Byfors, [5])

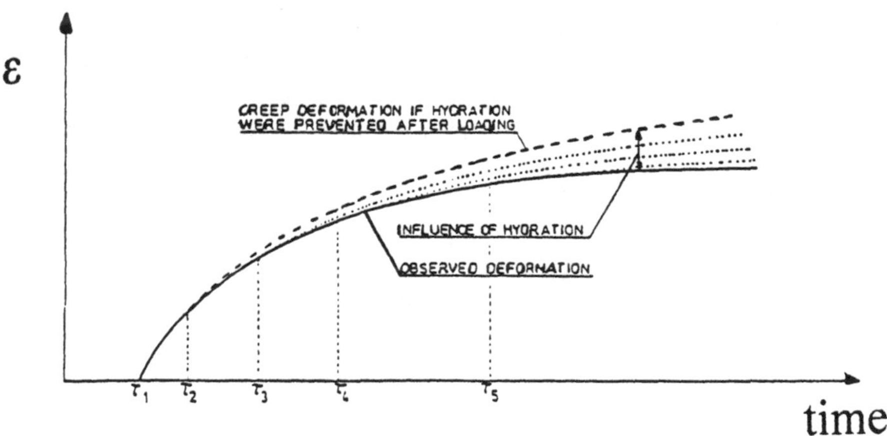

Fig 5.16 Creep deformation of hydrating concrete compared to creep of non-hydrating concrete, van Breugel [24].

Further, in the study of thermal stresses it is convenient to distinguish between basic creep and drying creep. In the mathematical models presented here, effects of drying (i.e. the Pickett effect) are not considered which may be a good approximation for early age concrete of medium-size and massive structures.

Finally, it is observed that, most of theoretical studies on thermal stress analysis have used creep properties for the modelling. This is because of the lack of data on stress relaxation at early ages. Due to the nature of restrained thermal dilatation and thermal stresses it is however considered more appropriate to employ relaxation function directly which has been obtained from relaxation tests. Some researchers have however studied relaxation experimentally e.g. Rostásy, Gutsch and Laube [58] and Morimoto and Koyanagi [67]. The latter authors reported results on both compressive and tensile stress relaxation, see Fig. 5.18. It is seen that tensile relaxation is smaller and terminates in a shorter period than the compressive relaxation. These results indicate the importance of more intensive studies of relaxation in future research.

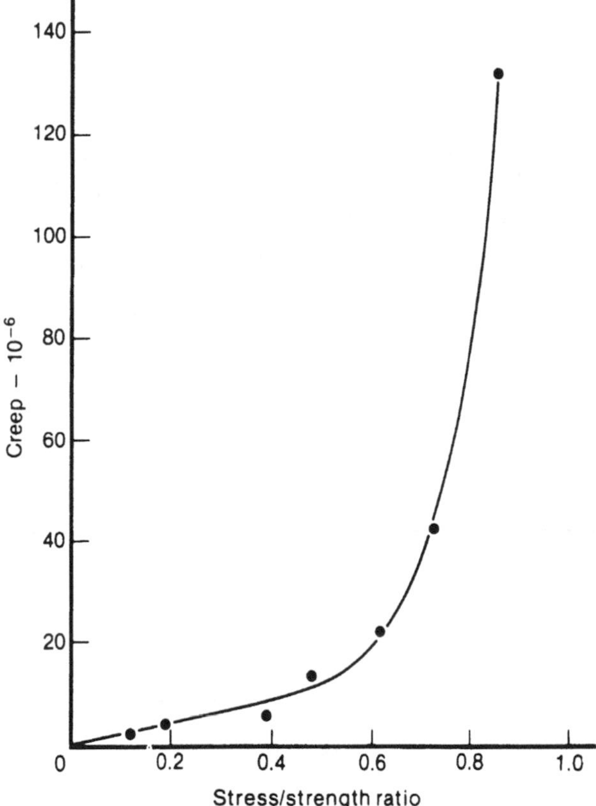

Fig. 5.17 Relation between creep after one minute under load to the stress/strength ratio for a five days old concrete based on results according to Jones and Richard (from Neville *et al.*, [19]).

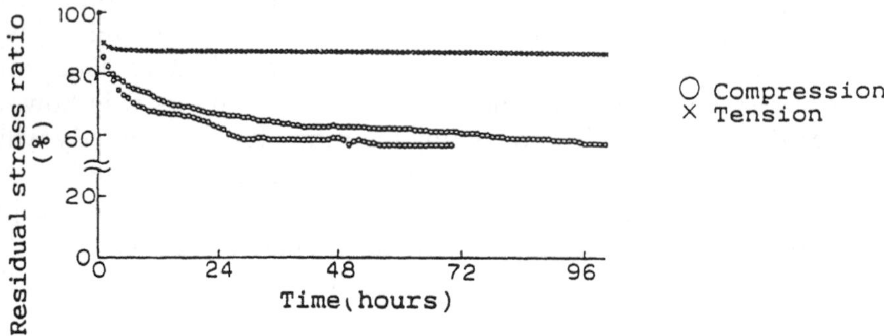

Fig. 5.18 Comparison of compressive and tensile relaxation (from Morimoto and Koyanagi [67]).

5.4.2 Constitutive equations

The viscoelastic behaviour of concrete may be introduced by a discussion of rheological models, among others the commonly used spring and dashpot models according to Kelvin-Voigt and Maxwell and the ones obtained when they are coupled in series or in parallel, see e.g. Fig 5.19.

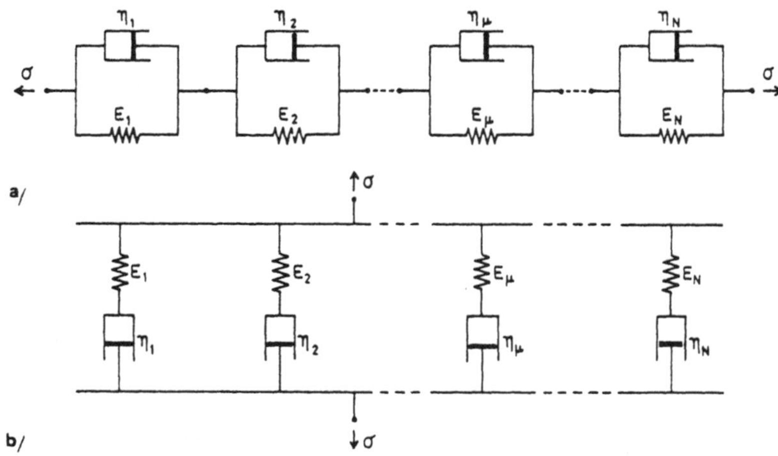

Fig. 5.19 Generalised spring and dashpot models: **(a)** Kelvin-Voigt Chain Model and **(b)** Maxwell Chain Models.

In this section, two ways of modelling creep and elasticity are discussed: the creep coefficient formulation and the creep compliance formulation. Both formulations may in some cases be related to these spring and dashpot models. Further, modelling of the viscoelastic response at varying stress history, such as the thermal stresses, is treated in the section. The section serve as a base for the discussion of methods and models for computation of thermal stresses in chapter 7.

Creep coefficient formulation

Use of the creep coefficient implies that the deformation at time t induced by a loading at time t' is subdivided into two parts: the instantaneous deformation and the creep. The total deformation can then be written

$$\varepsilon_{tot}(t,t') = \frac{\sigma(t')}{E(t')} + \varphi(t,t')\frac{\sigma(t')}{E(t')} \qquad (5.12)$$

where $\sigma(t')$ is the stress applied at time t', (Fig 5.20a)
$E(t')$ is the modulus of elasticity at time t'
$\varphi(t, t')$ is the creep function or the creep coefficient at t for a loading at t'

The first term of Eq (5.12) represents the instantaneous deformation and the second term is the creep. This creep law calls for a value of the modulus of elasticity at the time of application of load. However, it is difficult to define the true elastic deformation in a material test. The value of the E-modulus at a specific age shows a marked dependence of the rate of loading which is caused by creep phenomenon always present when loading the concrete. Thus, in analysis of test data, the value of the E-modulus is strongly influenced on how instantaneous deformation and creep are separated, see Fig 5.21. This means that errors are easily made when interpreting values of the E-modulus obtained by different authors which thus depends on different loading rates used in the tests. Such phenomenon is probably more evident at early ages.

Creep compliance formulation

With a compliance function, $J(t, t')$ the total time dependent deformation $\varepsilon(t, t')$ at time t from a loading with the stress $\sigma(t')$ at time t' can be expressed, see Fig 5.20a.

$$\varepsilon(t, t') = J(t, t')\,\sigma(t') \qquad (5.13)$$

It is seen that we do not need to define the instantaneous deformation in Eq (5.13). Several expression for the compliance function may be found in literature. The creep coefficient and effective E-modulus, E_{eff} , can (if the loading rate is exactly defined such as when both creep coefficient and E-modulus are picked from the same test) be expressed in terms of compliance function with the use of

$$J(t,t')=\frac{1+\varphi(t,t')}{E(t')}=\frac{1}{E_{eff}} \tag{5.14}$$

Principle of superposition - an integral type of viscoelastic law

With the superposition principle it is possible to evaluate the strain history caused by an arbitrary stress history (Fig 5.20b).

$$\varepsilon(t)=\int_0^t J(t,t')\,d\sigma(t')+\varepsilon_0(t) \tag{5.15}$$

in which $J(t, t')$ is the creep compliance at time t for a loading at time
$d\sigma(t')$ is the stress increment at time
$\varepsilon_0(t)$ is the stress-independent strain increment e.g. thermal strain and shrinkage

The superposition of the responses is permitted only if each response is purely linear with respect to stress, which implies that for concrete several restrictions should according to Bazant and Wittman [21] be fulfilled. It is concluded that it is doubtful if these restrictions hold for thermal stress analysis of young concrete, see also chapter 7. Further, with the superposition principle, the whole history of loading (of each Gauss point in a FE-calculation) must be stored for the calculation of each stress increment.

Relaxation function - principle of superposition

Similarly to above, the stress at a change of strain at time can be expressed with a relaxation function , see also Fig 7.4

$$\sigma(t, t') = R(t, t')\,\varepsilon(t') \tag{5.16}$$

where R(t, t') is the relaxation function of time t for a strain induced at time t'. Also, the variation of stress at a prescribed strain history may be written in the same way as Eq (5.15)

$$\sigma(t)=\int_{0}^{t}R(t,t')\left[d\varepsilon(t')-d\varepsilon_{0}(t')\right] \tag{5.17}$$

where $d\varepsilon(t')$ and $d\varepsilon_0(t')$ is the strain increment and stress independent strain increment respectively introduced at time t'. The compliance function may be converted into a relaxation function by solving the integral equation of Eq (5.15), (Emborg, [27], Bazant and Wittman, [21]). In the latter reference also an approximate formula this conversion is given, which has been used by authors like Gutsch and Rostásy [69] in thermal stress modelling.

Rate-type formulation

A rate-type formulation of the viscoelastic behaviour eliminates the need for a use of the complete history of stress or strain that was necessary with the integral-type of viscoelastic law. Further, with a rate-type formulation it is possible to model nonlinear effects due to e.g. high stress levels (microcracking), plastic flow and temperature effects in an more accurate way than with the integral type of viscoelastic law.

A modelling of concrete creep under varying stress history may be obtained with the Rate of Creep Method which was introduced to structural problems by Dischinger [71]. The method is based on the assumption that the rate of creep is independent of the age at application of load i.e. parallel creep curves are obtained for different loading ages. This leads to the following formula for the strain rate:

$$\frac{d\varepsilon_{tot}(T)}{dt}=\frac{\sigma(t)}{E}\cdot\frac{d\varphi(t)}{dt}+\frac{1}{E_c}\cdot\frac{d\sigma(t)}{dt} \tag{5.18}$$

where $\frac{d\varphi(t)}{dt}$ and $\frac{d\sigma(t)}{dt}$ are the rates of creep coefficient and stress respectively E_c is the modulus of elasticity. The Rate of Creep Method has been used by e.g. Bernander [72] in computation of early age thermal stresses, see chapter 7.

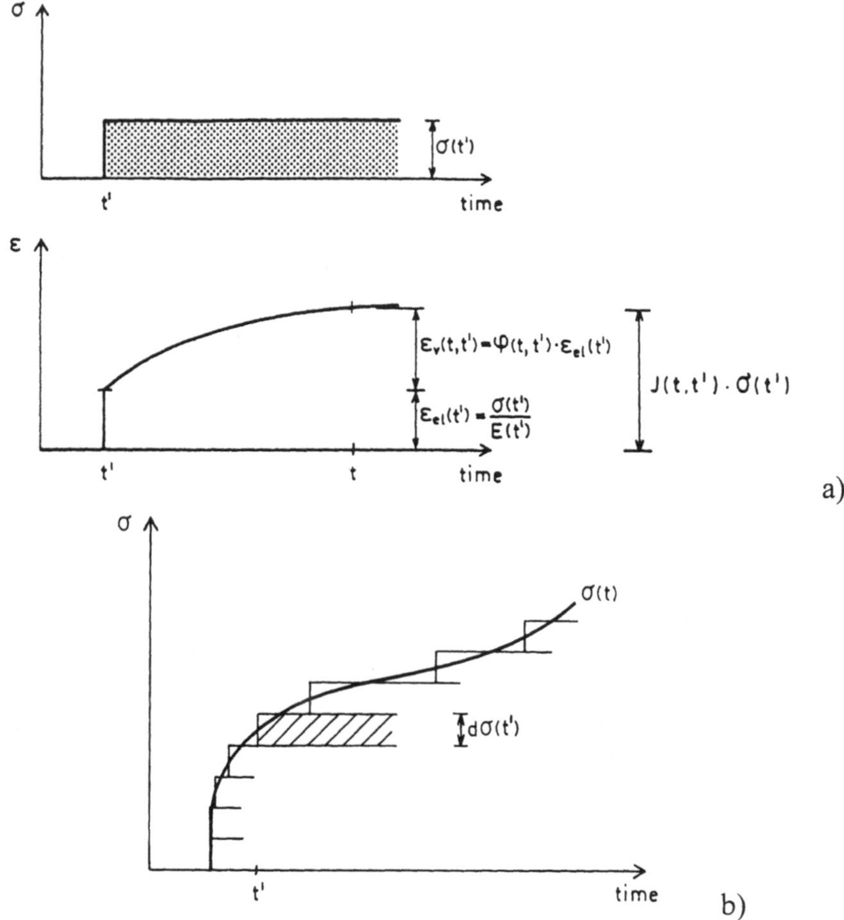

Fig. 5.20 a) Time-dependent deformation at time t for a loading at time t' expressed with a creep function φ(t, t') and compliance function J(t, t'). **b)** Representation of an arbitrary stress history by discrete stress increments dσ(t') which may be used in the integral type of viscoelastic law, Eq (5.15).

The rate-type of viscoelastic law implies a conversion of the response into differential forms of equations e.g. equations based on Kelvin Voigt Model or Maxwell Model (Fig 5.18) see e.g. Bazant and Wittman [21] and Anderson [73]. One example of differential-type of constitutive equation based on rheologic model is the one for the Maxwell Chain Model.

$$\dot{\sigma}_\mu + \frac{E_\mu(t)}{\eta_\mu(t)}\sigma_\mu = E_\mu(t)(\dot{\varepsilon} - \dot{\varepsilon}_0) \tag{5.19}$$

where $E_\mu(t)$ is the stiffness of the μ-th spring (Fig 6.18)
$\eta_\mu(t)$ is the viscosity of the μ–th dashpot
σ_μ is the stress and $\dot{\sigma}_\mu$ rate of stress of the μ-th element
$\dot{\varepsilon}$ and $\dot{\varepsilon}_0$ is rate of the total deformation and thermal deformation

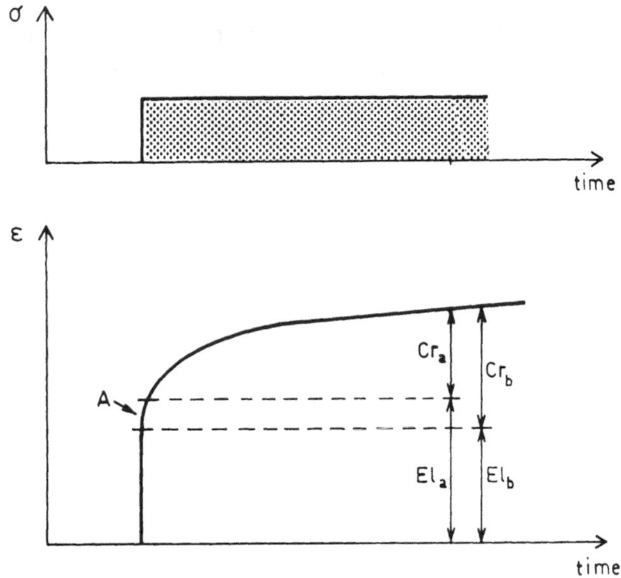

Fig 5.21 Inaccuracies related to separation of elastic and creep deformations . The instantaneous deformation represents a point on an almost vertical part of the response curve, point A in the figure. Therefore the subdivision into a creep component, C_r, and an elastic component, E_l, is difficult. Then errors indicated in the figure are easily obtained, specially after a long time and at early ages. (from Bazant, [64])

This rate-type of formulation has been adopted by Emborg [27] for thermal stress analysis obtaining good agreement with test results mainly dependent on the possibly to describe the nonlinear behaviour of the young concrete (low-stress nonlinearity at creep recovery and nonlinearity at high tensile stresses). Thus, computer programs were first used in order to convert creep data into relaxation data and to evaluate the age-dependent E-modulus E_μ in the Maxwell Chain model (see Bazant, [74], and Jonasson, [75]).

Rheologic models based on Kelvin-Voigt elements are used in the computation of thermal stresses by for instance Dahlblom [76] and Haugaard *et al.* [77].

We should also mention an incremental model including simplified formulation of creep strain rate $\dot{\varepsilon}_F$ is described in Acker *et al.* [78]. The model is based on the so-called "equivalent time method" according to the French Codes (BPEL 83, [79]), see also section 5.4.3.2. In Acker and Eymard [57] the incremental modelling is further developed by a nonlinear creep model.

Lokhorst and van Breugel [63] present a model that combines simple material laws and a description of the microstructure of hardening concrete. Used stress-strain relations are the following

$$\varepsilon_a = (\sigma/E_a) \quad \text{; aggregate}$$

$$\varepsilon_p = \sigma/E_p + a\,t^n\sigma \quad \text{; cement paste} \qquad (5.20)$$

where ε_a and ε_p is strain of aggregate and cement paste respectively
E_a and E_p is corresponding elastic modulus, $E_a = 6.10^4$ MPa, $E_p = 3.10^4$ MPa
σ is stress
t is time under load
a is basic creep rate; $a \approx 10^{-6}$
n $= 0.3$

Three coefficients (κ, λ and μ) are used to model relative volumes of the components of the model ; aggregate, bridges and clusters, see Fig. 5.22.

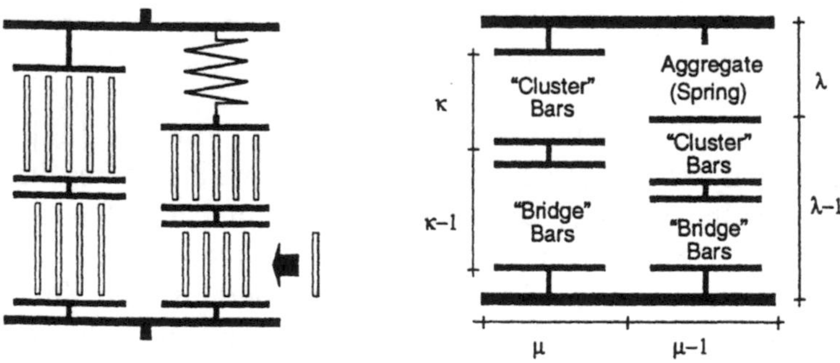

Fig 5.22 Rheologic model for young concrete, Lokhorst and van Breugel [63].

Other formulations

There are a number of other methods of creep analysis than the ones mentioned above that can be utilised in thermal stress computations. Most of such methods are based on the assumption that creep vary linearly with stress i.e. the principle of superposition is valid. In Dilger [80] the following methods of creep analysis are described: Effective modulus method (EM method), Rate of flow method (RF method), Improved Dischinger method (ID method), Rheological models (RM method) and Trost - Bazant method (TB method). Many of them are suitable for computing directly the strain under a varying stress or vice versa.

5.4.3 Mathematical models for elasticity and creep

5.4.3.1 Modulus of Elasticity

A number of test results were compiled by Byfors [5] indicating that the E-modulus of young concrete grows more rapidly than the compressive strength. It is also shown that the stress-strain relation at early ages has a marked nonlinear shape even at low stress levels. Further, it is shown that the E-modulus determined at lower load application rates (E_{stat}) is lower than the E-modulus determined by means of dynamic testing (i.e. E_{dyn}), due to creep phenomenon in the tests, an effect which probably is very strong at early ages. Table 5.5 shows some expressions for the E-modulus in relation to the compressive strength found in literature. Fig 5.23 compares the relation suggested by Byfors to test results and in Fig 5.24a compilation of the formulas in Table 5.5 has been made showing that there is, similar to Figs 5.10 and 5.12, a large scatter between expressions.

Further, a multiphase model for concrete based on a model of Christensen and Lu [81] has been developed by Paulini and Gratl [62] for description of elastic moduli as well as Poisson ratio and have been verified against tests.

5.4.3.2 Creep function (Creep coefficient)

The creep function may be formulated as a sum of two (or more) components (e.g. the delayed elastic and viscous flow in Fig 5.13) or as a product of age (t') and load duration functions (t - t'), see Table 5.6. Of course, there are many other functions for the creep coefficient than the ones above which are used for modelling of hardened concrete, see e.g. Neville *et al.* [19]. A primary question is, however, how accurate the modelling is for the very early age behaviour (t less than about 2-3 days and specially less than 1 day).

Table 5.5 E-modulus at early ages and for hardened concrete, some formulas

Reference	*Relation*	*Coefficients etc.*	*Comments*
CEB/FIP MC90 [36]	$E_c = 10^4 f_c^{0.333}$ $E_c(t) = \beta_E(t) E_c$ $\beta_E(t) = [\beta_{cc}(t)]^{0.5}$	E_c-28 day modulus $\beta_{cc}(t)$ is MC90 relative compressive stress gain, Table 6.1	Hardened concrete Young concrete
ACI (318-83) [53]	$E_c = 4730\, f_c^{0.5}$		Hardened concrete, (Young concrete - Oluokun *et al.*, [82])
Norwegian Code NS 3473 [34]	$E_c = 10^4 f_c^{0.3}$		Hardened concrete
Byfors [5]	$E_c = E_0\, f_{cc}(t)^a$ or $E_c = \dfrac{9.93 \cdot 10^3 \cdot f_{cc}^{2.657}}{1 + 1.370 \cdot f_{cc}^{2.204}}$	$f_{cc} \leqq 1$ MPa (prism): $E_o = 9.93 \,.\, 10^3$ $a = 2.675$ $f_{cc} > 1$ MPa: $E_o = 7.25 \,.\, 10^3$ $a = 0.471$	Dependence of w_o/c ratio (see Fig. 5.7) may be included $w_o/c < 0.40$ $\approx + 20\%$. $w_o/c > 0.60$ $\approx - 20\%$
Laube [40]	$\dfrac{E_c(\alpha)}{E_c(\alpha = 1)} = \dfrac{\left(1 + f_{ct}(\alpha = 1)^b\right) \cdot (rel \cdot f_{ct})^a}{1 + \left(f_{ct}(\alpha = 1) \cdot rel \cdot f_{ct}\right)^b}$	a=0.945 (-) b=0.823 (-)	rel . f_{ct} - see Table 5.4
Umehera, Uehara, Iisaka and Sugiyama [65]	$E_e(t) = \beta(t) \cdot 4.79 \cdot 10^3 \cdot \left(f_c^1(t)\right)^{0.5}$	$\beta(t) = 0.73$; $t \leq 3$ d $= 0.87$; $3\text{ d} < t \leq 4\text{ d}$ $= 1.0$; $4\text{ d} < t$	$f_t^1(t)$ compressive strength

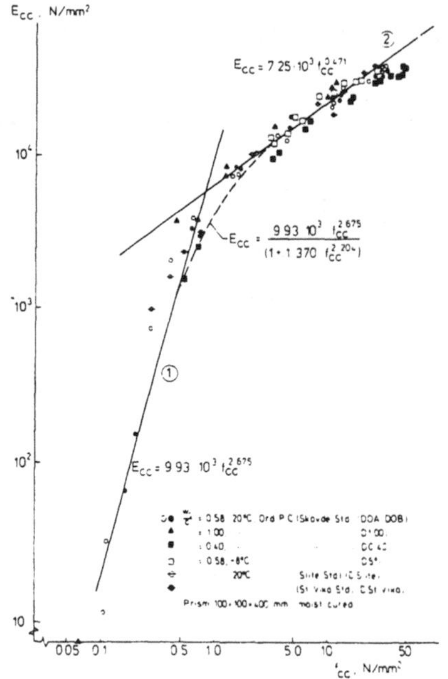

Fig 5.23 Example of experimental and theoretical relationships between early age E-modulus and compressive strength (prism 100 x 100 x 400 mm), Byfors [5].

Further in Guénot, Torrenti and Laplante [68] total strains with MEXO program have been evaluated with three different linear creep laws that showed good agreement with experimental data:
CEB Code [36]:

$$\varphi_c(t,\tau) = \frac{1}{E_{28}} \cdot \left(\frac{16.8}{f_{c\,28}^{0.5}}\right) \cdot \left(\frac{1}{0.1+\tau^{0.2}}\right) \cdot \left(\frac{t-\tau}{1500+t-\tau}\right)^{0.3} \tag{5.21}$$

Laplante [83]:

$$\varphi_c(t_e,\tau_e) = C_c(\tau_e)\frac{(t_e-\tau_e)^{A_c}}{B_c(\tau_c)+(t_e-\tau_e)^{A_c}} \tag{5.22}$$

Le Roy [84]:

$$\varphi_c(t,\tau) = \frac{131e^{-6}}{f_c(\tau)^{0.5}} \cdot \frac{(t-\tau)^{0.5}}{1.47\exp\left(3.4\dfrac{f_c(\tau)}{f_{c28}}\right)+(t-\tau)^{0.5}} \tag{5.23}$$

Table 5.6 Examples of expressions for creep coefficient at early ages reported in literature.

Reference	*Relation*	*Comments, coefficients etc.*
CEB/FIP [51]	$\varphi(t,t') = \varphi_{28}(t,t')\dfrac{E_c(t')}{E_c(28)}$ $\varphi_{28}(t,t') = \beta_d(t') + \varphi_d\beta_d(t-t') + \varphi_f[\beta_f(t) - \beta_f(t')]$	$E_c(t')$, $E_c(28)$ modulus at age of application of load and at 28d. $\beta_a(t')$, $\varphi_d\beta_d$, $\varphi_f\beta_f(t)$ see the code.
CEB/FIP MC90 [36]	$\varphi(t, t') = \varphi_0\beta_c(t - t')$	φ_0 = notional creep, effects of age and shrinkage, see the code. β_c = development of creep after loading also cement type and temperature effects may be modelled.
ACI [53]	$\varphi(t,t') = \dfrac{(t-t')^{0.6}}{10+(t-t')^{0.6}}\varphi_\infty(t')$ $\varphi_\infty(t,t') = 2.35\kappa'_1\,\kappa'_2\,\kappa'_3\,\kappa'_4\,\kappa'_5\,\kappa'_6\,\kappa'_7$	κ'_1 - κ'_7 age at application of load, concrete mixture etc.
Laube [40]	$\varphi(t,t') = P_1(t')\left[\dfrac{t-t'}{t_k}\right]^{P_2(t')}$	$P_1(t')$, $P_2(t')$ dependence of age $t_k = 1$ (h)
Swedish Code [49]	$\varphi(t, t') = \varphi_0 \cdot \varphi_h \cdot \varphi_{t'} \cdot \varphi_{t-t'}$	Basic creep φ influence of humidity, age at load application, load duration
Byfors [5]	$\varphi_{t'} = \left[\dfrac{\dfrac{f_{cc}(t')}{f_{cc}(28d)} + a}{1+a}\right]^{-2/3}$	Age dependency, a=0.17 for tests on Swedish Std Portland Cements.
Pfefferle [85]	$\varphi_{t-t'} = 1 - \sum_{i=1}^{\eta} a_i \exp\left(-b_i\sqrt{t-t'}\right)$	Creep under load a_i, b_i constants, see Emborg [86]

Reference	*Relation*	*Comments, coefficients etc.*
Bernander Gustafsson [55]	$\varphi(t,t') = \left[\sum_{i=1}^{\eta} \beta_n\, e^{-q_n t'}\left(1 - e^{-q_n(t-t')}\right)\right]\left[1 + \kappa\left(\frac{\sigma(t)}{\sigma_M(t)}\right)^m\right]$	$\sigma(t)$, $\sigma_M(t)$ is the compressive tensile strength at time t i.e. a consideration of nonlinear creep. β_n, q_n (1/day), m and κ are constants. Dischinger type of equation used in Rate of Creep Method.
BPEL 83 [79]	$\varphi(t,t') = \left(0.40 + \frac{k_1(h,r)}{1+0.01\cdot t'} \cdot \frac{\sqrt{t-t'}}{k_2(r)+\sqrt{t-t'}}\right)$	$k_1(h, r)$, $k_2(r)$ material functions
Acker, Eymard [78]	$\varepsilon_V = \sigma \cdot F\left(\frac{\varepsilon_V}{\sigma}, \mu, T, w\right)$ $\mu = \int_0^{t_0} \exp\left(-\frac{k}{T(\tau)} + \frac{k}{T_{ref}}\right) d\tau$	ε_v - creep deformation, T - temperature, w - water content, μ - maturity function, F by derivation of creep function φ according to BPEL 83

5.4.3.3 Compliance function

As mentioned earlier the compliance function may be expressed from the creep function. The perhaps most well known compliance function is the Double-Power Law (Bazant and Panula) by which creep and elasticity is described by power curves for loading age and by inverse power curves for age at loading

$$J(t,t') = \frac{1}{E_0} + \frac{\varphi_1}{E_0}(t'^{-m} + \alpha)(t-t')^n \tag{5.24}$$

where E_o, φ_1, m, α and n are material parameters, see Bazant and Panula [87], Bazant and Wittman [21], Kanstad [35] and Emborg [27].

Bazant and Chern [88] have extended the Double Power Law to the Triple Power Law in which the long-time creep is described more accurately

$$J(t,t') = \frac{1}{E_0} + \frac{\varphi_1}{E_0}(t'^{-m} + \alpha)\left[(t-t')^n - B(t,t',n)\right] \tag{5.25}$$

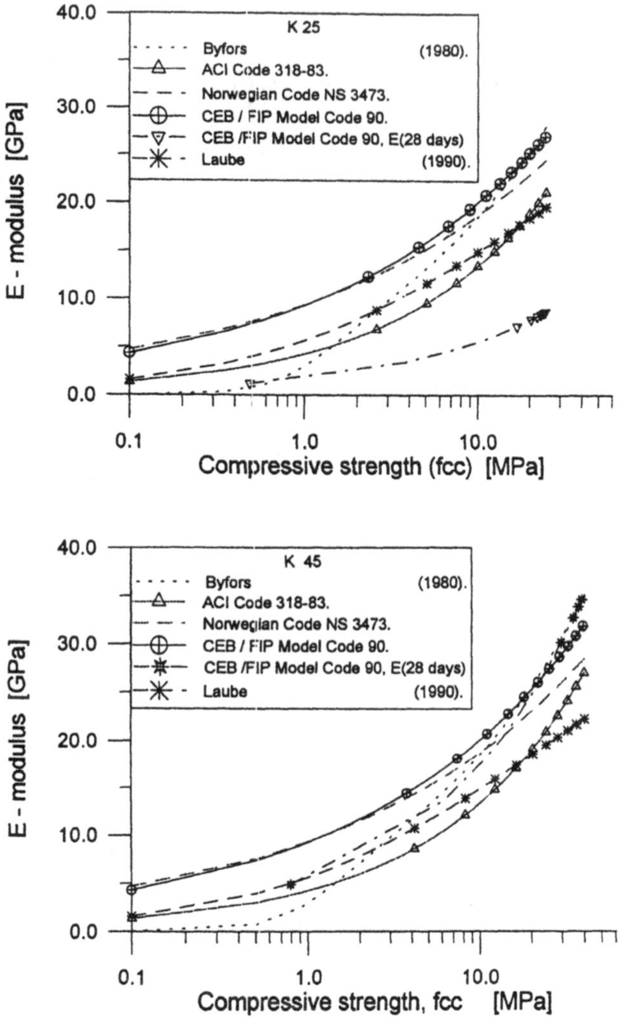

Fig 5.24 Comparisons between some of E-f_{cc} relationships treated in Table 5.5. Concrete: grade K45 (see Fig 5.10).

The Triple Power and Double Power Law are not valid for creep at very early ages (t less than about 1.5 day) and has therefore by Emborg [27] been supplemented with additional functions G(t') and H(t, t') of exponential type:

$$J(t,t') = \frac{1}{E_0} + \frac{\varphi_1}{E_0}(t'^{-m} + \alpha)\left[(t-t')^n - B(t,t',n)\right] + \frac{G(t')}{E_0} + \frac{H(t,t')}{E_0} \tag{5.26}$$

G(t') models the strong age-dependence of the instantaneous deformation (load duration 1.4 minutes) and H(t, t') models the increase of early age creep when the load has been applied, see Fig 5.25. Further, in Emborg [89], the evaluation of constants E_o, φ_1, m, n and α has been simplified compared with original equations in Bazant and Panula [87]. Thus, satisfactory agreements have been obtained with creep tests on normal strength concretes with Swedish cements and also other cements.

Westman [66, 90] has further refined Eq (5.26) for modelling creep of e.g. high strength concrete by conversion of functions $\varphi(t')$ and H(t, t') into functions $\psi_1(t')$ and $\psi_2(t, t')$ (parameters - see the reference):

$$\psi_1(t') = \gamma_1 \left(\frac{t - t'}{t_1 - t_0} \right)^{a_1} \tag{5.27a}$$

$$\psi_2(t, t') = \gamma_2 \left[1 - \exp\left(-\frac{t - t'}{t_2} \right)^{a_2} \right] \left(\frac{b_3 - t'}{t_3 - t_0} \right)^{a_3} \tag{5.27b}$$

In addition, Bazant and his collaborators have proposed new expressions for Double Power and Triple Power Laws, see e.g. Bazant, Kim and Panula [91] and Bazant and Kim [92], whose applicability to thermal stress evaluations can be examined in the future.

The basic creep with a formula where the times t and t' have been transformed into the variables u and u' has been described by Wilson [93].

$$J(u, u') = \frac{1}{E_c} \left[\frac{1}{\psi(u')} + \varphi_0 (u - u') \right] \tag{5.28}$$

where E_c is the E-modulus at 28 days. The function $\psi(u')$ expresses the influence of loading age and φ_0 is the basic creep value, see the reference and also Emborg [86].

By applying a rheological model to results of creep tests, Umehara, Uehara, Iisaka and Sugiyama [65], proposed the following model for time dependent deformation in compression

$$J'_c = [26.96\,(1 - e^{-24.74\,t}) + 71.99\,(1 - e^{-0.575\,t})] \cdot 10^{-6} \tag{5.29}$$

which is represented by a 4-element rheological model.

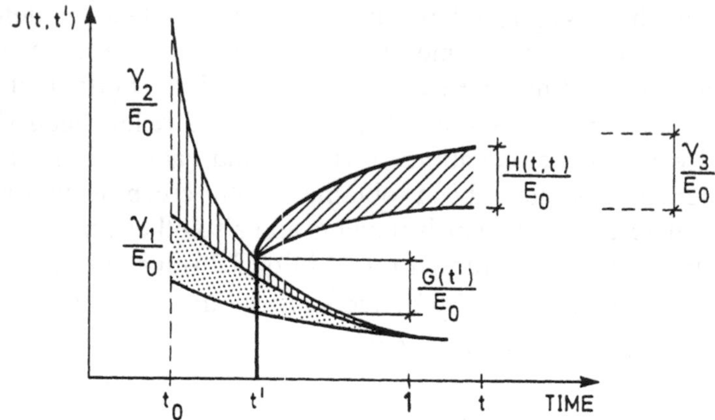

Fig 5.25 Additional functions G(t') and H(t, t') for early-age creep response shown schematically (from Emborg, [27]). (Compare also with Westman, [66, 90])

Also, a formula for creep in tension is suggested by Umehara *et al.* [65]:

$$J'_t = [28.74\,(1\text{-}e^{-0.801\,t}) + 8.13\,(1\text{-}e^{-45.38\,t}) + 4.468\,t]\,.\,10^{-6} \tag{5.30}$$

i.e. a slight difference from Eq. (5.29).

5.4.4 Temperature effects

Creep is affected in two ways if temperature changes: a direct effect on creep by an enhanced creep rate and an indirect effect on the creep due to the influence on the rate of ageing of the concrete.

The effect of temperature on ageing is taken into account by introducing the maturity concept into the creep laws i.e. the concrete ages t' and t are replaced by the equivalent ages t'_e and t_e.

The effect on the creep rate due to constant temperature in the time domain has been expressed in literature by various graphs and explicit formulas. For example in MC 90 (CEB/FIP, [36]) the effect of temperature is taken into account using temperature dependent coefficients $\beta_{H,\,T}$ and $\varphi_{RH,\,T}$. Umehara *et al.* [65] propose the following expression for temperature effect which is multiplied to the compliance function *i.e.*

$$J(t, t', T) = \Phi_c(T)\,.\,J(t, t') \tag{5.31a}$$

$$\Phi_c(T) = 0.112\,T + 0.552 \tag{5.31b}$$

Further, Bazant and Panula [87] models the temperature effect by changing φ_1 and n (Double Power Law) into φ_T and n_T respectively which are functions of curing temperature. The change of creep rate due to temperatures varying with time, as in thermal stress conditions, can, according to Bazant and Wittman [21] and others, only be done approximately and empirically for an integral-type law (Eq (5.15)). This is because of the inability of the history dependent integral-type law to model the effects of previous temperatures on present creep rates.

For the history-independent rate-type laws (e.g. Eq (5.19)) these difficulties do not arise. The change of present creep rate due to temperature may here be described with a modification of the parameter $n_\mu(t)$ in Eq (5.19), e.g. the viscosity of the m-th Maxwell element. This is done by writing, (Bazant *et al.*, [23]).

$$\dot{y}_\mu(t,T) = \varphi_T \, \dot{y}_\mu(t) \tag{5.32}$$

where $\dot{y}_\mu(t) = E_\mu(t)/\eta_\mu(t)$ and

$$\varphi_T = \exp\left[\frac{U_c}{R}\left(\frac{1}{T_0} - \frac{1}{T(t)}\right)\right] \tag{5.33}$$

T_0 is the reference temperature (20°C). R is the gas constant (8.31 KJ/mol) and U_c is the activating energy of creep; $U_c/R = 5000$ K. Thus, it seems that the activating energy concept may be applicable to creep as have been found also in other studies e.g. at Delft University.

The effect of temperature on maturity of the concrete is in the Rate-type formula modelled with the relaxation modulus $E_\mu(t')$ depending on the equivalent hardening period. For further description of temperature dependence of young concrete see e.g. Emborg [27].

5.4.5 Ageing effects

The influences of ageing of concrete under load, *i.e.* the consequence of volume growth of load-bearing hydrated cement (Fig. 5.15), have been studied by different authors. For instance Bazant and Prasannan [94] express the effect in a Kelvin Chain model with age-dependent moduli and viscosities. Further, van Breugel [24] models the phenomenon in a creep model based on a single Kelvin element and in a relaxation model.

5.4.6 Nonlinear creep

5.4.6.1 Nonlinear effects in compression, viscoplasticity

As earlier been mentioned, there exist two types of nonlinearity phenomena in creep when a concrete is subjected to compressive stresses: a) the low-stress nonlinearity (plastic flow) often denoted "adaption", leading to a stiffening behaviour at further loading or deloading and b) the high-stress nonlinearity (cracking) causing weakening of the response. A correct modelling of these nonlinear effects is easiest done in a rate-type creep formula. For a creep law based on the principle of superposition, nonlinear effects imply that restriction for use of this principle does not hold.

The *low-stress nonlinearity* of creep in concrete under compression, indicated in several investigations, is probably present in a thermal stress situation during the growth of thermal compressive stresses at the heating and at the subsequent deloading. Bazant and Kim [64] consider the phenomenon with a stress dependent coefficient in the maturity function β_T and an adaption parameter affecting the stiffness of m-th element for the rate-type model according to Maxwell Chain Model.

$$dt_e = \beta_T \, \beta_\sigma \, dt \tag{5.34}$$

$$E^*_\mu(t) = [1 + a(t)] \, E_\mu(t) \tag{5.35}$$

where β_T is the maturity function
β_σ is the stress dependent coefficient
$E_\mu(t)$ is the stiffness of m-th element in Maxwell Chain Model
$a(t)$ is the adaption parameter

The *high-stress nonlinearity* at compression is a more typical feature of inelastic behaviour of concrete. Its mechanism is microcracking mainly occurring at high hydrostatic pressures. For the Maxwell Chain Model, Bazant and Kim [64] have modelled this high-stress nonlinearity (compare with Eq (5.19))

$$\dot{\sigma}_\mu + \frac{E^*_\mu}{\eta^*_\mu}\sigma_\mu = E^*_\mu(t)(\dot{\varepsilon} - \dot{\varepsilon}_f - \dot{\varepsilon}_0) \tag{5.36}$$

where $E^*_\mu(t)$ and $\eta^*_\mu(t)$ are stress-dependent stiffness and viscosity and have been replaced instead of $E_\mu(t)$ and $\eta_\mu(t)$. Further, $\dot{\varepsilon}_f$ is the rate of flow which is expressed as

$$\dot{\varepsilon}_f(t) = \frac{\sigma(t) - \alpha(t)}{E_0} f[\sigma(t)]\dot{\phi}(t) \tag{5.37}$$

For description of the constants in Eq (5.37), see the reference. We may note, that for most cases of thermal stress situations, high-stress non-lineary in compression is not present.

5.4.6.2 Nonlinear creep in tension

Knowledge of nonlinear creep behaviour at high tensile stresses is of paramount importance when determining the risk of cracking in the concrete. Like nonlinear creep in compression nonlinear effects on creep in tension are very difficult to study experimentally for the early ages.

In Neville *et al.* [19] an existence of a fracture limit envelope is shown. Tensile stresses beyond the limit, which was found to be in the range of 0.60 to 0.85 of the tensile strength, induce tertiary creep and, hence, creep failure. The influence on high stress levels leading to tertiary creep is also shown in Wittman *et al.* [50], see section 5.3.4. Further, Bernander [72] found that the only way to simulate test results (thermal stresses at restrained temperature deformations) was to correct a creep function used for nonlinear behaviour at high stress levels in tension.

Cracking, both microcracking and at a later stage-visible cracks, have a dominating effect on tensile nonlinear creep. Cracking results in significant additional strains and a realistic model of calculating tensile creep is only complete if a model of cracking is considered. This will be discussed in section 5.5

5.4.7 Concluding remarks - comparisons of models

Section 5.4 deals with the modelling of the viscoelastic response at early ages and expressions for creep coefficients, compliance functions and E-modulus are given.

As mentioned earlier, the use of a creep coefficient formulation calls for a value of the E-modulus at the age of application of load. As the E-modulus is dependent on the rate of loading at tests, errors easily are made when interpreting results from different test series (Fig 5.21). These errors do not arise when using the compliance function.

It was stated that there exist various creep and compliance functions in literature which primary are used for modelling of hardened concrete. Of course, it is here of interest to see how accurate the modelling is for early age behaviour. An example of comparisons of different creep and compliance functions with results from creep tests is shown in Fig 5.26a. Creep coefficients φ(t, t') are used together with an E-modulus that has been evaluated from the tests for a load duration of 0.001 day. Thus, fairly good agreement with tests are obtained. However, if we use one of the general formulas for the growth of E-modulus described earlier in the section 5.4.3.1 together with the creep coefficient, it is observed how agreements with tests are far worse, Fig 5.26b. This demonstrates the effects of mixing formulas for j and E. (Fig 5.21)

Which one of the constitutive models (Rate of Creep Method, Superposition Principle etc.) that is most suitable for thermal stress computations depends on application and requirement on level of accuracy.

The rate-type formulation seems to be the most accurate way to consider temperature effects, nonlinear effects e t c which always are present in early-age thermal stress phenomenon. However, if a Maxwell Chain Model is used, relaxation data is required, directly from tests or by a conversion from creep tests. Although computer programs for this conversion may be obtained, Bazant [74] and Jonasson [75], the procedure may be considered as complex and time-consuming. Alternatively, the Kelvin Chain Model may be used in which creep data can be implemented directly.

For practicizing engineers the rate-type formulations may be considered too complicated and methods based on superposition principle can suite better due to e.g. simpler implementation in computer programs. The main disadvantage of the superposition principle is that the real material behaviour cannot be modelled correctly, thus implying errors in the calculations.

The Rate-of-Creep Method has, as far as the author knows, not thoroughly been tested for the application to thermal stresses in young concrete. Some experiences with varying agreements with test results were made in Bernander [72] and Emborg [86].

As a final remark on creep we may refer a conclusion raised by Dilger [80]: From observations and discussions of different types of creep functions and compliance functions he means that there exists <u>no</u> method that can be considered to be exact. One reason for this is, according to Dilger, the observed nonlinear behaviour of concrete particularly shortly after loading which hardly can be modelled mathematically.

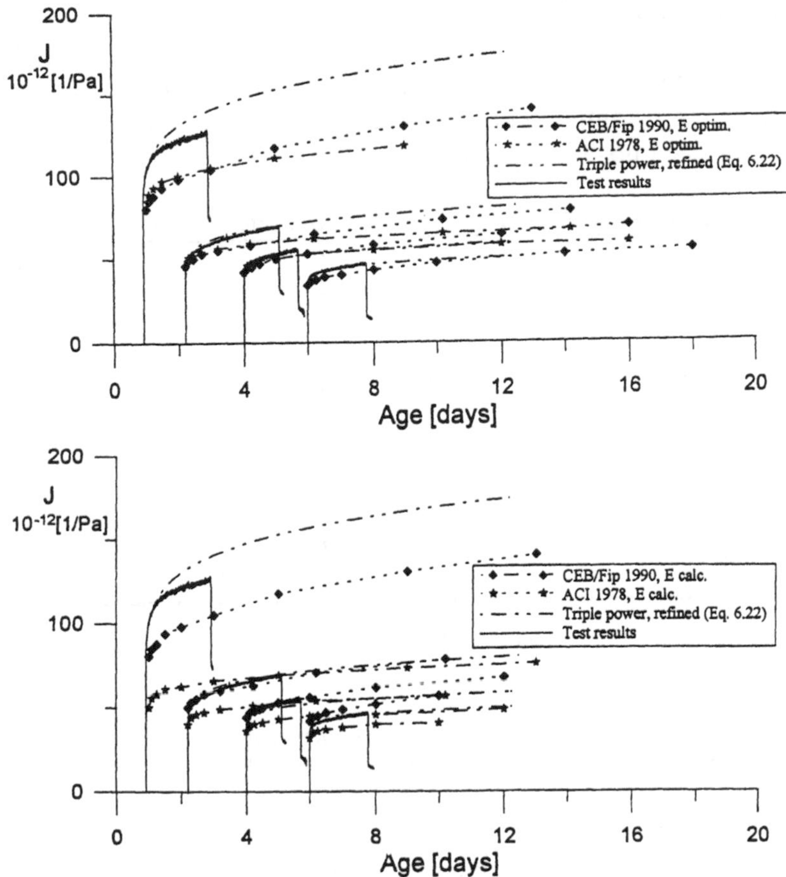

Fig 5.26 Comparison of different creep and compliance functions with results from creep tests. E-modulus is used with creep coefficient that is optimised from the tests (for a time under load t - t' = 0.001d) **(a)** or evaluated theoretically according to Table 5.5 **(b)** (CEB/FIP MC90 [36] and ACI 318-83 [53] resp.). Concrete: Swedish Standard Portland Cement type Degerhamn (ASTM Type II), w_o/c =0.4, cube strength at 1,2.5, 6 and 28 days: 8.5, 22, 28, 32 and 55 MPa resp. The specimens are unloaded after about 2 days.

5.5 FRACTURE MECHANICS BEHAVIOUR

5.5.1 General

The age-dependence of parameters governing fracture mechanics behaviour (e.g. tensile strength, modulus of elasticity, fracture energy, characteristic length) have been studied by e.g. Peterson [95] and rather extensively by Brameshuber and Hilsdorf [96] and Brameshuber [97]. A slight increase of the fracture energy G_f and a considerable decrease of characteristic length l_{ch}, implying a more ductile behaviour, have been found at early ages, see Figs 5.27 and 5.28.

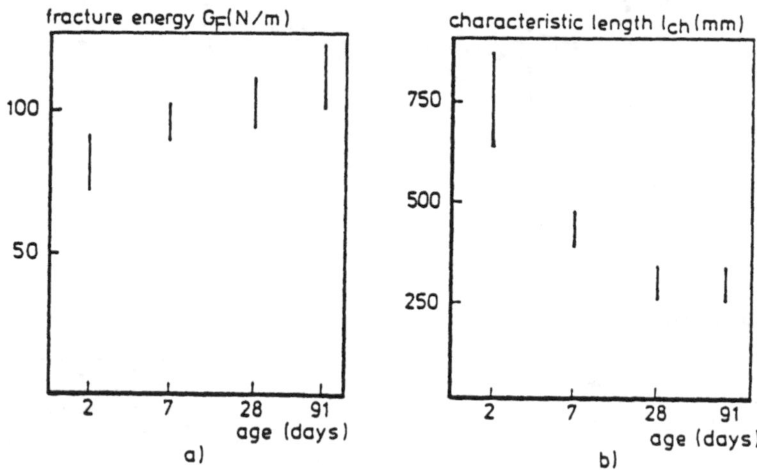

Fig. 5.27 Relationship between the age of the concrete and fracture energy **(a)** and characteristic length **(b)** respectively (from Peterson, [95])

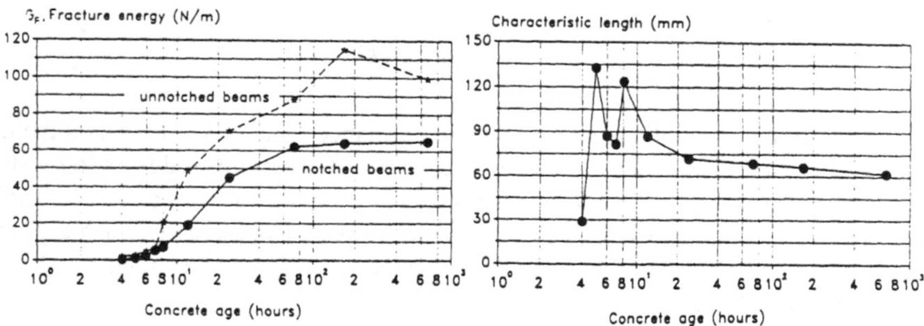

Fig. 5.28 Development of fracture energy and characteristic length in young concrete (from Brameshuber and Hilsdorf, [96])

The influence of temperature on fracture mechanics parameters has been examined for low and high temperatures, see e.g. Körmeling [98], Elices *et al.* [99], Ohlsson *et al.* [100] and Bazant and Prat [101]. The fracture energy has found to be much higher for tests at very low temperatures (-170°C). At higher temperatures than 20°C, the fracture energy was reduced, Fig 5.29.

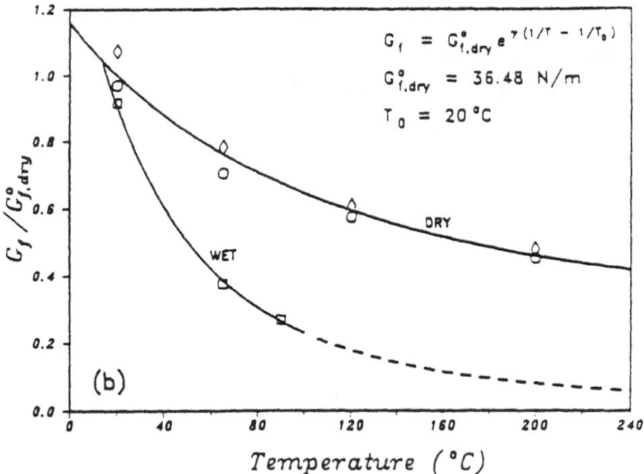

Fig. 5.29 Temperature dependence of fracture energy of dry and wet concrete. $G^{o}_{f,dry}$ is the fracture energy of dry concrete at 20°C. (The figure is based on test results of Bazant and Prat, [101])

The consideration of time-dependent behaviour in a fracture mechanics model may entail a substantial improvement towards more realistic models for the calculation of early-age thermal cracking. Time-dependent behaviour of the fracture mechanics parameters have been observed in experiments, see e.g. Hoordijk *et al.* [102], and have been discussed by means of a theoretical modelling in Modeer [103]. Further, the ultimate strain (failure strain) of very young concrete 0.5-2 days at uniaxial tensile tests have been found to be about 50 percent of the one of hardened concrete, see Byfors [5], Kasai *et al.* [1] and others. At very early ages (< 0.5 day) the ultimate tensile strain is very high.

However, there is a serious lack of material data for the fracture mechanics behaviour at early ages. To carry out e.g. bending tests with notched beams (according to RILEM recommendations) or uniaxial tensile test on cylinders for different loading rates, temperatures, concrete mixtures etc., is an important task for future work.

5.5.2 Modelling

The transferred stress in a crack is often assumed to be defined by the crack opening

$$\sigma = f(w) \tag{5.38}$$

where f(w) describes the softening behaviour of the material, often named softening function or softening curve, Fig 5.30. Function f(w) is a material function. The characteristic length l_{ch} and characteristic crack opening w_{ch} are defined from fracture energy G_F, tensile strength, f_t and modulus of elasticity (Hillerborg *et al.*, [104])

$$w_{ch} = G_F/f_t \tag{5.39}$$

$$l_{ch} = E\ G_F/f_t^2 \tag{5.40}$$

The parameters w_{ch} and l_{ch} are used in the characterisation of the structural brittleness numbers which, thus, describe how brittle failure is in a specific member.

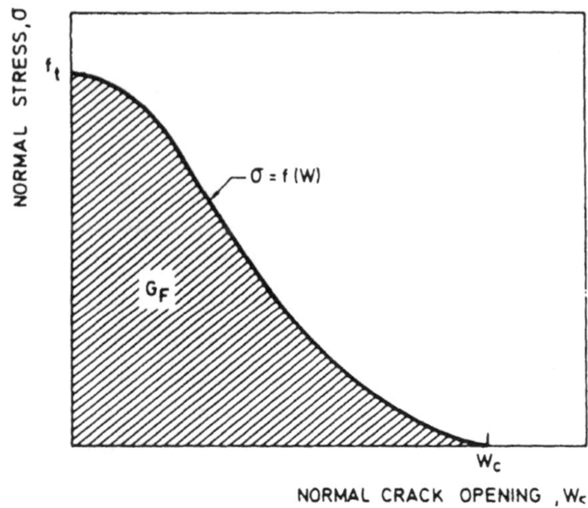

Fig. 5.30 Softening curve (from Elices and Planas, [99]).

Several models for the fracture mechanics behaviour of hardened concrete have been proposed in literature. Thus, two main concepts have been outlined: a discrete crack concept and a smeared crack concept, see e.g. Elfgren [105].

One basical model for describing the behaviour of concrete in a discrete crack concept is the "fictitious crack" model introduced by Hillerborg [106]. The general description of stress deformation properties is given by means of two curves according to Fig 5.31.

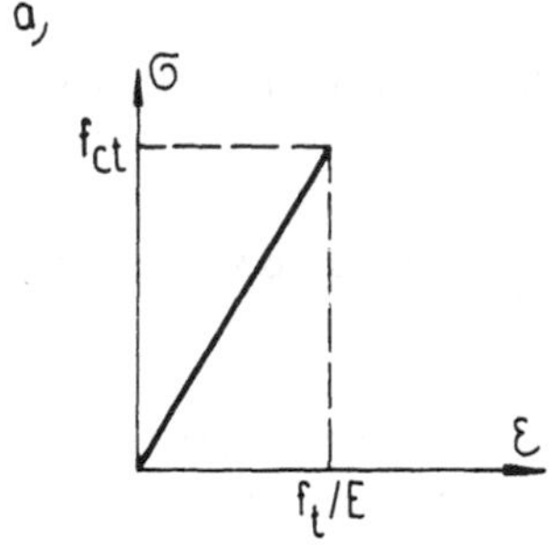

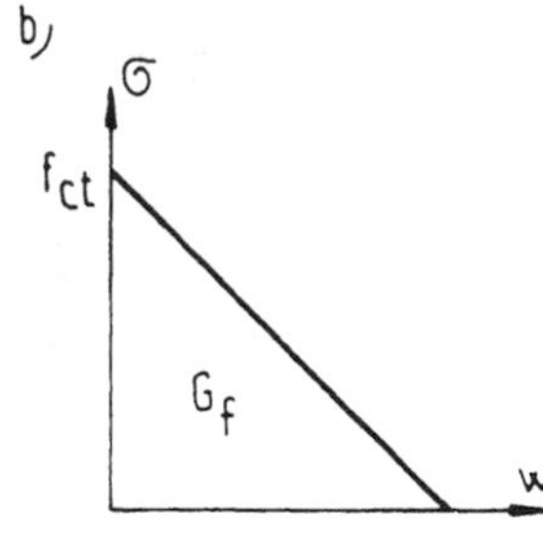

Fig. 5.31 Schematic characteristic curve for an element (a) before and (b) after cracking according to Hillerborg [106] (from Nagy, [108]).

A smeared crack relation that can describe the tensile strain softening with or without simultaneous creep (and shrinkage) has been proposed by Bazant and Chern [107]. In the relation, the strain due to strain softening is considered as additive to the strain due to creep, shrinkage, temperature and elastic deformation, see Figs 5.32 and 5.33

$$\varepsilon = \varepsilon_{ec} + \xi + \varepsilon^0 \tag{5.41}$$

in which ε, ε_{ec}, ξ and ε^0 are respectively the total strain, the strain due to elastic deformation and creep, the strain due to strain softening (fracturing strain), and the strain due to shrinkage and thermal deformation. The behaviour of the strain-softening element implies a relation between the stress σ and the fracturing strain ξ of the element

$$\sigma = C(\xi)\xi \tag{5.42}$$

where $C(\xi)$ is the secant modulus. Basic formulas of the strain-softening element (material parameters and behaviour at loading, unloading and reloading) are found in Bazant & Chern [107] and Emborg [27].

No matter of which model that is used for fracture mechanics behaviour in thermal stress evaluations it is most important to consider the behaviour of young concrete in the modelling. This can be done by introducing the development of parameters $f_t(t)$, $E(t)$ and $G_F(t)$ in formulas such as Eqs. (5.39) and (5.40). As an example, Emborg [27] shows an evaluation of early age fracture energy $G_F(t)$ based on formulas according to Bazant and Oh [109].

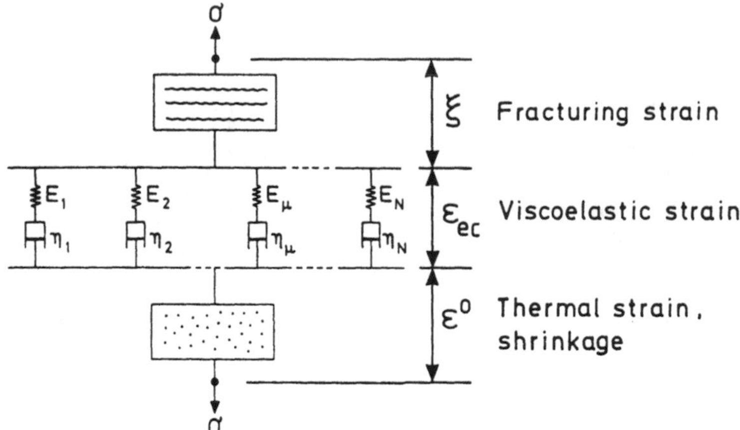

Fig. 5.32 Rheologic model describing deformation due to strain softening, elastic deformation, creep strain, shrinkage and thermal deformation (after Bazant and Chern, [107]).

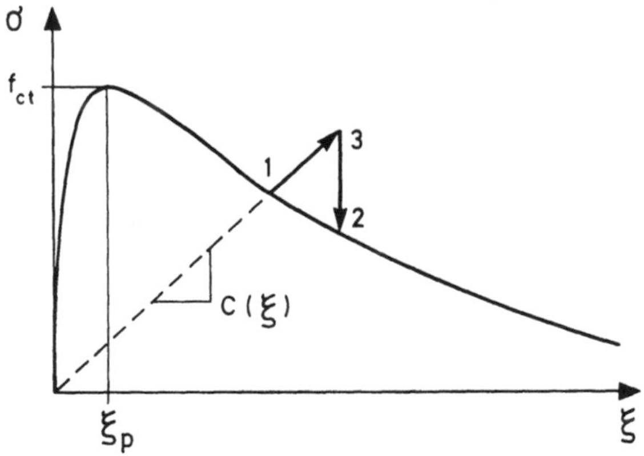

Fig. 5.33 Stress-strain relation with strain softening curve (after Bazant and Chern, [107]), notations: f_{ct}-peak stress (ultimate tensile strength), ξ_p - fracturing strain (nonlinear strain) at peak load, C(ξ) - secant modulus.

5.6 THERMAL EXPANSION AND CONTRACTION

The coefficient of thermal expansion for young concrete has been examined by some authors, e.g. Löfqvist [110], Alexandersson [111], Weigler and Karl [112], Emborg [27] and Hedlund [113]. A vital question is whether the coefficient of thermal expansion varies during the very early age heating and if it has a different value in the subsequent cooling phase.

For instance, Löfqvist [110] observed for the early-age heating values of the expansion coefficient up to 12 microstrain /°C, while for the cooling phase, values between 6-8 microstrain /°C were measured. When the concrete subsequently was exposed to repeated heating and cooling cycles, almost the same values (6-8 microstrain/°C) were observed.

Nolting [32] studied the development of thermal expansion coefficient at early ages. Among other things the influence of cement components was examined. He reports very high values of the coefficient during the first 5-8 h curing (up to 20 microstrain /°C). Thereafter the coefficient decreases rapidly to about 10-12 microstrain /°C. Between ages of 2h to 36h a small increase of the value was observed.

Emborg [27] reports results from tests where free thermal volume change has been studied for early-age heating and cooling cycles, see Fig. 5.34. It can be seen from the tests that coefficient of thermal expansion and contraction are - similarly to Löfquist's results - significantly different during the first heating cycle, see also Table 5.7. Thus, it seems appropriate to define thermal deformations not only by a coefficient of thermal contraction, α_e, but also by a coefficient of thermal contraction, α_c. In second and third heating cycles, the differences between the coefficients become smaller in the tests performed, and it seems likely that for older concretes values of α_e and α_c are almost the same.

Important studies on thermal expansion and contraction coefficient for high performance concrete - specially experimental methodologies - has been made by Hedlund [113]. For instance it was found that it is necessary to separate deformations due to early age shrinkage from the thermal dilation in order to find the "true" thermal coefficients, see Fig 5.35.

Regarding the thermal volume change, a phenomenon of stress-dependent thermal expansion has been observed in tests for temperatures in the interval 20-800°C. This phenomenon has been discussed by Thelandersson [114, 115] and Bazant & Chern [116]. They mean that deformations induced by temperature changes are not simply additive to the deformations caused by mechanical action (e.g. creep). According to Thelandersson and Bazant & Chern, the change of the thermal strain (or strain rate) should be modelled as a function of both the change of temperature (or the rate of temperature change) and the current state of stress. For a Maxwell element in uniaxial stress state the modelling may be done as follows, Thelandersson [115].

$$\frac{\dot{\sigma}}{E}+\frac{\sigma}{\eta_0}=\dot{e}-\gamma_0\sigma\dot{T} \tag{5.43}$$

where $\dot{T}$ is the rate of temperature change

σ and $\dot{\sigma}$ are the current stress and current stress rate respectively is the strain rate

η_o and E are the viscosity of the dashpot and the E-modulus of the spring respectively

γ_o is a function of temperature

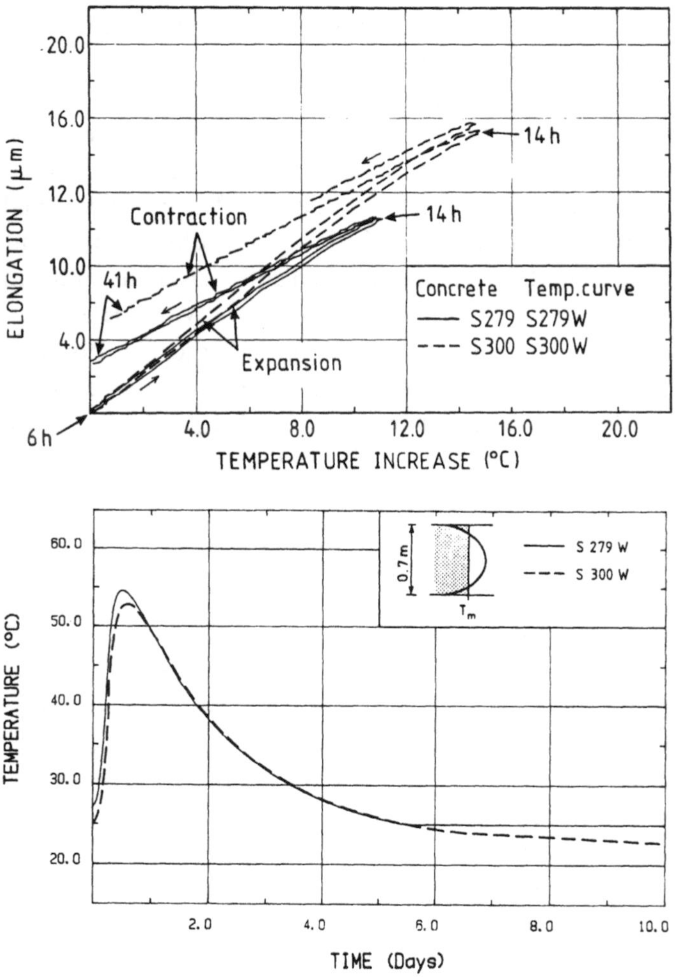

Fig. 5.34 Results from free thermal expansion and contraction tests on newly cast concrete specimen. (a): The specimens are heated directly after pouring by water with temperature curves according to e.g. Fig. (b). Measurement length = 100 mm. Concrete Standard Portland Cement S279: c = 279 kg/m³; w_o/c = 0.61, S300: c = 300 kg/m³, w_o/c = 0.54. (Emborg, [27]).

Table 5.7 Variation of coefficients of thermal expansion α_e and thermal contraction α_c at early-age heating for different water cement ratios in tests reported by Emborg [27] and Hedlund [113]

Cement type	w_o/c	α_e $[10^{-6}/°C]$	α_c $[10^{-6}/°C]$
Degerhamn (Type II) Standard Portland	0.25-0.35	9-12.5	6.0-8.3
Slite (ASTM Type I)	0.4-0.5	9.5-12.5	7.0-8.5
Standard Portland	0.5-0.7	9.5-10.5	7.0-8.0
	0.7-0.8	9.0-10.0	7.0-8.0
Degerhamn (Type II) Standard Portland	0.4-0.8	9.5-11.0	7.0-9.0

This equation based on the stress-dependent factor of the thermal coefficient has been found to be able to simulate observed experimental behaviour of hardened concrete in a satisfactory manner. The phenomenon of stress induced effects on thermal strain may affect the thermal volume change during the early-age heating and cooling of the constrained concrete. However, it seems to be very difficult to quantify experimentally and to model theoretically the effects of the phenomenon for very young concrete and no verification of Eq (5.43) has been found in literature.

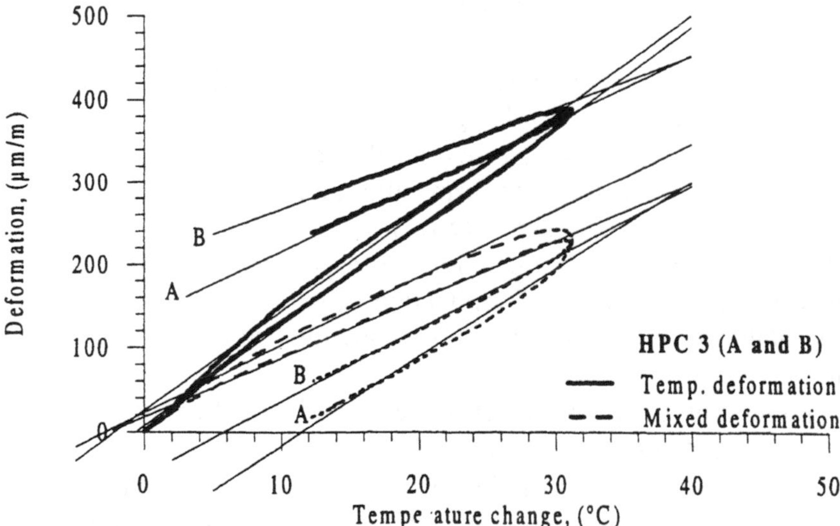

Fig. 5.35 Free thermal expansion and contraction tests (test A and B) of high performance concrete. The test started at the age of 9.5 hours. Observe the differences between recorded mixed deformation and the true temperature deformation which has been obtained by subtracting the deformation due to early age shrinkage (autogenous shrinkage). Hedlund [113].

5.7 CONCLUSIONS

It is of utmost importance, for an accurate thermal stress analysis, that a correct modelling of the mechanical behaviour is performed. The modelling of the mechanical properties at early ages is however difficult as the concrete in a few days changes from a nearly plastic state to a hardened matrix. Also, there is, depending on difficulties at tests, a considerable lack of test data on the behaviour of young concrete. This may be one explanation to the big scatter in the calculations of strength and E-modulus shown in the text.

Regarding the lack of experimental data, there exists, as an example, only few reported values on thermal volume change. This is remarkable as changes in coefficient of thermal expansion and contraction have a direct effect on the thermal stresses, see Fig 5.4.

Further, non-linear behaviour in compression (creep recovery) and in tension (non-linear creep an cracking) ought to be included in to modelling. Computations with linear models give a rather good description of the compressive stress gain but implies far too high tensile stresses, thus overestimating the cracking risks, see Chapter 7.

Chapter 5 may be concluded with some advises for the treating of mechanical properties in the thermal stress analysis:

- Check that formulas proposed are valid for the whole equivalent time of interest - specially at early ages.
- Check that the formulas are valid for the concrete mix, cement type etc. that is used. If necessary, perform calibration of material constants (e.g. by plotting a "strength gain spectra").
- Avoid separation of elastic and creep responses. If a separation is necessary: check that rates of application of load are identical when picking values of creep function and E-modulus from different laboratory tests.
- If preferably, use a non-linear model including softening due to cracking at high stress levels in tension.
- Be extra careful in evaluation of input data of parameters that have important effects on stresses e.g. input of coefficients defining thermal volume change.

APPENDIX A Temperature effects on final strength and effects of temperature at testing

For a cold concrete i.e. for an older concrete that have been cooled down after a temperature heating, the influence of previous high temperature on strength may be modelled in different ways. Jonasson [15] expresses a correction of the strength at a curing temperature other than 20°C as follows

$$f_{cc}^{corr} = f_{cc} - \Delta f_{cc} \tag{A.1}$$

where f_{cc} = strength at curing temperature 20°C
Δf_{cc} = loss of strength due to high temperature

The loss of strength Δf_{cc} is expressed as a value at 28 day equivalent time, Δf_{HT}, multiplied with a time-function:

$$\Delta f_{cc} = e^{-\left(\frac{t_2}{t_e}\right)^{\kappa_2}} \cdot \Delta f_{HT} \tag{A.2}$$

The time function is a step function between 0 and 1 models the fact that the influence is not present until about 2 days of equivalent age t_e. t_2 and κ_2 see Table A.1. Δf_{HT} is dependent on magnitude of higher curing temperatures and the time from pouring.

$$\Delta f_{HT} = \nu_2\, \Delta f_{max} \quad \text{with} \quad t_{HT} = \int_0^t \beta_{HT}\beta_{add} \quad \text{and} \quad \nu_2 = \begin{cases} (t_{HT}/t_3 & \text{for} \quad t_{HT} \le t_3 \\ 1 & \text{for} \quad t_{HT} > t_3 \end{cases} \tag{A.3}$$

Δf_{max} is maximum strength and t_{HT} is an integral for temperature-time effects. β_{HT} is a parameter including temperature sensibility at higher temperatures than T_{HT} which may be considered as the temperature above which the strength loss is present. Thus, β_{HT} may be expressed as

$$\beta_{HT} = \begin{cases} [\beta_T(T) - \beta(T_{HT})]^2 & \text{for} \quad T > T_{HT} \\ 0 & \text{for} \quad T \le T_{HT} \end{cases} \tag{A.4}$$

β_{add} is a factor for admixtures (accelerating additives: $\beta_{add} > 1$, retarding additives: $\beta_{add} < 1$).

Adaptations to strength tests at elevated temperatures for Swedish concretes have been performed leading to coefficients according to Table A.1.

Eqs (A.1) - (A.4) is one suggestion of taking account the temperature effect. However, due to the limited material data, the coefficients in Table A.1 may be considered as preliminary.

Table A.1 Parameters for cube strength loss at high curing temperatures. Swedish Standard Portland cements. Cubes tested wet and cold.

Cement type	T_{HT} *[°C]*	t_2 *[days]*	κ_2	t_3 *[days]*	Δf_{max} *[MPa]*
Std Portland, Slite (ASTM Type I)	28	72	2.0	60	5
Rapid Hardening, Skövde (Type)	28	72	1.4	60	6
Std Portland, Degerhamn (Type II)	28	240	2.0	60	4

An alternative way of modelling the thermal loss due to high temperatures have been proposed by Kjellsen and Detwiler [14]. They show, in support of several references, that different maturity functions are inadequate for predictions of later-age strength and propose a modified maturity model intended to take account the retardation effect of high temperatures on the strength development.

The modified maturity (equivalent age) τ_e, mod is according to Kjellsen and Detwiler expressed as

$$\tau_e, \text{mod} = \int_0^t f(T, f_{cr}) \cdot dt \tag{A.5}$$

where $f(T, f_{cr})$ is a relative temperature function (compare the Arrhenius concept)

$$f(T, f_{cr}) = \exp\left[\frac{E(T, f_{cr})}{R}\left(\frac{1}{T_{ref}} - \frac{1}{T}\right)\right] \tag{A.6}$$

and

$$\frac{E(T, f_{cr})}{R} = \ln\left[\frac{\tau_e}{t_T}\right]\left[\frac{1}{T_{ref}} - \frac{1}{T}\right]^{-1} \tag{A.7}$$

where τ_e = equivalent age to reach f_{cr}=0.2 at the reference temperature T_{ref}
t_T = age to reach f_{cr} = 0.2 at the particular temperature T.

Further, f_{cr} is the relative compressive strength based on standard trend curves. The following strength-time relation have been used by the authors (according to Freiesleben-Hansen and Pedersen).

$$f_c = P_1 \exp(-(P_2 / t)^{P_3}) \tag{A.8}$$

where f_c is the estimated compressive strength and P_1-P_3 are model parameters of which numerical values for various concrete mixes are given in the reference.

The high temperature effect on strength is by Emborg [27] corrected by using a factor G_c which is dependent on temperature, water cement ratio and cement type. Thus, the reduction of the compressive strength growth at reference temperature 20o C, Δf_{cc} (t_e, 20), during a time-step may be expressed as a corrected strength growth

$$\Delta f_{cc}^{*}(t_e, T) = G_c \, . \, \Delta f_{cc}(t_e, 20) \text{ for } f_{cc}(t_e, 20) \geq 0.54 \, . \, f_{cc}(28d, 20) \tag{A.9}$$

where G_c is written

$$G_c = \exp\,[-0.08(T - 20)] + a_c \text{ with } G_c^{max} = 1.0 \tag{A.10}$$

in which T is the mean temperature (°C) during the time step and a_c is a correction factor for the w/c ratio, see the reference. It is seen that only strength growth over 54 per cent of the 28-day strength is corrected for. The function G_c is shown in Fig A.1 in comparison with results from tests on warm cubes.

Laube [40] has also proposed a function for the reduction of strength growth which is based on the same test results as Eqs (A.9) and (A.10), see Fig A.2.

$$\gamma_o(t, \tau_{max}) = 1 - \frac{T_{max} - 20}{80} \cdot \frac{t}{t_{max}} \leq 1 \tag{A.11}$$

where T_{max} and t_{max} is the maximum temperature in a structure and time for this maximum respectively.

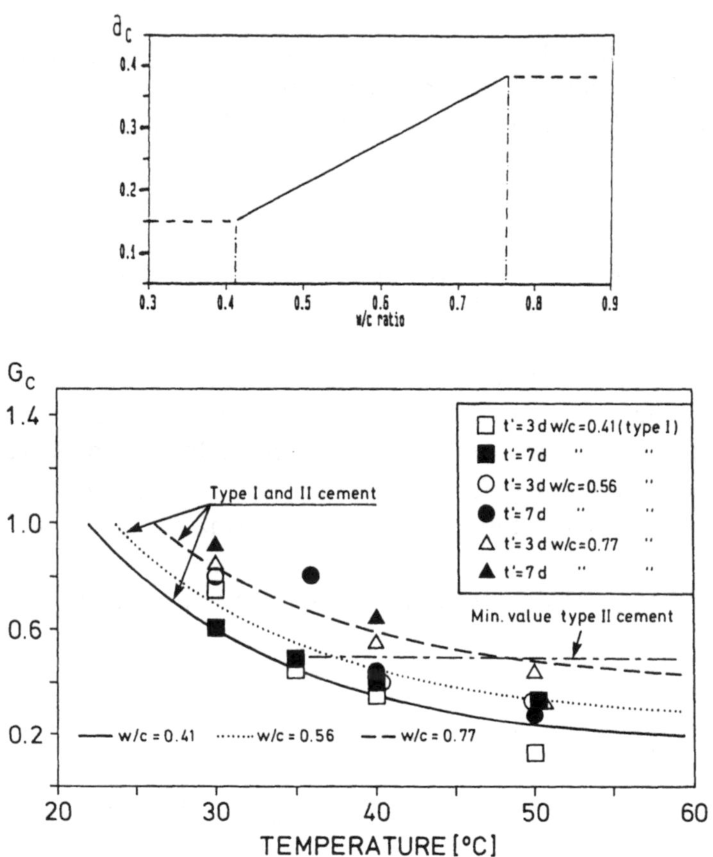

Fig A.1 The correction factor G_c for reduction of strength growth and factor a_c as functions of the curing temperature and of the w/c ratio (Type I and Type II cements). Tests on 150 mm cubes, water cured, tested wet and warm (Emborg, [27]).

The effect of the temperature of the specimen at the moment of the strength test, mentioned earlier, may also be modelled in different ways. One mathematical adaption of this effect has been suggested by Emborg [27].

$$f_{cc}(t_e, T) = R_c(T) \, . \, f_{cc}(t_e, 20) \tag{A.12}$$

in which $f_{cc}(t_e, 20)$ is the strength at 20°C testing temperature. $R_c(T)$ is a linear function: $0.8 \leq R_c(T) \leq 1.0$, see Fig A.3b.

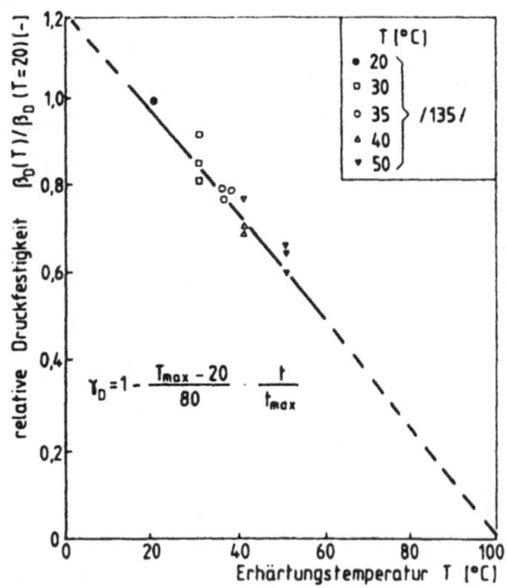

Fig A.2 Influence of higher curing temperatures on the cube strength (Laube, [40]).

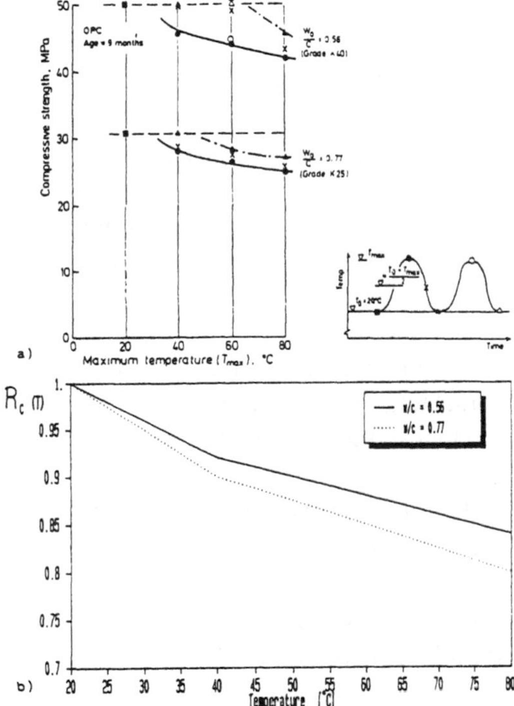

Fig. A.3 **a)** Measured strengths of hardened concrete at elevated temperatures at the moment of testing 150 mm cubes, water cured, tested wet (Jonasson, [48]). **b)** Function $R_c(T)$ for the reduction of strength due to the temperature at the testing.

The effect of temperature at the moment of strength test is included in the results in Fig A.1 *i.e.* the cubes were tested in a warm state. This implies that the values in these figures represent the true in-situ strength of the concrete of a warm section, which is of interest in many cases. Therefore, we do not need to use the corrections according to Eq (A.12) in Eq (A.9).

If a warm specimen is *cooled prior to testing* (which normally occurs), *the compressive strength obtained actually should be reduced* e.g. *according to Eq (A.12)* for an accurate describing of the *in-situ* strength. This means when a determination of strength of warm concrete is carried out with models proposed by Jonasson and by Kjellsen and Detwiler, Eq (A.12) should be used.

Further, we may also note that often when temperature effects on compressive strengths are treated in literature and in experimental studies, the strengths have been based on *cubes that have been cooled down* prior to testing.

REFERENCES

[1] Kasai, Yokoyama, Matsui: Tensile properties of early-age concrete. Proceedings of the 1971 International Conference on Mechanical Behaviour of Materials, Vol IV, pp 288-299.

[2] ACI Committee No 207: Effect of restraint, volume changes and reinforcement on cracking of massive concrete, *ACI Journal*, July 1973, pp 445-470.

[3] Weigler, Karl: Junger Beton - Beanspruchung - Festigkeit - Verformung. Forschungsberichte aus dem Institut für Massivbau der Technischen Hochschule Darmstadt, 1974, Nr 20, 55 pp.

[4] Freiesleben-Hansen, Pedersen: Måleinstrument til kontrol av betons haerding, *Nordisk Betong*, No 1, Stockholm 1977, pp 21-25.

[5] Byfors: Plain concrete at early ages. Swedish Cement and concrete Institute, Fo 3:80, Stockholm 1980.

[6] Horden W. C. and Reinhardt H.: Temperatur en spanningen im tetrapuds tijdens de verharding, *Cement*, No 12, 1986, pp 49-52.

[7] Jonasson J.-E.: Slipform construction - calculations for assessing protection against early freezing. Swedish Cement and Concrete Research Institute, Fo 4:84, Stockholm 1984, 70 p.

[8] Maatjes E. and van Breugel K.: Temperatur - ein Sterkteontwikkeling a verhardend beton (in Dutch), *Cement*, No 5, 1986, pp 44-52.

[9] Helland S.: Temperatur og fasthetsutvikling i betong med v/c lavere enn 0,40. (Temperature and strength development in concrete with w/c ratios lower than 0,40. In Norwegian). *Nordisk Betong*, No 5, Stockholm 1987, pp 26-29.

[10] van Breugel, K.: Simulation of hydration and formation of structure in hardening cement-based materials, Technical University of Delft, Faculty of Civil Engineering, Dissertation, Delft, 1991, 295 pp.

[11] Kasai Y., Hiraya and Yokoyama K.: Initial strength of concrete at variable curing temperature. Japan Cement Engineering Association, 18th General Meeting, 1964, pp 123-128, (quoted from Byfors, [5]).

[12] Sadgrove B. M.: The early development of strength in concrete, CIRIA, Technical Note 12, 1970, 43 pp (quoted from Byfors, [5]).

[13] Haugland, Hofsøy, Garborg, Fjeldstad: Tidiligfastheter for beton. Sammandrag Fältforsøk, Laboratorieforsøk. (Early strength of concrete. Field tests, laboratory tests. In Norwegian). FCB, Report 76-1, 2, 3, 4, 5, Trondheim 1976. (Quoted from Byfors, [5]).

[14] Kjellsen K. and Detwiler R.: Later-age strength prediction by a modified maturity model, submitted for publication in *ACI Journal of Materials*.

[15] Jonasson J.-E.: Modelling of temperature, moisture and stresses in young concrete. Division of Structural Engineering, Luleå University of Technology, Doctoral Thesis 1994:153D, 225 pp.

[16] Hellman: Beziehungen zwischen Zug- und Druckfestigkeit des Betons. *beton*, No 2, 1969, pp 68-70, (quoted from Byfors, [5]).

[17] Bellander U.: Hållfasthet i färdig konstruktion, Del 1. Förstörande metoder, Rimliga kravnivåer (Strength in completed structures, part 1, in Swedish), Swedish Cement and Concrete Research Institute, Research 13:76, Stockholm 1976, (quoted from Byfors, [5])

[18] Neville A.M.: *Properties of concrete.* Pitman Publishing Limited, London, 685 p.

[19] Neville A.M., Dilger W.H., Brooks J.: *Creep of plain and structural concrete*, Construction Press, Longman Group Limited, New York, 1983, 361 pp.

[20] Bazant Z.P.: Viscoelastic of solidifying porous material - concrete - Swedish Cement and Concrete Research Institute, Fo 5.77, Stockholm (quoted from Byfors [5]).

[21] Bazant Z.P. and Wittman F.H. (editors): *Creep and shrinkage in concrete structures.* Wiley & Sons, New York 1982, pp 363.

[22] Rostásy F.S. and Budelmann H.: Creep of concrete with variable moisture cement at elevated temepratures up to 90°C. Creep and shrinkage of concrete. Mathematical modelling (RILEM), Technological Institute. Northwestern University, Evenston, 1986, pp 581-590

[23] Bazant Z.P. *et al.*: Material models for structural creep analysis, Chapter 2 in Preprints "RILEM Symposium on Creep and Shrinkage of concrete structures: Mathematical modelling" held in Evanston, August 26-29, 1986 (Editor Z P Bazant), Evanston, Illinois 1986, pp 80-232.

[24] van Breugel, K.: Relaxation of young concrete. Department of Structural Concrete, Faculty of Civil Engineering. Delft University of Technology, Research Report 5-80-D8, Delft 1980, 140 pp.

[25] Bazant Z.P. and Carol I. (editors): *Creep and Shrinkage of Concrete*, Proceedings of the Fifth International RILEM Symposium, Barcelona, September 6-9, 1993, Chapman & Hall, London, 1993, 934 pp.

[26] Tanabe T. and Ishikawa: Time dependent behaviour of concrete at early ages and its modelling, see Bazant & Carol [25], pp 435-452.

[27] Emborg M.: Thermal stresses in concrete structures at early ages, Division of Structural Engineering, Luleå University of Technology, Doctoral Thesis 1989:73D, 280 pp.

[28] Parrot L.: Effect of loading at early ages upon creep and relaxation of concrete. RILEM Comittee 42-CEA Internal report UK5, 1978 (quoted from Byfors, [5]).

[29] Browne: Thermal movements of concrete. Current practical sheet. *Concrete* No 3PC/06/1, November 1972, pp 51-53.

[30] Emanuel and Hulsey: Prediction of thermal coefficient of expansion of concrete, *ACI Journal*, April 1977, pp 149-154.

[31] Harrison T.A.: Early-age thermal crack control in concrete. Construction Industry Research and Information Association (CIRIA), Report No 91, 1981, 48 pp.

[32] Nolting E.: Zur Frage der Entwicklung lastunabhängiger Verformungen und Wärmedehnzahlen junger Betone, Dissertation, Institut für Baustoffkunde und Materialprüfung, Technische Universität Hannover, Heft 56, Hannover 1989, 206 pp.

[33] Graf O., Albrecht W. and Schafflet H.: *Die Eigenschaften des Betons*, Zweite neuarbeitete Auflage, Springer-Verlag, 1960.

[34] NS 3473: Prosjektering av betongkonstruksjoner. Beregnings- og konstruktsjonsregler, (Design of concrete structures) Norges Standardiseringsforbund, 1988.

[35] Kanstad T.: Nonlinear analysis considering timedependent deformations and capacity of reinforced and prestressed concrete, Dissertation, Division of Concrete Structures, The Norwegian Institute of Technology, The University of Trondheim, May 1990, 360 pp.

[36] CEB-FIP Model Code: Comité Euro-International du Beton, Bulletin D'information No 213/214, Tomas Telford, London, May 1993, 473 pp.

[37] Hamfler H.: Berechnung von Temperatur-, Feuchte- und Verschiebungsfeldern in erhärtenden Betonbauteilen nach der Methode der Finiten Elemente, Institut für Massivbau, Hannover, 1988, 159 pp.

[38] Wesche, K.: *Baustoffe für tragende Bauteile*, Band 2: Beton. Bauverlag, Wiesbaden-Berlin 1981 (quoted from Hamfler, [37]).

[39] Carino N.J.: Maturity functions for concrete, Proceeding from "International Conference on concrete at Early Ages" (RILEM) Paris 6-8, April 1982, Vol I, Paris 1982, pp 123-128.

[40] Laube M.: Werkstoffmodell zur Berechnung von Temperaturspannungen in massigen Betonbauteilen im jungen Alter. Doctoral Thesis, Universität Braunschweig, 1990, 251 pp.

[41] van Breugel, K.: Development of temperature and properties of concrete as a function of the degree of hydration. Proceedings from "RILEM International Conference on Concrete at Early Ages", Paris, April 6-8, 1982, Vol 1 Paris, pp 179-185.

[42] Powers & Brownyard: Studies of physical properties of hardened portland cement paste. Research Laboratories PCA, Bulletin 22-1948 (quoted from Fagerlund, [43]).

[43] Fagerlund G.: Relation between the strength and the degree of hydration or porosity of cement paste, cement mortar and concrete, Seminar on hydration of cement, Copenhagen, Nov 20th 1987. Nordisk Betonforbund/Aalborg Portland, Aalborg,

[44] Fagerlund G.: Tendenskurvor för våra cement, Trend curves for our cements, Cementa AB Internal report, January 1991.

[45] Taplin J.M.,: A method for following the hydration reaction in portland cement paste. *Australian Journal of Applied Science*, 10 ,3 (quoted from Fagerlund, [43]).

[46] Locher: *Beton*, Heft 7, 1976, pp 247-249 and No. 8 1976, pp 283-285 (quoted from van Breugel, [10]).

[47] van Breugel, K.: Numerical simulation of the effect of curing temperature on the maximum strength of cement-based materials, *Thermal Cracking in Concrete at Early Ages*, Proceedings of the RILEM International Symposium, edited by R. Springenschmid, E & FN Spon, London, 1995, pp 127-134.

[48] Jonasson J.-E.: Early strength growth in concrete - preliminary test results concerning hardening at elevated temperatures. Int. symp. on Winter Concreting (RILEM), Espoo 1985, Technical Researc Center of Finland (VTT), Espoo 1985, pp 249-254.

[49] Swedish Concrete Handbook - Material: (In Swedish) Svensk Byggtjänst, Stockholm, 1994 (in print).

[50] Wittman F.H., Roelfstra P.E., Mihashi H., Huang Y.-J., Zhang X.-H., Nomura N.: Influence of age at loading, water-cement ration and rate of loading on fracture energy of concrete, *Materials and Structures*, **20**, No 116, 1987, pp 103-110.

[51] CEB/FIP: Model code for concrete structures. Comité Européen du Béton - Fédération Internationale de la Précontrainte, CEB Bulletin No 124/125-E, Paris 1978, 348 pp.

[52] DIN 1045: Beton und Stahlbeton, Bemessung und Ausführung. Dez 1978 (quoted from Hamfler, [37]).

[53] ACI Committee No 318: Building Code Requirements for Reinforced concrete, American Concrete Institute, Detroit, 1983, 111 pp.

[54] Oluokun F., Burdette E., Deatherage H: Splitting Strength and Compressive Strength Relationship at Early Ages. *ACI Materials Journal*, **88**, No 2, March-April 1991, pp 115-121.

[55] Bernander S. and Gustafsson S.: Egenspänningar i ung betong p g a temperaturförioppet under hydrationen (Temperature stresses in early age concrete due to hydration. In Swedish with English Summary). *Nordisk Betong*, No 2, Stockholm 1981, pp 25-31.

[56] Wierig H-J: Kriechen and Relaxation. Verhalten von jungem Beton, *beton*, no 1, 1988, pp 17-20.

[57] Acker P. and Eymard R.: Thermal and hygral effects in concrete structures: How to account for creep? In Proceedings of Third International RILEM Workshop on Behaviour of concrete elements under thermal and hygral gradients, Weimar 1992, p 191-197.

[58] Rostásy F.S., Gutsch A. and Laube M.: Creep and relaxation of concrete at early ages - experiments and mathematical modelling, see Bazant & Carol [25], pp 453-458.

[59] Dameron R.A., Dunham R.S., James R.J. and Rashid Y.R.: Constitutive Modelling of Early time effects in concrete, see Bazant & Carol [25], pp 459-464.

[60] Ohshita H., Ishikawa Y. and Tanabe T.: Creep mechanism of early age concrete modelling with two phase porous material, see Bazant & Carol [25], pp 465-470.

[61] Serra I., Marri A.R., Martinez-Abella F. & Lopez-Almanza F.: Experimental measurement of creep and shrinkage of early-demolded reinforced concrete beams, see Bazant & Carol [25], pp 471-477.

[62] Paulini P. and Gratl N.: Stiffness formation of early age concrete, *Thermal Cracking in Concrete at Early Ages*, Proceedings of the RILEM International Symposium, edited by R. Springenschmid, E & FN Spon, London, 1995, pp 63-70.

[63] Lokhorst S.J. and van Breugel K.: From microstructural development towards prediction of macro stresses in hardening concrete, *Thermal Cracking in Concrete at Early Ages*, Proceedings of the RILEM International Symposium, edited by R. Springenschmid, E & FN Spon, London, 1995, pp 71-78.

[64] Bazant Z.P. and Kim S.S.: Nonlinear creep of concrete - adaption and flow, *J. Eng. Mech. Div.* (ASCE), **105**, 1979, pp 419-446.

[65] Umehara H., Uehara T., Iisaka T. and Sugiyama A.: Effect of creep in concrete at early ages on thermal stresses, *Thermal Cracking in Concrete at Early Ages*, Proceedings of the RILEM International Symposium, edited by R. Springenschmid, E & FN Spon, London, 1995, pp 79-86.

[66] Westman G.: Basic creep and relaxation of young concrete, *Thermal Cracking in Concrete at Early Ages*, Proceedings of the RILEM International Symposium, edited by R. Springenschmid, E & FN Spon, London, 1995, pp 87-94.

[67] Morimoto H., Koyangi W.: Estimation of stress relaxation in concrete at early ages, *Thermal Cracking in Concrete at Early Ages*, Proceedings of the RILEM International Symposium, edited by R. Springenschmid, E & FN Spon, London, 1995, pp 95-102.

[68] Guénot I., Torrenti J.-M., Laplante P.: Stresses in concrete at early ages: comparison of different creep models, *Thermal Cracking in Concrete at Early Ages*, Proceedings of the RILEM International Symposium, edited by R. Springenschmid, E & FN Spon, London, 1995, pp 103-110.

[69] Gutsch A., Rostásy F.S.: Young concrete under high tensile stresses - creep, relaxation and cracking, *Thermal Cracking in Concrete at Early Ages*, Proceedings of the RILEM International Symposium, edited by R. Springenschmid, E & FN Spon, London, 1995, pp 111-118.

[70] Wierig H.-J.: The creep behaviour of heat-treated concretes loaded at an early age. *Betonwerk + Fertigteil-Technik*, Heft 6, 1985, pp 369-373.

[71] Dischinger: Elastische und plastische Verformung der Eisenbetontragwerke. *Der Bauingenieur*, **20**, No 5/6, 21/22, 31/32, 47/48, Berlin 1939, pp 55-63, 286-294, 426-437 and 563-572.

[72] Bernander S.: Temperature stresses in early age concrete due to hydration. Proceedings from "International Conference on Concrete at Early Ages", (RILEM, Paris), April 6-8, 1982 ,Vol II, pp 218-221.

[73] Anderson C.A.: Numerical creep analysis of structures. Chapter 8 in *Creep and shrinkage in concrete structures* (Edited by Bazant Z.P. and Wittman F.H, [21]), Wiley & Sons, New York 1982, pp 259-303.

[74] Bazant Z.P.: Input of creep and shrinkage characteristics for a structural analysis program. Materials and Structures (RILEM, Paris), Vol 15, No 88, pp 283-290.

[75] Jonasson J.-E.: Analysis of creep and shrinkage in concrete and its applications to concrete top layers. *Cement and Concrete Res.*, **8**, New-York 1978, pp 441-454.

[76] Dahlblom O.: HACON-S-A program for simulation of stress in hardening concrete. Vattenfall Hydro Power Generation, Vällingby, 1992, 77 pp.

[77] Haugaard M., Berrig A., Fredriksen J.O.: Curing technology - A 2 dimensional simulation program, Proceedings from the Nordic Concrete Res. Meeting, Gothenburg, Aug 1993, The Nordic Concr. Fed., Oslo, 1993, pp 222-224.

[78] Acker P., Eymard R., Piau J.-M.: Structural Analysis of Concrete Structures: Accounting for thermal and hygral effects. Laboratoire Central des Ponts et Chaussées, Paris, internal report, 11 pp.

[79] BPEL 83: Règles techniques de conception et de calcul des ouvrages et constructions en béton précontraint par la méthode des états-limites, Bull. off. fasc. spec. no 83-31 bis, Paris (quoted from Acker *et al.*).

[80] Dilger, W.: Methods fo structural creep analysis, chapter 9 in *Creep and shrinkage in concrete structures*, see Bazant and Wittman [21],pp 305-340.

[81] Christensen R.M., Lo K.H. (no title), *Journal of Mechanic Physical Solids*, **27**, No 4, 1979, pp 315-329. (quoted from Paulini and Gratl, [62]).

[82] Oluokun F., Burdette E., Deatherage H.: Elastic Modulus, Poisson's Ratio and Compressive Strength Relationships at Early Ages, *ACI Materials Journal*, **88** No 1, January-February 1991, pp3-10

[83] Laplante, P.: Propriétés mécaniques des bétons durcissants: analyse comparée des bétons classiques et à très hautes performances. Doctoral Thesis, Ecole Nationale des Ponts et Chaussées, Paris, 1993, published in Rapport de recherche des laboratoires des Ponts et Chaussées OA13.

[84] Le Roy, R.: Déformations instantanées et différées des bétons à hautes performances. Doctoral Thesis, Ecole Nationale des Ponts et Chaussées, Paris, 1995.

[85] Pfefferle R.: Das Kriechen des Betons, eine kritisch gedämpfte Schwingung, *Beton- und Stahlbetonbau*, Heft 12, 1979, pp 269-301.

[86] Emborg M: Temperature stresses in massive concrete structures. Viscoelastic models and laboratory tests. Division of Structural Engineering, Luleå University of Technology, Licentiate Thesis 1985:011L, 143 pp.

[87] Bazant Z.P. and Panula L.: Practical predictions of time dependent deformations of concrete. Materials and structures (RILEM) Vol 11, 1978 (part I-IV): pp 307-316, 317-328, 415-423, 424-434, Vol 12, 1979 (part V-VI), pp 169-174, 175-183.

[88] Bazant Z.P. and Chern J.C.: Triple Power Law for concrete creep. *J. Eng. Mat.*, **3**, No 4, Jan 1985, pp 63-83.

[89] Emborg M.: Creep of concrete at very early ages - adjustment of Triple Power Law. Division of Structural Engineering, Internal report, 1994:x, (to be published).

[90] Westman G.: Thermal cracking in high performance concrete. Luleå University of Technology, Division of Structural Engineering, Licentiate Thesis, 1995:27L, 122 pp.

[91] Bazant Z.P., Kim J.-K. and Panula L.: Improved prediction model for time dependent deformations in concrete: Part 1-Shrinkage, *Materials and Structures*, **24**, 1991, pp 327-345.

[92] Bazant Z.P. and Kim J.-K. and: Improved prediction model for time-dependent deformations in concrete. *Materials and Structures*: **24**, 1991 (part 2), pp 409-421, **25**, 1992 (part 3), pp 21-28, (part 4), pp 84-94, (part 5), pp 163-169.

[93] Wilson R.: Further development of the Improved Dischinger Method, preprints, Int. symp. Creep and Shrinkage of Concrete: Mathematical Modelling, August 26-29, Evanston 1986, (Ed. Bazant Z.P.), Evanston 1986, pp 873-878.

[94] Bazant Z.P. and Prasannan.: Solid falsation theory for concrete creep. Center for concrete and Geomaterials, The Technological Institute, Northwestern University, Evanston, Report No 87-12/4985, 35 pp, also: Cement and Concrete Research, Vol 18, 1988, pp 923-932.

[95] Peterson P.-E.: Crack growth and development of fracture zones in plain concrete and similar material. Lund, Report TvBM-1006, Lund 1981.

[96] Brameshuber W. and Hilsdorf H.: Development of strength and deformability of very young concrete. Proceedings from "RILEM International Conference on Facture of Concrete and Rock", held in Houston 1987, (Editors Shah S.P. and Schwartz S.E.), pp 655-667.

[97] Brameshuber W.: Bruchmechanische Eigenschaften von jungem Beton, Dissertation, Institut für Massivbau und Baustofftechnologie, Karlsruhe, Heft 5, Karlsruhe 1988, 283 pp.

[98] Körmeling H.A.: Strain rate and temperature behaviour of steel fibre concrete in tension, Dissertation, Delft University of Technology, 1986.

[99] Elices M., Planas J., Maturana P.: Fracture of concrete at cryogenic temperatures. Proceeding from "International Conference on Fracture of concrete and Rock", Houston, June 17-19, 1987 (RILEM). (Editors Shah S.P. and Schwartz S.E.), pp 159-169.

[100] Ohlsson U., Daerga P.-A., Elfgren L.: Fracture energy and fatigue strength of unreinforced concrete beams at normal and low temperatures. Paper presented at the International Conference on concrete and damage of concrete and rock at the Technical University of Vienna, July 1988. Also published in Engineering Fracture Mechanics (London. 8 pp.

[101] Bazant Z.P. and Prat P.C.: Effects of temperature and humidity on fracture energy of concrete, Report No 86-121 6922, Center for Concrete and Geomaterials, Northwestern University, Evanston, Illinois, 1986.

[102] Hordijk D., van Mier J. and Reinhardt H.-W.: Material properties, Chapter 4 in "RILEM Technical Committee 90 FMA", see Elfgren [105], pp 67-127.

[103] Modeer M.: Time dependent behaviour, section 6.4 in "RILEM Technical Committee 90-FMA", see Elfgren [105], pp 162-171.

[104] Hillerborg A., Modeer M. and Peterson P.-E.: Analysis of crack formation and crack growth in concrete by means of fracture mechanics and finite elements. *Cem. Concr. Res.*, 6, pp 773-782.

[105] Elfgren L. (editor): Fracture mechanics of concrete - From theory to applications, A State-of-the-art report prepared by RILEM Technical Committee 90-FMA, Chapman & Hall, London, 1989.

[106] Hillerborg A.: A model for fracture analysis. Report TVBM-3005. Div. of Build Mat., Lund Institute of Technology, Sweden 1978 (quoted from Nagy, [108])

[107] Bazant Z.P. and Chern J.C.: Strain Softening with creep and exponential algorithm, *J. Eng. Mech. Div.* (ASCE), **111**, No 3, 1985, pp 391-415.

[108] Nagy A.: Cracking of concrete structures due to early thermal deformations. Department of Structural Engineering, Lund Institute of Technology, Licentiate Thesis Report TVBK-1009, 110 pp.

[109] Bazant Z.P. and Oh B.H.: Strain rate effect in rapid triaxial loading of concrete. *J. Eng. Mech. Div.*, ASCE, **108**, Oct 1982, pp 764-782.

[110] Löfqvist B.: Temperatureffekter i hårdnande betong (Temperature effects in hardening concrete, in Swedish), Dissertation, Kungliga Vattenfallsstyrelsen, Tekniskt meddelande No 22, Stockholm 1946, 195 p.

[111] Alexandersson J.: Strength losses in heat cured concrete. Swedish Cement and Concrete Research Institute. Handlingar No 43, Stockholm 1972, 135 pp.

[112] Weigler H. and Karl S.: Creep of concrete under early loading - Influence of the rate of hardening of the cement, *Betonwerk + Fertigteil-Technik*, Heft 9, 1981, pp 519-522.

[113] Hedlund H.: Stresses in high performance concrete due to temperature and moisture variations at early ages. Luleå University of Technology, Division of Structural Engineering, Licentiate Thesis, 1996:38L, 238 pp.

[114] Thelandersson S.: On the multiaxial behaviour of concrete exposed to high temperatur. *Journal of Nucl. Eng. Des.*, **75**, No 2, pp 271-282 (quoted from Thelandersson, [115]).

[115] Thelandersson S.: Modelling of combined thermal and mechanical action in concrete. *J. Eng. Mech. Div.* (ASCE), **113**, No 6, June 1987, pp 893-906.

[116] Bazant Z.P. and Chern J.C.: Concrete creep at variable humidity, *Materials and Structures*, **18**, No 103, 1985, pp 1-19.

6

Assessment of External Restraint

F. S. Rostásy, T. Tanabe and M. Laube

6.1 INTRODUCTION AND SCOPE

The occurrence and control of cracks at an early age of concrete are an important practical problem. Especially in massive concrete members cracks are predominantly caused by thermal stresses. These stresses arise due to the heat of hydration and its flow to the surrounding air, to the subsoil or to older concrete members connected with the newly cast concrete, etc.. Drying shrinkage only negligibly contributes to early age cracks in the first weeks after casting. For HPC chemical shrinkage is important.

The thermal stresses are usually interpreted as the sum of two stress components:

a) One component comprises the self-equilibrating stresses (so-called eigenstresses) due to the non-uniformity of temperature across the member's section. These stresses represent the internal restraint.
b) The other component is due to external restraint. External restraint occurs if the free thermal deformation of the concrete member is impeded by adjacent older concrete elements, by the subsoil, etc.. The stresses due to external restraint are not self-equilibrating. They lead to inner actions and to external reactions. Considering the restrained element itself, the reactions self-equilibrate.

Internal restraint alone occurs only in the totally unrestrained concrete member. However, all structural concrete elements are more or less externally restrained. Hence, internal and external restraint nearly always coexist. For practical purposes, it is convenient to distinguish between these two types of restraint and to determine their stress responses sepa-

Prevention of Thermal Cracking in Concrete at Early Ages. Edited by R. Springenschmid. RILEM Report 15. Published in 1998 by E & FN Spon, 11 New Fetter Lane, London EC4P 4EE, UK. ISBN 0 419 22310 X

rately. However, such procedure is a simplification because of the non-linear behaviour of concrete. In spite of this fact this simplification can be up-held in many cases, especially in estimating the restraint actions which may lead to macrocracking of the member. The same applies to the hardened, older concrete member which during its service life may be subjected to temperature changes in the course of climatic variations, etc..

In view of the fact that external restraint may cause severe cracks, their control is pre-eminent. Avoidance of cracking in various ways is described in this State-of-Art Report (STAR) and in various papers of [17]. To control the effectiveness of a chosen method (e.g. cooling of concrete member), the designer has to estimate the thermal restraint stresses for the specific case in the design stage. The purpose of this report is to present the most important aspects of the practical assessment of the stresses caused by external restraint.

6.2 SYMBOLS

t, Δt	age, time; age or time increment
t_{ef}	effective age
α_T	coefficient of linear thermal expansion
T	temperature in °C
ΔT	temperature difference in K
ε, κ	strain and curvature, general
ε_0, κ_0	free thermal strain and curvature
f_c	strength of concrete, general
f_{cm}, f_{ck}	mean and characteristic cylinder compressive strength
f_{ctm}	mean axial tensile strength
E_{ctm}	mean tensile Young's modulus (≈ compressive)
$E_{m,ef}$	effective tensile Young's modulus (≈ compressive)
ψ	relaxation function
ρ_c	density of concrete
σ	stress, general
F, M	force, moment
k_R, R_N, R_M	restraint factors
f_{yk}	characteristic yield strength of reinforcement

6.3 PRINCIPAL CAUSE OF RESTRAINT STRESSES

6.3.1 Relevant differences of temperature and thermal strain

The principal cause for internal and external restraint stresses is the time-dependent rise and fall of concrete temperature and the subsequent generation of the free thermal strains ε_0 within the young concrete element. Section 6.5 of the STAR describes the methods to determine the fields $T(x_j, t)$ and $\varepsilon_0(x_j, t)$ dependent on the relevant parameters.

The thermal stresses shall be described with an example. Figure 6.1 shows the time-dependence of concrete temperature for a wall at a certain height z above foundation.

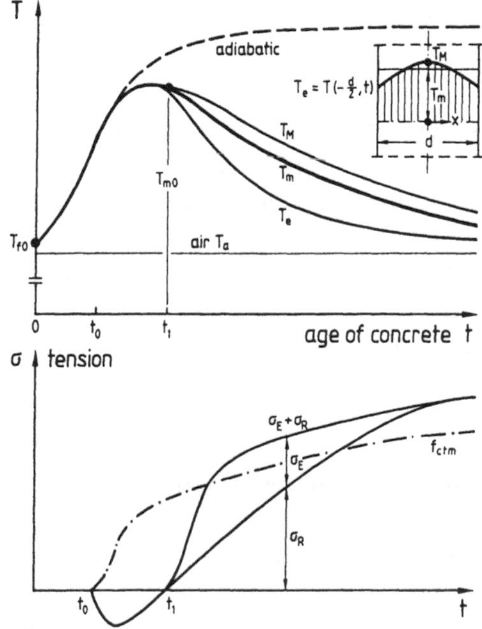

Fig. 6.1 Temperatures and stresses in a restrained wall in the uncracked stage (schematic).

Several simplifications are assumed. With the mean free thermal strain

$$\varepsilon_0(z,t) = \alpha_T \, \Delta T(z,t) \tag{6.1}$$

the concrete tensile stress due to external restraint is caused by the difference of the mean temperatures:

$$\sigma_R(t) \approx -k_R \alpha_T [T_m(t) - T_{m0}] E_{m,ef} \tag{6.2}$$

The restraint stress increases from zero to a maximum value. It is subjected to long-term relaxation, but if it reaches the tensile strength of concrete, macrocracking will occur. At the age t_0 the concrete has become solid matter, responding with stress to imposed strain. The effective modulus also takes into account the relaxation of concrete.

Internal restraint is caused by the non-uniformity of temperature across the section. The maximum tensile stress at the wall's surface can be expressed by:

$$\sigma_E\left[\frac{d}{2}, t\right] \approx -\alpha_T \left[T\left(\frac{d}{2}, t\right) - T_m(t)\right] E_{m,ef} \tag{6.3}$$

The eigenstress tends from zero to a maximum and vanishes thereafter. It is transient and subject to relaxation. If it reaches the surface tensile strength, it will break down, leaving shallow and narrow surface cracks behind.

As Fig. 6.1 shows, both stresses superimpose. Hence, the first surface cracks will occur if the sum of stresses $\sigma_E(d/2, t) + \sigma_R(t)$ reaches the momentaneous axial tensile strength of the concrete. Thereby, the thickness of members and the temperature distribution play a decisive role.

6.3.2 Determination of thermal dilatation

The determination of the restraint stress requires the knowledge of the field of temperature and thermal strain to calculate the unrestrained thermal dilatations of the structural member. Besides that several material properties must be known.

Several reports on methods exist for the calculation of the concrete temperature ([1], [2], [3]). In the sequel, the principles of determination of the thermal dilatations will be shown. Reference is made to the Chapters 2 and 5 of the STAR.

In the case of a large concrete member (wall, slab, etc.) the distribution of the thermal strain ε_0 at a specific location and for the age t is depicted in Fig. 6.2.

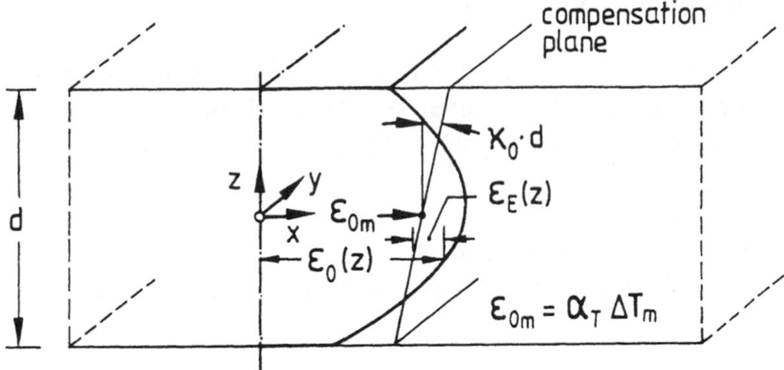

Fig. 6.2 Tensile stress-deformation relationship in accordance with MC 90.

One-dimensional heat-flow in z-direction may occur. The free thermal dilatations are characterised by the mean strain ε_{0m} and mean curvature κ_0. Both are found by the introduction of the compensation plane and are determined with the 0th- and 1st-order moments. If the member is modelled e.g. by equal finite strips as shown in Fig. 6.8 we obtain:

$$\varepsilon_{0m}(t) = \frac{\sum_{1}^{m} \varepsilon_{0k}}{m} \tag{6.4}$$

$$\kappa_0(t) = \frac{\sum_{1}^{m} \varepsilon_{0k} \cdot z_k}{\sum_{1}^{m} z_k^2} \tag{6.5}$$

In the general case of two- and three-dimensional heat flow the mean strain and curvature will have to be determined accordingly.

6.4 TYPES OF EXTERNAL RESTRAINT

6.4.1 Definition

If the free deformations are hindered, restraint will arise. The magnitude of hindrance is often described by a restraint factor.

6.4.2 Common types of restraint

Nearly all structural members are restrained. The degree of restraint varies, it depends on several parameters which will be dealt with later. One may distinguish between the following main types in practice:

End restraint
Restraint may arise at the end and at inner supports of beams, slabs, frames, etc., if the deformation is impeded there by neighbouring members.

Continuous restraint
If a slab or foundation is cast directly on to the subsoil, restraint actions are built-up from the ends of the member. Especially on rock, the restraint may become high. On clayey soil and gravel, the shear resistance of the soil and other factors influence the magnitude of restraint.

A frequent case of continuous restraint is a wall on foundation. Cracks in the wall may arise because the stiffness of the foundation will exceed the stiffness of the wall. In practice combinations of end and continuous restraint occur. Abrupt change of cross-section within a structural member also lead to restraint.

6.5 RESTRAINT ACTIONS

6.5.1 Introductory remarks

In contrast to load-induced actions, the estimation of restraint actions is hampered by additional uncertainties regarding

- the time-dependent viscoelastic and viscoplastic behaviour and strength of the young, rapidly ageing concrete;
- the realistic forecast of the field of temperature, thermal strain and degree of hydration;
- the realistic modelling of the interaction of the restraining effects.

Practice deals with restraint on different levels. For r/c-elements with small to medium dimensions (max $d \leq 0.8$ m), which under service loads presupposedly will be in the cracked stage usually little effort is made to realistically estimate the restraint. A minimum restraint reinforcement is required by most codes to control the crack width. Thereby usually full restraint is assumed. This crude approach is often acceptable, but will lead to excessive reinforcement especially if water tightness is required and if the thickness of a member increases. Hence, for massive members the control of early age cracks is also economically paramount.

In the sequel, several ways to realistically estimate the restraint actions at different levels of complexity will be presented. It must be stressed that the realistic assessment of stresses and crack risk not only requires mechanical modelling and numerical work but above all test data on the young concrete's properties.

6.5.2 Relevant material properties

For the realistic forecast of restraint actions, the tensile Young's modulus E_{ct}, the axial tensile strength f_{ct} and the creep or relaxation must be known with reasonable accuracy. Chapter 5 of the STAR gives an overview on known relationships.

If test data for the concrete to be used for the cast of the member are not available, code values might assist. Some codes such as ACI [1], CEB-FIP MC 90 [4], EC 2, pt 1 [5] and JSCE [11] present values of f_{ct} and E_{ct}, mostly in relation to the characteristic compressive strength f_{ck}. Usually such relationships are valid for standard curing of concrete specimens at 20°C and for the age of 28 days. However, for a massive member, the dependence of the mechanical properties on the effective age must be known. Hence, in conclusion, with code values the true behaviour of concrete in the structure can only crudely be described. In spite of this fact, several formulae are presented (values in MPa).

The mean tensile modulus at 28 days can be estimated with MC 90 as follows

$$E_{ctm,28} \approx 21500\,\alpha_E \left(\frac{f_{ck}+8}{10}\right)^{0.333} \qquad (6.6)$$

with e.g. $\alpha_E \approx 1.0$ for quartzitic aggregate. The mean axial tensile strength is accordingly formulated:

$$f_{ctm,28} \approx 1.4 \left(\frac{f_{ck}}{10}\right)^{0.667} \qquad (6.7)$$

In Fig. 6.3 the relationships between stress and strain and crack-opening, resp., are depicted for hardened concrete (t ≥ 28 d).

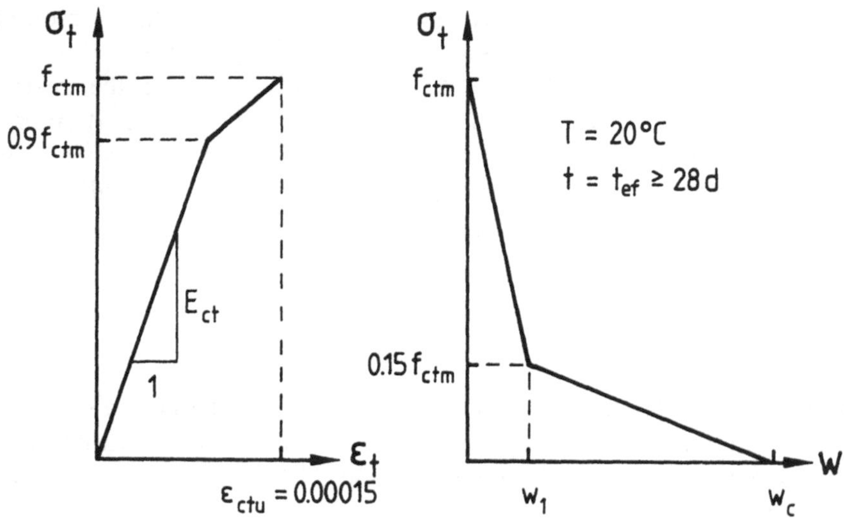

Fig. 6.3 Age-dependence of tensile strength and Young's modulus in accordance with MC 90.

The knowledge of the values f_{ctm} and E_{ctm} at 28 days and for curing at 20°C is insufficient. Also their dependence and of the relaxation on age, type of cement, curing condition T(t), etc. should be known. The accelerating effect of elevated temperature T(t) ≥ 20°C can be described by the effective age t_{ef} (for CEM I and II cements, [5]):

$$t_{ef} = \sum_{1}^{n} \Delta t_i \exp\left[13.65 - \frac{4000}{273 + T(\Delta t_i)/T_0}\right] \tag{6.8}$$

where: T_0 = 1°C; Δt_i ... time interval; $T(\Delta t_i)$ ... mean temperature within time interval. If T(t) ≥ 20°C, then $t_{ef} \geq t$. The effective age is the maturity-equivalent time if hardening would occur at a constant temperature of 20°C. For the evaluation of Eq. (6.8), the time function of the mean concrete temperature $T_m(t)$ must be known; e.g. on basis of the calculation field $T(x_j, t)$.

In MC 90 the age-dependence of the cylinder compressive strength and Young's modulus is described. Figure 6.4 shows the relationship between the hardening factor β_{cc} of the compressive strength and the effective age. With β_{cc} the age-functions of the compressive strength and of the Young's modulus can be expressed:

$$f_{cm}(t,T(t)) = f_{cm}(t_{ef}) \approx \beta_{cc} f_{cm,28} \tag{6.9a}$$

$$E_{ctm}(t,T(t)) = E_{ctm}(t_{ef}) \approx \beta_{cc}^{1/2} E_{ctm,28} \tag{6.9b}$$

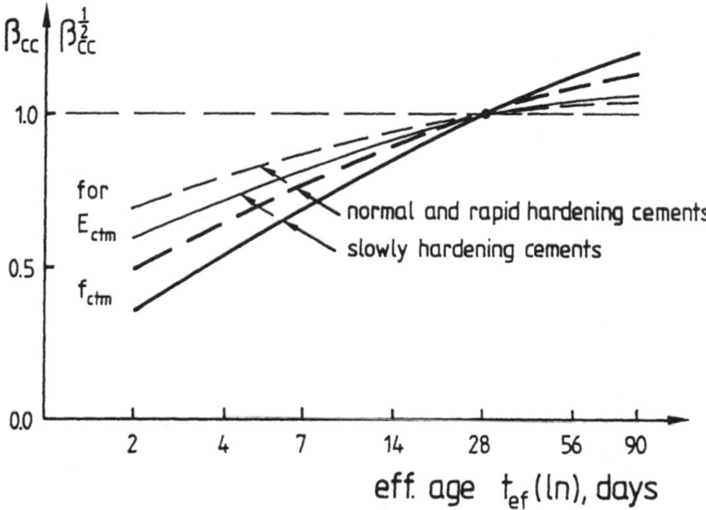

Fig. 6.4 Free thermal dilatations in a wall or slab.

MC does not give the age-dependence of the tensile strength. Eq. (6.10) presents an approximation, though the early age rate of strength development will be overestimated by it [6]:

$$f_{ctm}(t,T(t)) = f_{ctm}(t_{ef}) \approx \beta_{cc}^{2/3} f_{ctm,28} \tag{6.10}$$

If the tensile strength is forecast in accordance with Chapter 6 or with the Eq. (6.7) and (6.10) or even if it is determined by tests for the concrete of the cast, it will differ from the effective tensile strength $f_{ctm,ef}$ of the concrete in the structure. For this fact several causes are responsible. In [6] is shown that it can be estimated with Eq. (6.10):

$$f_{ctm,ef} \approx (0.75 \div 0.80)\ f_{ctm} \tag{6.11}$$

because of the simultaneous presence of tensile thermal stress and other reasons. Eq. (6.11) represents the relevant cracking resistance of concrete.

The realistic assessment of creep and relaxation at early age is difficult. In Chapter 6 of STAR several relationships are presented. Well verified by tests is the relaxation function of [7]:

$$\psi(t,t_i) = \frac{\sigma(t,t_i)}{\sigma(t_i)} = \exp\left[-P_1(t_{ef,i})\left(\frac{t-t_i}{t_k}\right)^{P_2(t_{ef,i})}\right] \tag{6.12}$$

with: $t_k = 1$ h; P_1 and P_2 are parameters depending on the effective age $t_{ef,i}$ at first loading; $t - t_i$ time under load.

Relations of the material values based on codes or other sources which were not calibrated by tests with the specific concrete are insecure. Hence, for massive structures tests to calibrate the relationships are essential. Reference [8] shows which tests are necessary.

6.5.3 Determination of restraint actions

Provided the thermal deformations and the material properties were realistically assessed, the interaction between the restrained and the restraining elements has to be modelled. This is a statical problem which can be treated on such level of complexity as the problem warrants. In the sequel several typical problems are treated, beginning with an ideal case.

6.5.3.1 Ideal axial restraint

In Fig. 6.5 the longitudinal section of a member (e.g. slab) between supporting elements and the thermal deformations ε_{0m} and κ_0 are depicted. The connection between the member and the restraining elements may be modelled by a hinge at the left hand end and by an extensional spring at the right hand end. Hence, restraint is not total but resilient. Warping is assumed to be unhindered. The restraint force is then:

$$F_R = -\alpha_T\,\Delta T_m\,E_{m,ef}\,A\frac{1}{1+\dfrac{E_{m,ef}A}{c\,L}} = -\alpha_T\,\Delta T_m\,E_{m,ef}\,A\,k_R \tag{6.13}$$

with: A ... cross-section; $E_{m,ef}$ ... mean effective modulus; c [N/mm] ... spring constant. If the slab were connected in such a way that the thermal curvature is impeded, then also a restraint moment would arise. A rotational spring would have to be introduced for modelling.

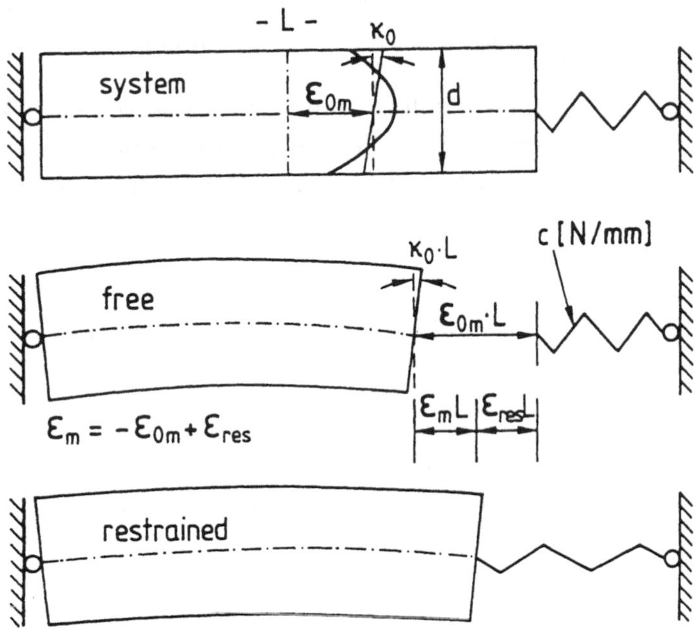

Fig. 6.5 Axial resilient restraint.

The intensity of restraint can be described by the restraint factor:

$$k_R = \frac{1}{1 + \frac{E_{m,ef} A}{c\,L}}; \quad 0 \le k_R \le 1 \tag{6.14}$$

with: $c \to \infty$, $k_R \to 1$, total restraint; and with $c = 0$, $k_R = 0$, no restraint. Eq. (6.14) shows that the restraint is influenced by a stiffness ratio which in this case is defined as

$$S_D = \frac{E_{m,ef} A}{c\,L}; \quad 0 \le S_D \le \infty \tag{6.15}$$

Both k_R and S_D are age-dependent because the mean effective modulus depends on the age and includes the viscoelasticity of young concrete. This fact can be taken into account by means of the effective modulus

$$E_{m,ef} \approx E_m \Psi_m \tag{6.16}$$

with E_m and ψ_m being representative mean values within e.g. the first 10 days of a gradually increasing restraint. In [3] values were given for ψ_m: 0.65 for $t_{ef,0} \approx 2$ to 3 d, and 0.75 for $t_{ef,0} > 5$ d, with $t_{ef,0}$ the effective age at the on-set of restraint.

In Fig. 6.6 the force-strain line of the member according to Eq. (6.13) is depicted. The member with resilient restraint responds to a strain ($-\varepsilon_{0m}$) with a lower force than with rigid restraint. If the cracking force

$$F_{R,cr} = f_{ctm,ef} A \tag{6.17}$$

is reached, restraint vanishes for the unreinforced member.

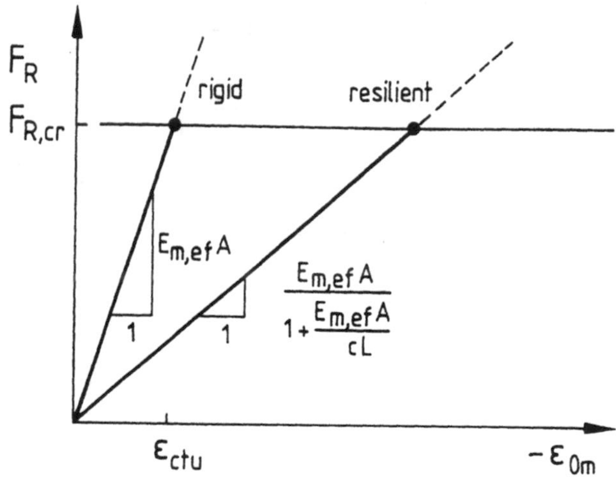

Fig. 6.6 Rigid and resilient restraint; cracking force.

6.5.3.2 General Approach

Massive concrete members can often be modelled by a representative longitudinal cut-out from the centre part of member. This is, e.g., shown in Fig. 6.7 for a foundation slab. One - or two - dimensional heat-flow and plain strains are assumed. Figure 6.8 shows the strain distribution in such a strip. The plane which equilibrates the strains is called compensation plane. The magnitude of local or end restraint conditions can be expressed by:

$$\varepsilon_{res} = \varepsilon_{0m} + \varepsilon_m \tag{6.18}$$

$$\kappa_{res} = \kappa_0 + \kappa \tag{6.19}$$

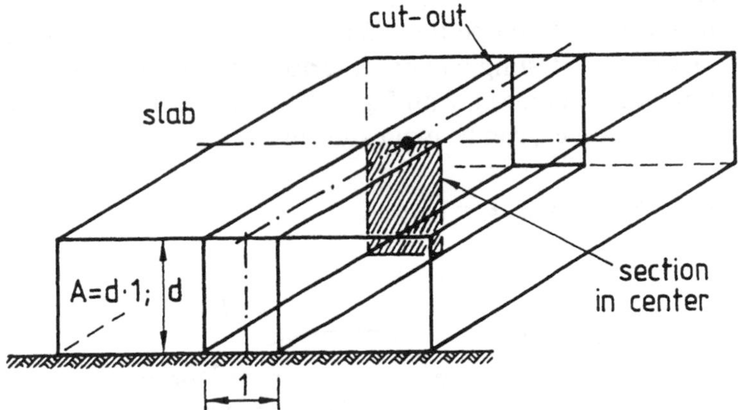

Fig. 6.7 Slab on ground, representative cut-out and section.

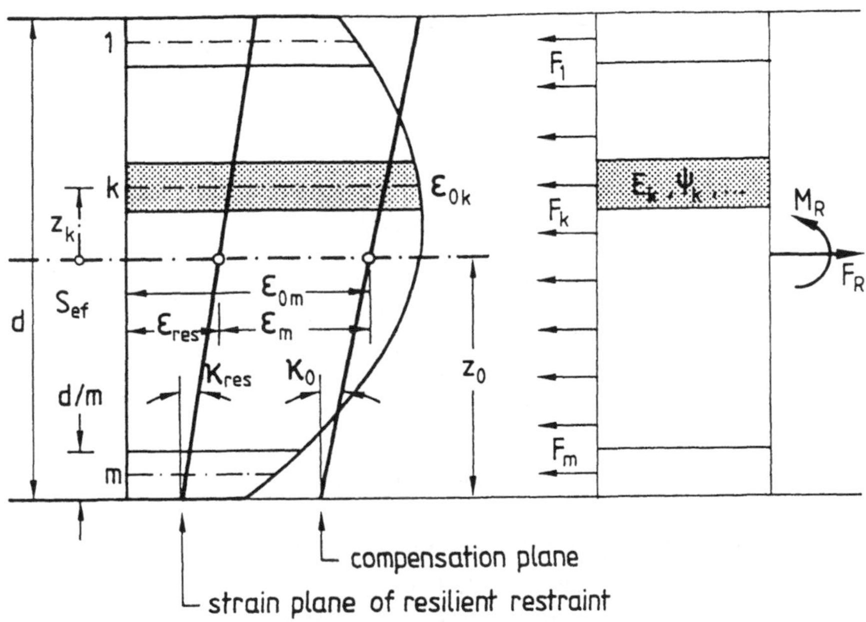

Fig. 6.8 Finite layer model.

The residual deformations ε_{res}, κ_{res} describe the boundary conditions; they are the unknowns. The resultant restrained deformations are ε_m and κ. One recognises immediately: $\varepsilon_{res} = \kappa_{res} = 0$ means total restraint; $\varepsilon_{res} - \varepsilon_{0m} = 0$, $\kappa_{res} - \kappa_0 = 0$ means zero restraint.

Methods to assess the restraint actions on a practical level are reported in the literature (e.g. [8] to [11]). For a practical approach the section can be modelled by m finite and equidistant layers with age-dependent mechanical properties. With a modified stress-strain-line of e.g. Fig. 6.3 or from testing and by describing the strain history by incremental strain steps at times t_i with Fig. 6.9, the compatibility condition for the layer k at the time t_i can be expressed by:

$$\Delta\varepsilon_{el,ki} + \Delta\varepsilon_{0,ki} - \Delta\varepsilon_{res,i} - \Delta\kappa_{res,i}\, z_k = 0 \tag{6.20}$$

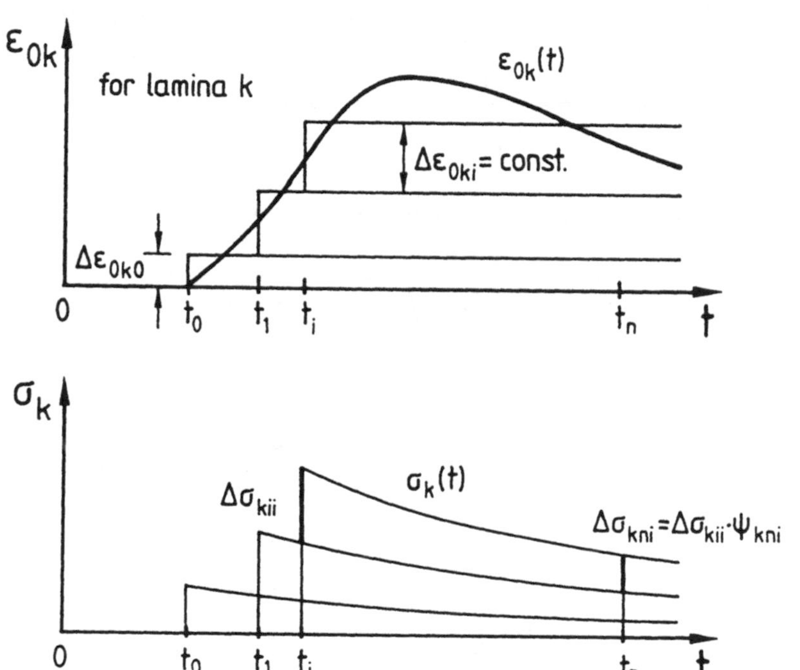

Fig. 6.9 Incremental strain history and stress response.

Each strain step causes a spontaneous stress response which diminishes in course of time due to relaxation.

For the on-set of restraint at time t_0 (i = 0), the force and stress of the layer k at time t_n can be expressed by:

$$F_{kn} = -\frac{d}{m}\sum_{i=0}^{n-1}\left[\Delta\varepsilon_{0,ki} - \Delta\varepsilon_{res,i} - \Delta\kappa_{res,i}\, z_k\right] E_{ki}\,\psi_{kni} \tag{6.21}$$

$$\sigma_{kn} = \frac{m}{d} F_{kn} \tag{6.22}$$

The resultant restraint actions are:

$$F_{Rn} = \sum_{1}^{m} F_{kn} \tag{6.23}$$

$$M_{Rn} = \sum_{1}^{m} F_{kn} \cdot z_k \tag{6.24}$$

For hardened concrete the free thermal deformations ε_{0m} and κ_0 can be calculated in accordance with Eq. (6.4) and (6.5) to determine the compensation plane. For young concrete this would lead to gross errors, because the viscoelastic and viscoplastic properties vary across the section and with time. This calls for the modification of the compensation plane. Hence, Eq. (6.4) and (6.5) have to be modified. We obtain incrementally for any time t_i of strain step (i = 0; 1 ... n-1) at time t_n:

$$\Delta\varepsilon_{0m,ni} = \frac{\sum_{1}^{m} \Delta\varepsilon_{0ki}\, E_{ki}\, \psi_{kni}}{\sum_{1}^{m} E_{ki}\, \psi_{kni}} \tag{6.25}$$

$$\Delta\kappa_{0,ni} = \frac{\sum_{1}^{m} \Delta\varepsilon_{0ki}\, E_{ki}\, \psi_{kni}\, z_k}{\sum_{1}^{m} E_{ki}\, \psi_{kni}\, z_k^2} \tag{6.26}$$

Clearly, at $t_i = t_n$ the relaxation function is $\psi_{kni} = 1$. For the elastic material, $\psi_{kni} = 1 = \text{const.}$, and if the Young's modulus does not vary across the section the Eq. (6.4) and (6.5) are obtained. The origin of the ordinate z is the position of the neutral axis of section S_{ef}.

For zero external restraint, $\varepsilon_m = 0$, $\kappa = 0$, the actions F_{Rn} and M_{Rn} vanish. However, the internal restraint stays on. It is in balance. For full restraint, we have to set: $\varepsilon_{res} = \kappa_{res} = 0$. These relations can be applied to the case of sec. 6.5.3.1. The total axial restraint force is then:

$$F_{Rn} = -\frac{d}{m} \sum_{1}^{m} \sum_{0}^{n-1} \left[\Delta\varepsilon_{0ki} - \Delta\varepsilon_{res,i}\right] E_{ki}\, \psi_{kni} \tag{6.27}$$

For comparison, $E_k\psi_{kni}$ is substituted by the mean effective $E_{m.ef}$ modulus.

Then we obtain with

$$F_{Rn} = -d\,E_m\,\psi_m\,(\varepsilon_{0mn} + \frac{F_{Rn}}{c\,L})$$

the result of Eq. (6.13):

$$F_{Rn} = -\varepsilon_{0mn}\,d\,E_{m,ef}\,\frac{1}{1+\dfrac{d\,E_{m,ef}}{c\,L}} \tag{6.28}$$

In the JSCE-standard [11] the so-called compensation plane method is presented. With the notations of sec. 6.2 the residual deformations ε_{res} and κ_{res} of Eq. (6.18) and (6.19) are expressed in terms of the axial restraint factor R_N and the moment restraint factor R_M:

$$\varepsilon_{res} = \varepsilon_{0m}(1-R_N) \tag{6.29}$$

$$\kappa_{res} = \kappa_0(1-R_M) \tag{6.30}$$

If these relations are introduced into Eq. (6.20), the increments of axial restraint force and of restraint moment at the time t_i are obtained:

$$\Delta F_{Ri} = -R_N\,E_{mi}\,d\,\Delta\varepsilon_{0mi} \tag{6.31}$$

$$\Delta M_{Ri} = -R_M\,E_{mi}\,\frac{d^3}{12}\kappa_{0i} \tag{6.32}$$

They lead to the increment of the local restraint stress at the time t_i:

$$\Delta\sigma_{Ri}(z) = R_N\,\frac{\Delta F_{Ri}}{d} + R_M\,\frac{\Delta M_{Ri}}{d^3}12\,z \tag{6.33}$$

The restraint actions and stresses at the time t_n are determined by summation over time. If we apply Eq. (6.33) to the case of sec. 6.5.3.1, we obtain the axial restraint force:

$$F_{Rn} = -\varepsilon_{0mn}\,d\,E_{m,ef}\,R_N \tag{6.34}$$

which coincides with Eq. (6.13) if $k_R = R_N$. The restraint factors vary with the time. They have been identified by FEM analysis, and are used in [11].

The most general approach to solve a restraint problem is the three-dimensional FEM. More important than a complex statical solution is the realistic modelling of the boundary conditions and interaction of members, resp..

6.5.3.3 Base restraint on rock and soil

In practice foundations, slabs, etc. are cast directly on rock, soil or pile caps. Before concreting the member, often a e.g. 10 cm thick concrete layer is cast on ground to provide a level bottom plane for the member. In other cases such a layer is covered by e.g. polymer sheets to break the bond.

If the concrete member is cast directly or via a concrete base layer on solid rock, several effects have to be taken into account. Because of the heat-flow, the rock will deform and import this deformation to the fresh concrete. Restraint occurs in all three directions. Modelling of the joint characteristics and restraint parameters becomes difficult.

One of the first reports on restraint of blocks on rocks is the one by ACI-committee 207 [1]. In this report only axial restraint is considered, the bending stiffness of the subbase is assumed to be infinite. Restraint depends on the ratio of extensional stiffness: $S_D = A_cE_c/A_FE_F$, with A_FE_F stiffness of rock and $A_F \approx 2.5\ A_c$. Figure 6.10 shows the dependence of the restraint factor on the height above base and on the slenderness ratio L/H.

The restraint stress is defined as:

$$\sigma_R\left(\frac{z}{H}\right) = -k_R\left(\frac{z}{H}\right) \alpha_T\ \Delta T_m\ E_c \tag{6.35}$$

with the restraint factor at the base:

$$k_R = \frac{1}{1+\dfrac{E_cA_c}{E_FA_F}} \tag{6.36}$$

Because of several arbitrary assumptions in the equations and other reasons, the results neither correspond with recent theoretical approaches nor with the results of stress-meter tests. The restraint by rock can be more suitably described by the methods of sec. 6.5.3.4.

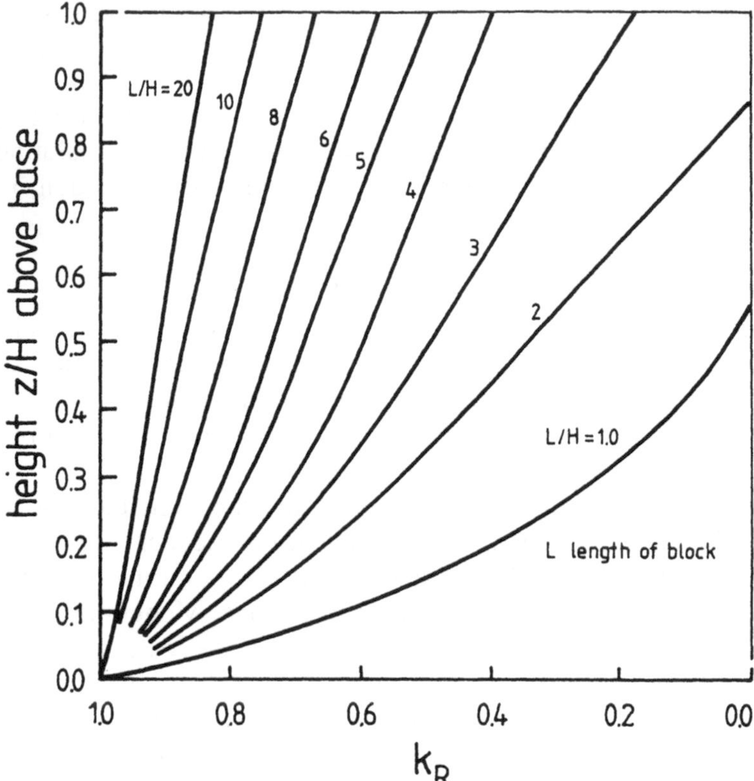

Fig. 6.10 Restraint factor in accordance with ACI 207 at centre section.

If the slab is cast on clay or gravel, the free deformations will be impeded by friction and/or by interlock with the subbase. Figure 6.11 shows for the cooling phase of concrete that the end region of the slab is subjected to friction bond stresses

$$\tau_{fr} = \mu_{fr}\,\sigma_V = \mu_{fr}\,\rho_c\,d \tag{6.37}$$

with μ_{fr} ... coefficient of sliding friction; and σ_v ... pressure on soil by dead-weight of slab. In the central region of long slabs the slip u between slab and soil may be very small. Hence, as free sliding cannot occur, bond stresses due to interlock will arise. This behaviour can be modelled as shown by Fig. 6.12 [18]. In Fig. 6.13 the coefficients of static friction μ_{st} and sliding friction μ_{fr} are plotted vs. the pressure σ_v. The lines pertain to sand and gravel subbase with different densities of about 0.4 and 0.6. With the coefficient μ_{st} the modulus of horizontal bedding c_x can be determined. Then the distributions of bond stress and longitudinal restraint stress $\sigma_R(x)$ can be assessed and compared with the tensile strength of concrete, Fig. 6.11.

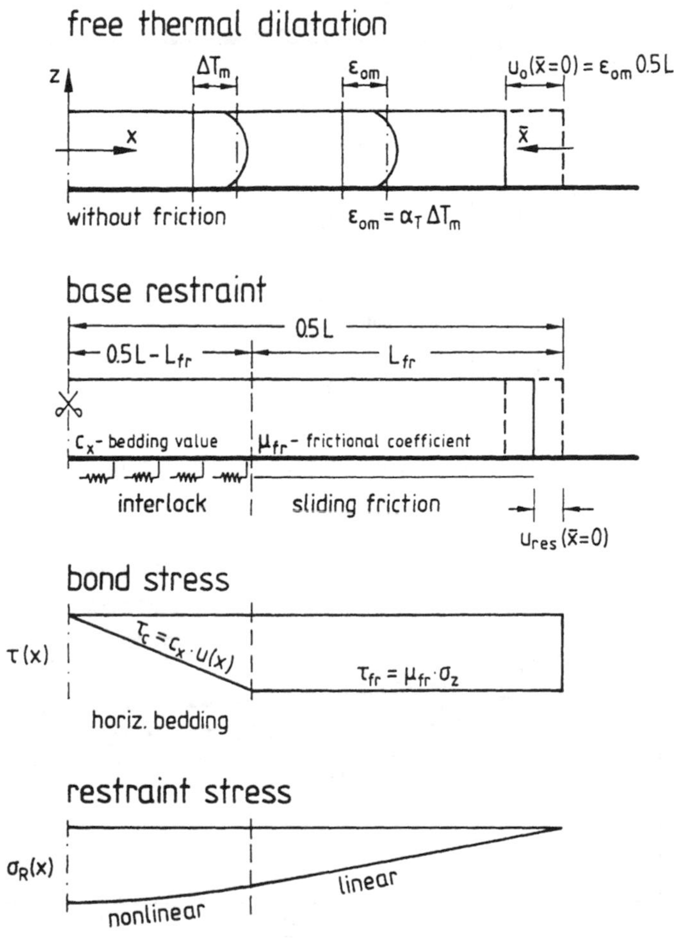

Fig. 6.11 Restraint due to friction on subbase.

Evaluation and experience prove that restraint by friction is only important for rather thin slabs d ≤ 40 cm.

Friction between the slab and the subbase can be reduced by a sliding layer placed between the concrete subbase and the slab. Several types of material are used, differing in effectiveness. Smoothness of the underside of the slab and top of the subbase has a great influence on friction.

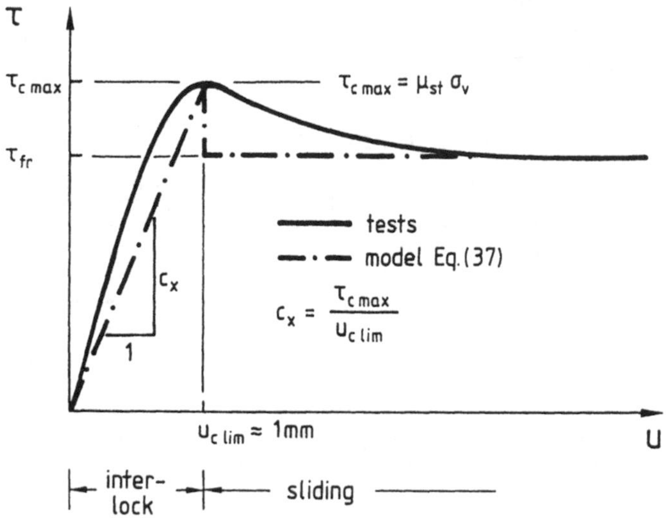

Fig. 6.12 Modelling of bond stresses due to interlock considering sliding.

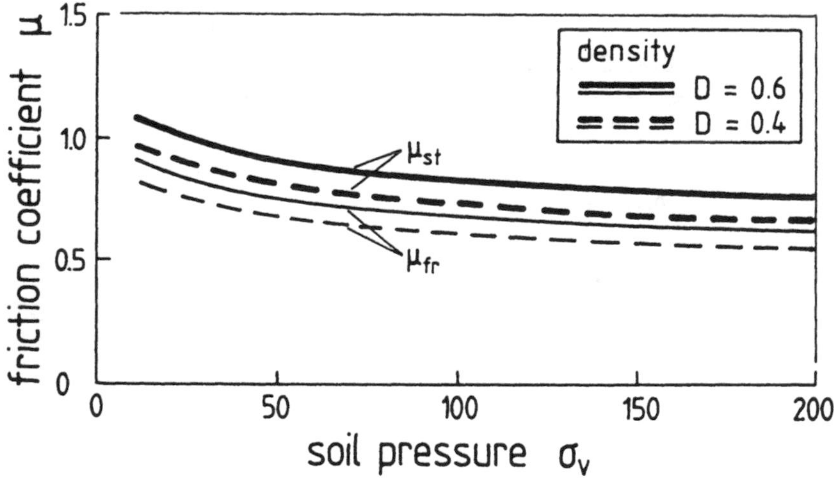

Fig. 6.13 Coefficients of static friction μ_{st} and sliding friction μ_{fr}.

6.5.3.4 Wall on foundation

The junction of a wall with an older slab or foundation strip is the most common type of restraint. As Fig. 6.14 shows the free dilatation of wall is impeded by the foundation. In course of restraint, both members - being monolithically connected along the working joint - will deform. Restraint is introduced into the wall on both ends within St. Venant's zones of the length ≈ H. Wall and foundation form a two-layer system. If it is bedded on soil and if $L/H < 10$, the influence of dead weight on restraint actions can be neglected [12], [14].

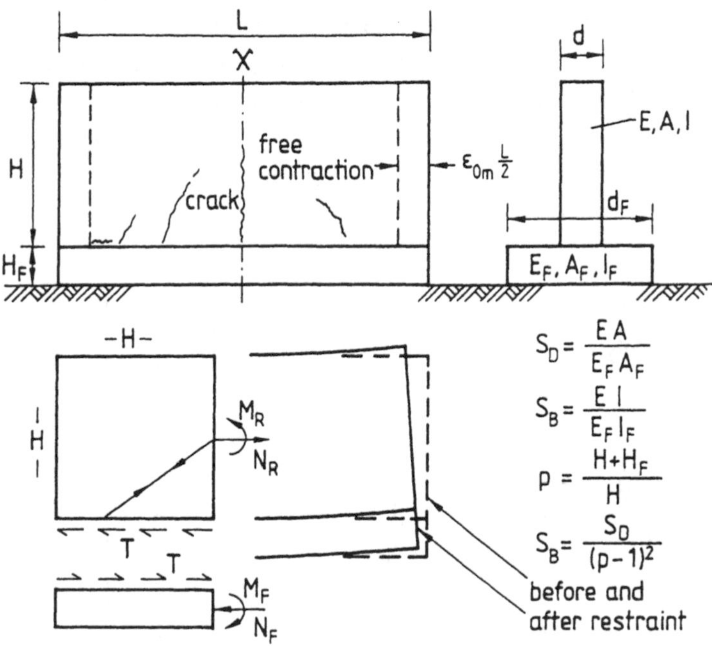

Fig. 6.14 Wall/foundation system, actions and stiffness ratios.

Types of cracks

In practice two types of throughcracks can be observed, Fig. 6.15. They differ with respect to their height above working joint and to width. Their occurrence depends primarily on geometrical relations. In walls with a slenderness of $L/H \leq 1$, only cracks occur which starting at the bottom only partially cut through the height of wall, type PCr. They are rather narrow ≤ 0.1 mm. The cracks, type TCr cut through the entire height of wall, they may be wide. They depend primarily on the relative stiffness of both wall and foundation relations and occur if $L/H \geq 2$ to 2.5. Control of cracking has to concentrate on the crack type TCr.

Two-layer solution with effective elasticity

Many reports deal with this problem either theoretically and/or experimentally (e.g. [12] to [15]). Approximate solutions were found by assuming that materials are elastic. This assumption is justified for the older concrete of the foundation, however, crude for the hardening concrete of the wall, even if its relaxation is taken into account by means of the effective Young's modulus, e.g. [16].

It was found that the restraint actions N_R, M_R primarily depend on the stiffness ratios S_D and S_B which are defined in Fig. 6.14 and less on the slenderness ratio L/H. For total restraint we obtain $S_D = S_B = 0$ and for zero restraint, $S_D = S_B \rightarrow \infty$. Evaluations of executed structures described in [3] and [12] prove that high restraint is seldom. The stiffness ratios are in practice in the range: $0.2 \leq S_D < 2$ and $2 \leq S_B < 20$. Hence, the realistic assessment of restraint is important. The assumption of full restraint is often grossly erroneous.

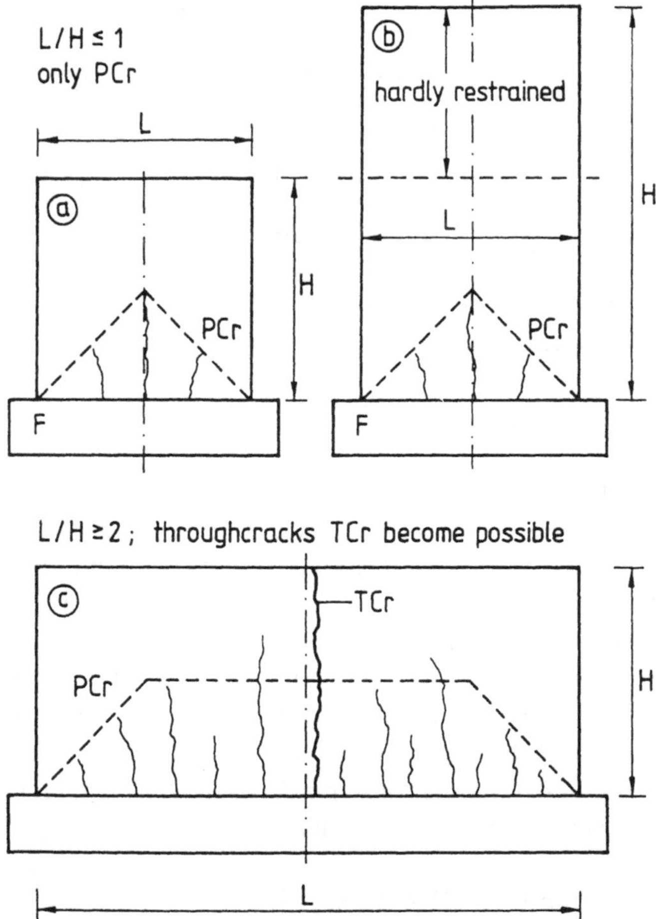

Fig. 6.15 Type of cracks in restrained walls.

On the assumption that the wall uniformly contracts by the free mean strain ε_{0m}, the restraint actions of the two-layer system are for L/H ≥ 2 [3]:

$$N_R \approx -\varepsilon_{0m} E_{m,ef} A \frac{1+S_B}{3S_B p^2 + (1+S_D)(1+S_B)} \tag{6.38}$$

$$M_R \approx -\varepsilon_{0m} E_{m,ef} A \frac{H}{6} \frac{3pS_B}{3S_B p^2 + (1+S_D)(1+S_B)} \tag{6.39}$$

With these equations, the stresses at the on-set of cracking can be estimated. However, these first cracks of the type PCr must not necessarily travel over the entire height of wall, as the restraint is somewhat relieved. Whether they will, depends on the stiffness parameters. This difficult problem was dealt with approximately in [3]. Fig. 6.16 shows the dependence of the critical thermal strain crit ε_{0m} for the transition of a crack PCr to a wall-high crack TCr dependent on stiffness parameters. If the corresponding thermal strain max ε_{0m} for a maximum drop of temperature max ΔT_m exceeds the critical value crit ε_{0m}, then the crack will run to the crown. For example: p = 1.25; $S_D = 0.5$; $\varepsilon_{ctu} \approx 0.000075$; max $\varepsilon_{0m} = 0.00045$; - max $\varepsilon_{0m}/\varepsilon_{ctu} = 6$; through-cracking is highly probable. As the stiffness ratios decrease, the likelihood for through-cracks increases. In [16] this problem was treated in terms of probability of crack occurrence. In [3] the described approach was extended to estimate the crack width and to determine the reinforcement for crack control.

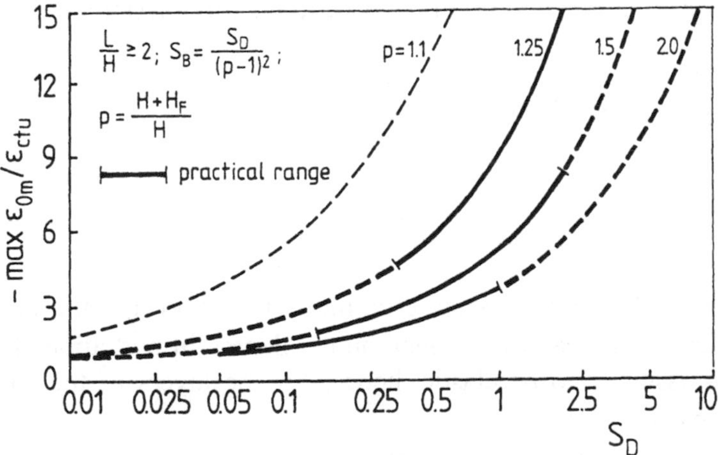

Fig. 6.16 Criterion for the formation of wall-high through-cracks.

Application of finite layer method

A more realistic estimation of the restraint actions can be achieved with the method of sec. 6.5.3.2. Fig. 6.17 shows the distribution of the thermal strains and the compensation plane.

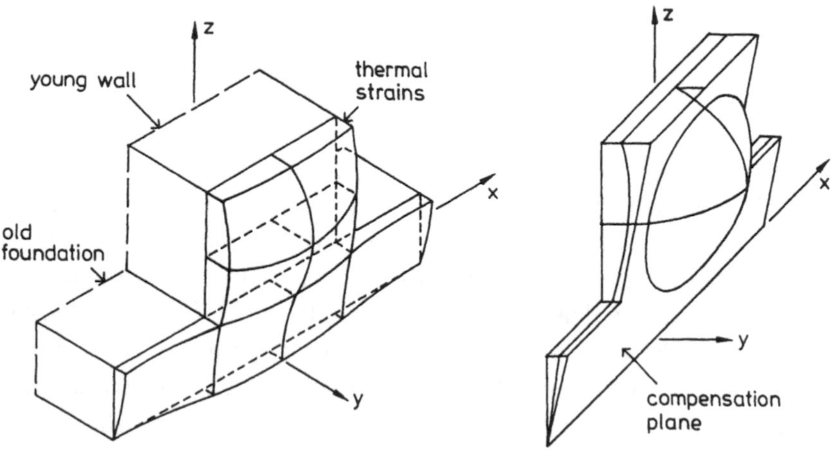

Fig. 6.17 Temperature strain distribution and compensation plane.

In Fig. 6.18 the two-layer system and its discretisation is depicted. With the afore-hand calculated temperature field the relevant mean thermal strain ε_{0mk} across section in layer k and the modified compensation plane, Eq. (6.25) and (6.26) can be calculated. Compatibility requires a common resultant plane of strains for the wall and foundation. Hence, we can describe the restraint actions of the wall/foundation-system.

$$F_{Rn} = -\sum_{k=1}^{m}\sum_{i=0}^{n-1}\left[\Delta\varepsilon_{0m,ki} - \Delta\varepsilon_{res,i} - \Delta\kappa_{res,i}\, z_k\right] E_{m,ki} \cdot A_{m,ki} \cdot \psi_{m,ki} \tag{6.40}$$

$$M_{Rn} = \sum_{k=1}^{m} F_{kn}\, z_k \tag{6.41}$$

where $A_{m,ki}$ refers to the area of each individual layer. For the young concrete of the wall E_k and ψ_k depend on the degree of hydration. The foundation is assumed to be elastic, hence, for the layers of the foundation $E_{F,k} = \text{const}$, $\psi_{F,k} = 1$.

With the equilibrium conditions the unknown deformations $\varepsilon_{res,n}$ and $k_{res,n}$ can be determined. Then, the local stress σ_{kn} due to external restraint can be calculated. The eigenstresses must be added.

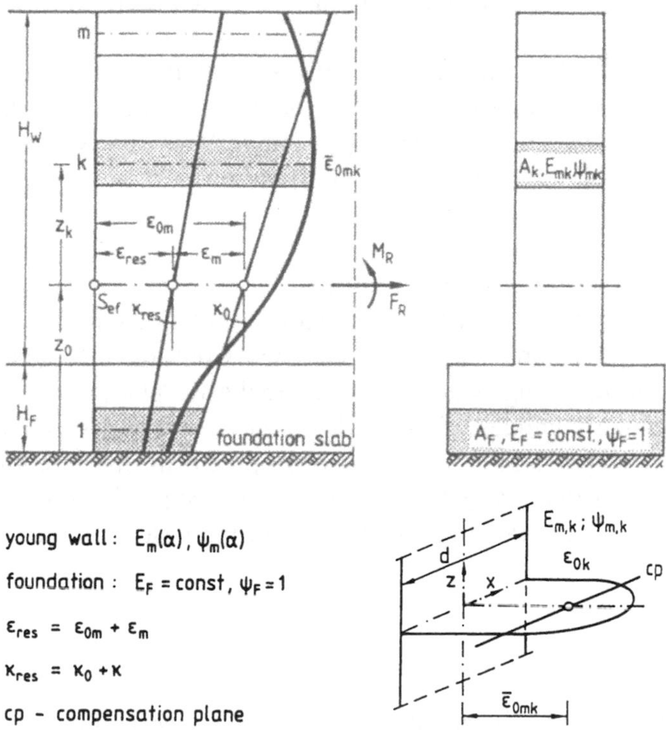

Fig. 6.18 Finite layer method for the wall-foundation system.

6.5.3.5 Improvement of restraint forecast by testing

The evidential quality of the models presented heavily relies on the realistic assessment of the material properties and restraint interaction parameters. Certainly, the numerical procedure chosen is of influence too. Assessment of the young concrete's properties with empirical formulae alone will lead to a crude forecast of stresses. The quality of forecast can be significantly improved by testing the properties of the specific concrete for the cast of member aforehand. The laboratory tests should comprise, dependent on effective age:

- adiabatic calorimetry,
- calibration of maturity functions t_{ef} and $\alpha(t_{ef})$,
- short-term σ-ε-lines in axial tension,
- isothermal creep and/or relaxation tests at varying temperature.

With such tests the material laws can be calibrated. Then the fields of T, α, ε_0, κ_0, of the material properties, and of the thermal stresses can be determined.

Final proof of actual quality is the structural member itself. Hence, the on-site monitoring of temperature, stress and strain - as described in Chapter 9 - becomes increasingly important for various reasons: improvement of forecast model, modification of concrete composition, curing, gain of experience, etc..

6.5.3.6 Stochastic nature of cracking

All important parameters which are incorporated in the numerical forecast of thermal stresses are scattering properties. Consequently, also the stress response and cracking occurrence are stochastic phenomena. Therefore, the deterministic models described so far should be supplemented by probabilistic aspects in the future. Development in this direction has not proceeded significantly yet. A promising approach is reported in [16]. There the so-called cracking index as scattering base variable

$$J_{cr} = \frac{f_t(x_j, t)}{\sigma(x_j, t)} \tag{6.42}$$

was introduced and calibrated by on-site crack observation.

6.6 CRACK CONTROL WITH REINFORCEMENT AND PRE-STRESSING

The limitation of the width of restraint cracks by means of reinforcement is dealt with in most codes following uniform principles. These principles will be elucidated for the case of sec. 6.5.3.1, Fig. 6.5. For general relationships, see MC 90, e.g..

Because the external restraint is accompanied by internal restraint, the influence of the latter on the formation of through-cracks is of interest, though yet in dispute. MC 90 presupposes that a through-crack spontaneously occurs if the sum of the stresses at the member's faces reaches the tensile strength. Hence, the mean restraint stress - acting on the entire section - can be expressed by a reduced tensile strength:

$$\sigma_R = k\, f_{ctm,ef} \tag{6.43}$$

with $0.8 \geq k \geq 0.5$ depending on the member's thickness.

A through-crack will only then occur if the imposed strain and the degree of restraint are high enough. This fact was already shown in Fig. 6.6. Behaviour after the first crack acc. to MC 90 is depicted in Fig. 6.19. In order to prevent wide cracks, the steel must not yield at first cracking:

$$F_{Rr} \leq F_{sy} = f_{yk}\, A_s \tag{6.44}$$

Eq. (6.44) leads to the lower boundary amount of the minimum reinforcement $A_{s,min}$. Control of crack width requires besides $\sigma_{sr} \leq f_{yk}$ and $A_s > A_{s,min}$ the fulfilment of requirements regarding diameter and spacing of bars. E.g. in MC 90 details are presented.

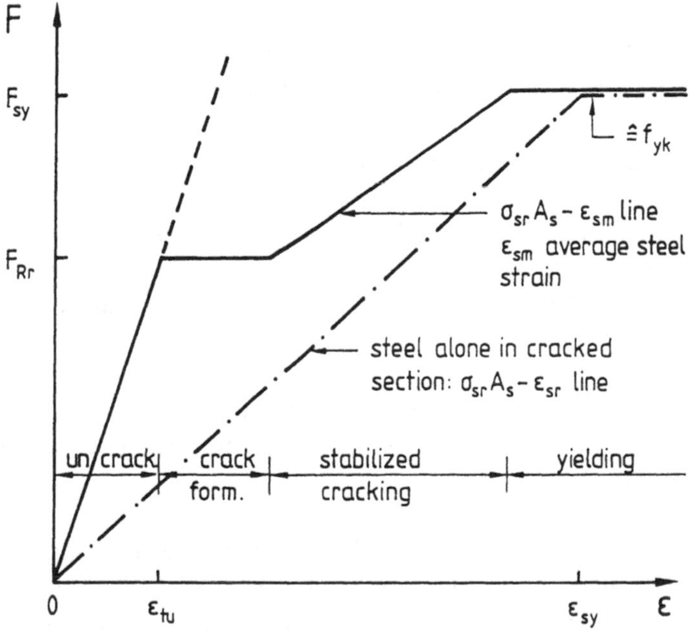

Fig. 6.19 Force - deformation line for the axially tensioned r/c-member.

By means of prestressing, avoidance and control of thermal and shrinkage cracks can be improved. Thereby it is essential that the desired precompression of the endangered regions of the member can be realised. E.g. for post-tensioned foundation slabs prestress must not be dissipated by friction or at deformation blocking inserts etc..

REFERENCES

[1] ACI Committee 207: Effect of restraint, volume change, and reinforcement on cracking of massive concrete. *ACI-Journal*, Proc. 70, **7**, July 1973, pp. 445 - 470.

[2] Breugel, K. van: Prediction of Temperature Development. Ibid [17], General Report.

[3] Rostásy, F.S. and Henning, W.: Zwang in Stahlbetonwänden auf Fundamenten. *Beton- und Stahlbetonbau*, **84**, (1989), H. 8, S. 208 - 214; H. 9, S. 232 - 238.

[4] CEB-FIP Model Code 1990; *CEB Bulletin d'Information* No. 213/214. Thomas Telford Serv. Ltd. London 1993.

[5] Eurocode 2, pt. 1: Design of Concrete Structures, 1992.

[6] Rostásy, F.S. and Onken, P.: Effective concrete tensile strength in the structure at early age thermal restraint. Res. Report, Institut für Baustoffe, Massivbau und Brandschutz, TU Braunschweig, April 1992.

[7] Laube, M: Werkstoffmodell zur Berechnung von Temperaturspannungen in massigen Betonbauteilen im jungen Alter. Doct. Thesis, TU Braunschweig, 1990.

[8] Tanabe, T.: Thermal stress analysis of massive structures by the proposed compensation plane method. *Proc. of First East Asian Conference on Structural Eng. and Construction*. Bangkok, pp. 778 - 785, 1986.

[9] Rostásy, F.S.; Laube, M. and Onken P.: Zur Kontrolle früher Temperaturrisse in Betonbauteilen. *Bauingenieur*, **68**, (1993), S. 5 - 14, Springer-Verlag.

[10] JCI Committee: Thermal stress of massive concrete structures. Report of the Japan Concrete Institute, November 1985.

[11] JSCE (Japan Society of Civil Engineers): Standard Specification for Design and Construction of Concrete Structures. Pt. 2, 1986.

[12] Stoffers, H.: Cracking due to shrinkage and temperature variation in walls, *Heron*, **23**, No. 3, 1978.

[13] CUR Rapport 85: Scheurvorming door krimp en temperatuurwisseling in wanden. Nov. 1978.

[14] Ishikawa, M.; Imaeda, Y.; Takatsuji, K. and Maeda, T.: An Effect of External Restraint in Mass Concrete. *Proc. JSCE*, No. 460, T-18, Febr. 1993.

[15] Emborg, M.: Thermal stresses in concrete structures at early age. Doct. Thesis, Luleå Techn. University, 1989.

[16] Kimura, K. and Ono, S.: Evaluation of thermal crack occurrence in massive concrete structures. *Concrete Library of JSCE*, No. 11, June 1988, pp. 59 - 81.

[17] *Thermal Cracking in Concrete at Early Ages*. RILEM-Proc. **25** of the Int. Symposium, Munich, Oct. 1994.

[18] Kolb, H.: Ermittlung der Sohlreibung von Gründungskörpern unter horizontalem kinematischen Zwang. Doct. Thesis, University Stuttgart, 1987.

7

Models and Methods for Computation of Thermal Stresses

M. Emborg

7.1 GENERAL

The fact that concrete, especially in massive concrete structures, tends to crack due to volume change during hydration is a problem that has been adressed by engineers for about 100 years. Measures to control cracking have almost been focused on restricting temperature differences within the pour and temperature rise in relation to older adjoining concrete and surrounding. Thus, research were concentrated on e g producing low-heat cement, lowering cement content, and to control the temperature rise. As a consequence, most efforts were made to develope methods for temperature predictions as well as for examining the thermal properties of the concrete. It is not until recently that engineers with the aid of modern laboratory techniques and the progress of computers have obtained more advanced understanding of the problems with thermal stresses thus including other influencing factors than temperature in analysis of thermal cracking.

One crucial point is to have reliable material models and methods for structural analysis by which the risk of cracking may be estimated and by which effects of different measures counteracting cracking can be quantified and compared in a correct way. Various methods of calculating cracking risks are presented in literature and some examples will be treated here.

First, we may look into the causes of thermal cracking. The most important parameters to be considered in a thermal stress analysis of a newly cast concrete element may thus be sketched as in Fig 7.1. Three main factors are identified in the figure: the temperature development in

Prevention of Thermal Cracking in Concrete at Early Ages. Edited by R. Springenschmid. RILEM Report 15. Published in 1998 by E & FN Spon, 11 New Fetter Lane, London EC4P 4EE, UK. ISBN 0 419 22310 X

the element being cast, the mechanical behavior of the young concrete and the degree of restraint imposed on the element. (An alternative sketch of influencing factors where the influence of adjoining structure specially is highlightened is shown by Bernander in Chapter 9.
In Fig 7.1 effects of shrinkage, superimposing the thermal effects, are not treated. It is wellknown that the temperature gradient at early ages in many cases is not the only cause of crack formation. Specially for concretes with low water to binder ratio, early age shrinkage (autogenious shrinkage and drying shrinkage) may induce considerable eigenstresses and cracking. The combined effects of temperature and shrinkage have gained in interest during the last decade and are treated in several reports see e.g. Roelfstra and Salet [1], Jonasson [2], Jonasson, Hedlund and Groth [3], Sellevold *et al.* [4], Hedlund [5] as well as in various other contributions of the Rilem Conference in Munich (Rilem, [52]). However, for the case of simplicity, the influence of shrinkage is only scarcely treated in this section.

The *temperature development* in the element is dependent on (a) thermal properties of the concrete (hydration heat etc), (b) environmental conditions (ambient air temperature, temperature of adjoining structures and subgrade, wind etc.), (c) geometry and dimension of the structure being cast and adjoining structures and (d) conditions at concreting (placing temperature, form work, insulation, cooling etc). Evaluation of temperature development in newly cast structures is treated by van Breugel in Chapter 4.

It has been found that the difference between placing temperature and temperatures of adjoining structures has a paramount effect on the temperature drop in the cooling phase, and subsequently the total contraction potential of the maturing concrete and thus cracking risk. It has also been found that for slender structures, the difference between initial temperature and the temperature to which the concrete is to reach is essential in the tensile stress built up, see Chapter 9.

The following properties, with regard to the *mechanical behavior* of the young concrete, are of great importance in the thermal stress analysis: (a) development of elasticity and creep (viscoelastic behavior) at normal and high stress levels and at elevated temperatures, (b) strength development, (c) thermal expansion/contraction, (d) non-linear behavior and fracture mechanics behavior and (e) plasticity all of them influenced by the temperature history of the concrete (i.e. by the maturity of the concrete). The mechanical properties are treated in Chapter 5.

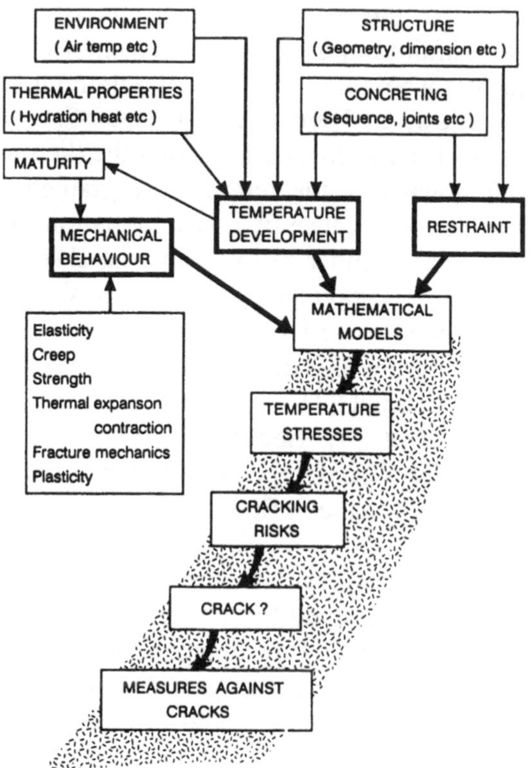

Fig 7.1 Analysis of early-age thermal stresses and risks of thermal cracking - influencing factors (from Emborg and Bernander, [6]).

The *restraint* is affected by the general structural configuration as such (stiffness of adjoining structures, dimensions, friction to underground etc.) and how the element is poured (casting sequences, casting joints etc.), see also Chapter 6 (Rostásy, Tanabe and Laube) for further descriptions and Chapter 9 for examples of the influence of restraint.

Traditionally, the risks of thermal cracking in hardening concrete have - as mentioned above - been estimated by means of temperature differences within cast section, between different cast or between concrete temperature and air temperature. Restrictions based on temperatures must however be considered as uncertain as they reflect only a fraction of the parameters in Fig 7.1.

Evaluation of cracking risks with due to all effects in the figure gives most accurate results where preferably tensile stresses or strains are compared with the tensile strength or ultimate tensile strain. For various situations of casting different maximum stress levels may then be accepted. For example, the consequence of cracking in a long unreinforced element is much more severe (leading to wider cracks) than in a short element with a large quantity of reinforcement. Therefore, a lower stress level may be allowed for the former case. (However, when repairing cracks, the experience is that wider cracks often are easier to inject than very thin cracks. This fact can complicate the judegment of the consequences of cracking).

With estimations of cracking risks based on stress computations, temperature restrictions may be defined for varying cases of casting as temperature measurements is the most suitable control method for practical use. Such stress related temperature restrictions for use in codes are in production in some countries, see section 7.8.

It is observed that temperature and stresses can be carried out separately although the hardening process of the concrete is a coupled thermo-mechanical problem. The methodology of separate stress and temperature computation has been chosen by almost all researchers and scientists all over the world.

Some methods used in crack risks evaluations are treated in next sections and it is convenient to divide them into methods for rough estimations, equations based on the superposition principle, differential types and other types of material models and methods for structural analysis.

7.2 METHODS FOR ROUGH ESTIMATIONS OF CRACKING RISKS

7.2.1 General feature

These types of methods are conveniently used when more approximate results are needed in e.g. bidding phase or at the follow up on site. The methods have some important general advantages compared to other more complicated methods, namely their high rapidity and the user friendly handling of input and output data. Hence, it is very easy to carry out sensibility analyses for certain part of a building to e.g. check efficiency of a chosen crack preventing measure.

7.2.2 Temperature related methods

The simplest way to estimate risks of thermal cracking of an element is to compute temperature development in the young concrete and to make comparisons with maximum values of absolute temperature in the concrete, temperature differences within pour and differences related to adjoining structures or to air temperature. For example, values between 10 - 25 °C are often found in prescriptions as maximum temperature difference in cast section and between pours.

Depending in the complexity and nature of the structure, 1D, 2D or even 3D computer programs may be used for the temperature estimations, see Chapter 4.

However as stated above, with use of temperature related models, only one of the parameters influencing on thermal stresses is taken into account. The error of using only temperature differences as crack criterion may be exemplified in Fig 7.2 where computed tensile stress levels of surface in walls of different depths are illustrated for different air temperatures and for varying initial concrete temperatures. The parameter on the horizontal axis indicates corresponding maximum temperature differences ΔT over the crossection. *We see that there exists no general relationship between ΔT and stress level.* The lines B-B and C-C can be said to represent the relationship between stress levels and temperature difference under respectively favorable and unfavorable conditions.

Several researches have tried to explain why in most cases so complete different cracking risks are obtained with temperature and stress related critera. For instance, Mangold and Springenschmid [7] mean that the uncertainty of temperature related criteria is due to the fact that the zero-stress temperature gradient (see the reference) is not taken into account.

The uncertainty of a temperature based crack risk estimation may also explained physically by studying Fig 7.2. Three scenarious with regard to initial and ambient temperature may be identified, the stress levels of which are shown in Table 7.1 derived from the graph of the figure. As can be seen the stress level in Case I are in fact lower than in Case II and Case III. Yet the temperature difference is markedly greater in Case I (24.5°C) than in Case II and Case III (18°C and 14.5°C respectively). Thus, the correlation factors in Table I are not only very different from unity but differs in the three cases by a factor up to 2 allowing for very little confidence in temperature-related criteria. What is the reason for this?

In Case I, with the low air temperature and the higher initial casting temperature, concrete maturity near the surface will be effectively retarded by heat losses to the cold environment while the maturity development in the core will be enhanced by the higher temperature there. Hence, much of the expansion phase in the central parts of the wall will take place while the concrete at the surface is in a more plastic stage. This condition is favourable as it effectively decreases the risk of early age surface cracks.

In Case III - on the contrary - the surface concrete will be leading the strength and maturity development across the section. Hence, when, eventually, the delayed rapid volume expansion takes place in the interior, high stresses develop in the stiffer surface concrete layers. As a result of advanced maturity, the stiffness of the surface is already high and its capacity to deform plastically is very restricted. Therefore, the risk of surface cracks increases appreciably.

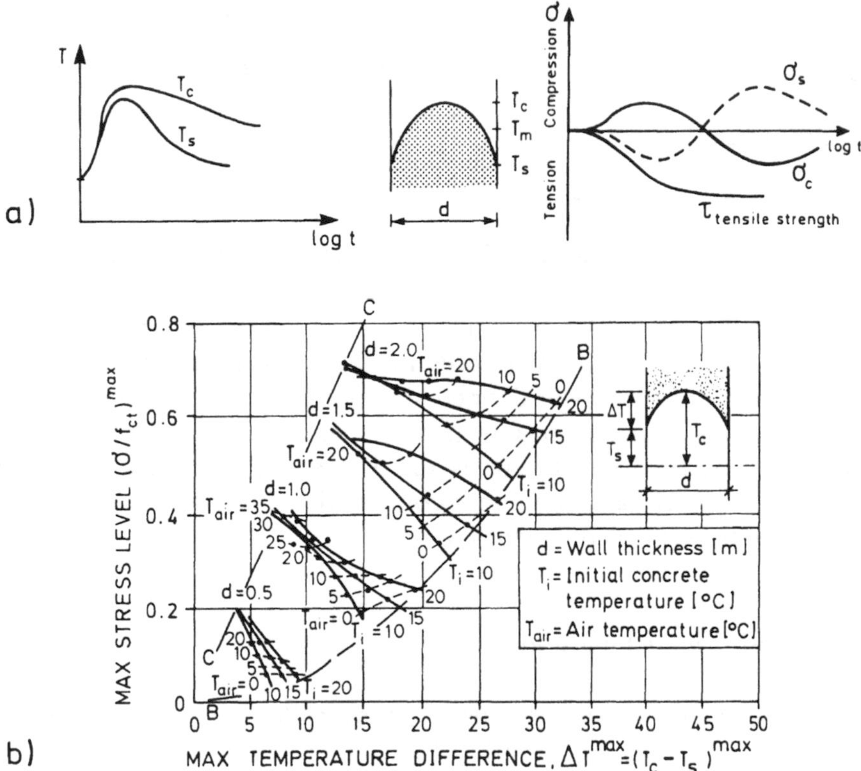

Fig 7.2 Risks of surface cracks due to temperature differentials. **a)** Due to differential temperature across a concrete section early age tensile stresses occur at surface while the center is in compression. At later ages the tensile stresses change into compression, thus closing possible cracks (schematic graphs, T_s, T_c temperatures at surface and center respectively, σ_s and σ_c corresponding stresses. **b)** Maximum tensile stress levels (see also section 7.7) as a function of the maximum temperature difference center - surface at different inital concrete temperatures T_i and air temperatures T_{air}. Theoretical computations. Concrete: Swedish standard Portland cement (ASTM Type I) C = 331 kg/m³, w/c = 0.55. (Emborg and Bernander, [6]).

Table 7.1 Three cases of surface cracking risks according to Fig 7.2 for d = 1.5 m

	Parameter	*Case I*	*Case II*	*Case III*
Initial concrete temperature	T_i	20°C	20°C	10°C
Ambient air temperature	T_{air}	5°C	20°C	20°C
Temperature difference cross section	ΔT^{max}	24.5°C	18°C	14.5°C
Cracking risk - max stress level	η^{max} (Eq 7.26)	0.45	0.53	0.54
Cracking risk - temperature diff. $\Delta T_{cr} = 20°C$	$\Delta T^{max}/\Delta T_{cr}$	1.225	0.90	0.752
Correlation factor	$\frac{\Delta T / \Delta T_{cr}}{\eta^{max}}$	2.72	1.69	1.39

It may be concluded trom Fig 7.2 and Table 7.1 that winter condition is more favourable than summer situations despite the temperature difference mostly is larger. The example shows that temperature related criteria in many typical cases is very uncertain and must be used carefully.

However, for some well defined situations of concreting such a criteria may give acceptable results if all other parameters of Fig 7.1 are defined and entered into the basis for the estimations. Hence, diagrams similar to Fig 7.2, based on stress computations, can be very useful and can also be implemented in codes of practice.

7.2.3 Strain related methods

One approach to model the tensile stress build-up in a member for a external end restraint situation has been reported by Harrison [8]. It was assumed that no stresses are induced before the maximum temperature is reached and that tensile stresses occur during the whole cooling phase. Harrison is aware of "small" compressive stresses that are present during heating phase which means that tensile stresses do not develop until some time after the peak temperature. Thus, the assumption of no stresses during heating leads to an overestimation of the tensile stresses and is according to Harrison therefore a safe basis for design. Using a strain criterion and including end restraint R the approach was expressed as

$$\varepsilon_t = \alpha_c (T_{max} - T_{air})\, R \tag{7.1a}$$

which was reformulated by Harrison [9] to

$$\varepsilon_t^* = \frac{f_t}{E_{eff}} = \alpha_c (T_{max} - T_{air}) R \tag{7.1b}$$

where ε_t^* is a creep modified tensile strain capacity
f_t is the tensile strength
E_{eff} is the effective E-modulus

The whole cooling is taking into account by temperature difference $T_{max} - T_{air}$. In Eq (7.1b) creep is modeled with the average effective modulus during cooling phase. Empirically, we may assume $\varepsilon_t^* \approx 2\varepsilon_t$ where ε_t is the strain capacity of the concrete (in the range of 10^{-4}). Restraint R is 1 at 100 % restraint and < 1 at flexible support. In Harrison [] various values of R for end restraint as well as for continous edge restraint etc are found.

Bernander [10] has suggested a strain related method for estimating cracking risks which takes into account that tensile stresses occurs only at the later part of the cooling phase. The method can be adopted for cases of external restraint and not too non uniform temperatures within the cast. With the method, the temperature $T(t_2)$ is estimated when concrete is stress free after it has been in compression during the heating phase i e when the elastic strain $\varepsilon_{el} = 0$ in Fig 7.3. $T(t_2)$ is often denoted the zero stress temperature. Remaining elastic strain ε_t is compared with the ultimate tensile strain ε_u. The method leads to the following equations (see also chapter 9):

$$\varepsilon_t = \alpha_c \left[T(t_2) - T_u \right] R \tag{7.2}$$

$$T(t_2) = T_{max} - \frac{\alpha_e}{\alpha_c} (T_{max} - T_c)(1 - k_o) \tag{7.3}$$

where T_{max} and T_c is maximum temperature and placing temperature respectively
T_u is air temperature i e the temperature at the end of cooling
α_e, α_c is coefficient of thermal expansion and contraction (section 5.6)
R is the end restraint
k_0 $= \varepsilon_v / \varepsilon_{tot}$

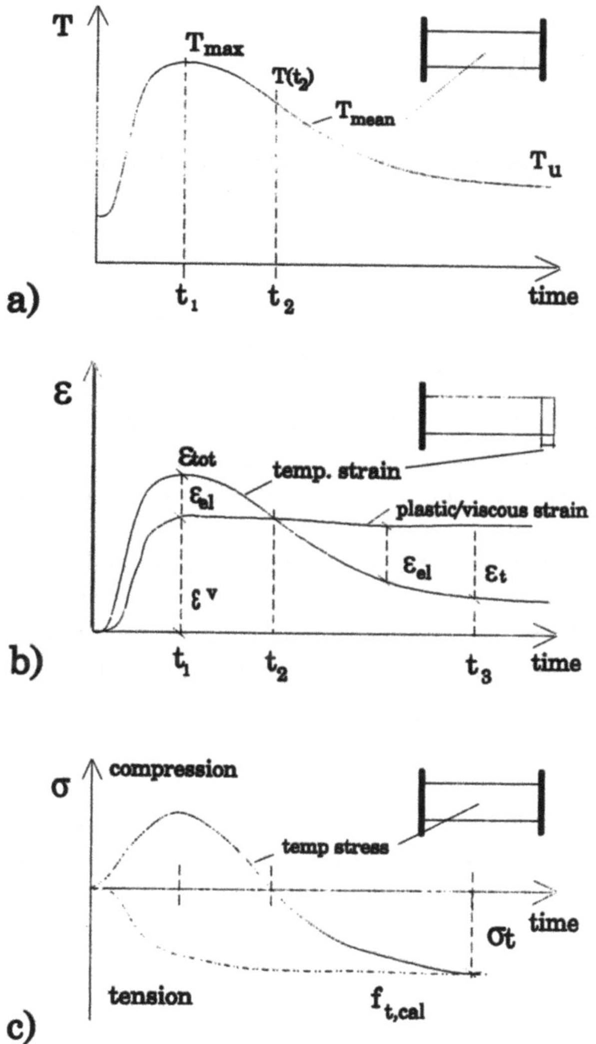

Fig 7.3 **a)** Mean temperature rise in a structure due to hydration **b)** restrained elastic and viscous strain and **c)** thermal stresses at inflexible supports

The deduction of Eq (7.3) makes use of the fact that very little viscous deformation takes place from time t_1 to t_2 due to deloading of the hardening concrete. The parameter k_0 expresses how much of restrained strain that is viscous (and plastic) compared to the total strain (see Fig 7.3b). Thus, k_0 is a measure of the non-elastic (irreversible) deformations while the concrete is in compression, see section 7.2.4.

The restraint R is defined from support with "stiffness" S (spring constant, [m/N])

$$R = \frac{1}{\left[1 + E_{eff} AS / L\right]} \qquad \text{or} \qquad R = \frac{1}{\left[1 + AS / \left(J(t,t') \cdot L\right)\right]} \tag{7.4}$$

With S → 0, R → 1 implying total restraint and S → ∞, R→ 0 equal no restraint. Alternatively, c=1/S [N/m] may be used according to chapter 6. E_{eff} is the effective modulus and J(t, t') is the compliance function (see section 5.4.2). The ultimate concrete strain ε_u varies with concrete quality, age of the concrete, loading rate, fracture mechanics behavior etc and can be difficult to obtain accurately in tests. Little information about the ultimate strain for young concrete is thus reported in literature. Values between 0.08 - 0.16 ‰ may however be found in e.g. Byfors [11], Gutsch and Rostásy [12]. For normal concrete qualities ε_u may approximately be set to 0.1 ‰.

7.2.4 Stress related models

For the case of external end restraint, the axial tensile stress has in chapter 6 of STAR been expressed as

$$\sigma_R(t) = -k_R \alpha_T \left[T_m(t) - T_{mo}\right] E_{m,eff} \tag{7.5}$$

where T_m (t) is mean temperature rise and $E_{m,eff}$ is effective E-modulus. Other parameters, see the chapter. Temperature differences are related to temperature at time t_0 when the concrete is responding for temperature changes after being in plastic state.
In chapter 5 an approximate expression was given also for tensile stress at surface of a member due to non-uniform temperatures over cross section

$$\sigma_E(d/2,t) \approx -\alpha_T \left[T(d/2,t) - T_m(t)\right] E_{eff} \tag{7.6}$$

where T(d/2,t), T(t) etc. see chapter 5.

Alternatively, Eq (7.2) may for external end restraint be expressed in terms of stresses (Emborg and Bernander, [13])

$$\sigma_t = \alpha_c \left[T(t_2) - T_o\right] \cdot E_{eff} \cdot R = \alpha_c \left[T(t_2) - T_o\right] \cdot \frac{1}{J(t,t')} \cdot R \tag{7.7}$$

Tensile stress is thus computed for the remaining cooling to time t_3 after that the zero-stress temperature has been reached (Fig 7.3c). With J(t, t') or E_{eff} reduction of stress due to creep is considered. It is also possible to adjust tensile stress σ_t with due to nonlinear behavior.

A forementioned parameter k_0 (= $\varepsilon_v / \varepsilon_{tot}$) has been calibrated for some concrete mixtures in Sweden to a more accurate method described in section 7.4 and a slight dependence of the placing temperature was found: $k_o = 0.64 + 0.003T_c$. Recent research has shown that the values of k_0 is in the range of 0.60 - 0.80. However, the coefficient should be further investigated considering different concrete qualities, cement types etc.

Evaluated tensile stress σ_t is compared with the failure stress $f_{t,cal}$ of concrete which is related to the tensile strength as

$$f_{t,cal} = k_3 f_{ct} \tag{7.8}$$

With k_3, reduction of strength on account of tertiary creep related to slow loading rates is done (i e creep failure, see section 5.3.4), observed in several laboratory tests. $f_{t,cal}$ is thus considered as a "true" failure strength at time t_3. For loading rates normally occurring at cooling phase, k_3 may according to Fig 5.11 be set to 0.6 - 0.8.

The uniaxial tensile strength is preferably computed from the compressive strength. It is thus important to include effects of high hydration temperatures, see section 5.2.3. Further, early high tensile stresses (thin sections, low air temperature etc.) imply low compressive and tensile strengths and vice versa, i e an estimation of the strength must be done with due to the age t_3.

A stress level $\sigma_t(t)/f_{t,cal}(t)$ less than 0.6-0.7 has been found to give reasonable low level of cracking risk. Different values of J(t, t') (or E_{eff}) are used depending on concrete quality, early/late occurence of tensile stresses, rapid/slow cooling etc. Table 7.2 was evaluated for Swedish Handbook of Concrete whose values may be a base for calibration to other cement types, admixtures, aggregates etc.

As variations of k_0 and J(t, t′) have been found to have considerable effects on computed tensile stresses and cracking risks, it is recommended to check possible variations of the parameters within reasonable ranges.

Table 7.2 Examples of values of creep compliance and effective E-modulus used at calculation with approximate thermal stress method (Jonasson *et al.*, [14])

concrete[d)]	*occurence tensile stress*	t_2[a)] [days]	*d*[b)] [m]	*J(t, t')* $[10^{-12}$ 1/Pa]	E_{eff}[c)] [GPa]
Swedish	early	2-4	≈0.3 - 0.7	47 (40-55)[e)]	21 (25-18)[e)]
Std Portland	normal	4-12	≈0.5 - 1.5	42 (35-50)	24 (29-20)
Type I	late	10-15	≈1.0 - 2.0	37 (30-45)	27 (33-22)
C=330 kg/m³	very late	14-24	≈1.5 - 3.0	32 (25-40)	31 (40-25)
Swedish	early	2-3	≈0.3 - 0.7	57 (50-65)	18 (20-15)
Std Portland	normal	3-12	≈0.5 - 1.5	52 (45-60)	19 (22-17)
Type II	late	10-14	≈1.0 - 2.0	47 (40-55)	21 (25-18)
C=330 kg/m³	very late	14-24	≈1.5 - 3.0	42 (35-50)	24 (29-20)

(a) t_2 see Fig 7.3, (b) Corresponding wall thickness at air temperature of 10-20°C, placing temperature 10-20°C and with 12 mm plywood form (not removed), (c) E_{eff} is directly computed from J(t, t′); E_{eff}=1/J(t, t'), (d) For higher cement content than 330 kg/m³ (≈ Grade K30, cube strength) the values of J(t, t') may be reduced by about 5% for C=360 kg/m³ (≈ Grade K35) by about 10% for C=390-400 kg/m³ (≈ Grade K40-45) by about 15%. In the same way the values may be increased for corresponding lower concrete qualities. (e) Mean values are given. The scatter is within bracket.

7.3 CONSTITUTIVE EQUATIONS BASED ON THE PRINCIPLE OF SUPERPOSITION

A viscoelastic law of integral type was in section 5.4.2 expressed as

$$\varepsilon(t) = \int_0^t J(t,t')\,d\sigma(t') + \varepsilon^{o}(t) \tag{7.9}$$

where J(t, t') is compliance function at time t for a loading at time t', see Fig 7.4 (the Effective E-modulus may also be entered)
$d\sigma(t')$ is stress increment at time t'
$\varepsilon^0(t)$ is stress independent strains at time t, i. e. temperature strain, shrinkage

The variation of stress at prescribed strain history may be written in the same way as above

$$\sigma(t) = \int_0^t R(t,t')\left[d\varepsilon(t') - d\varepsilon^{o}(t')\right] \tag{7.10}$$

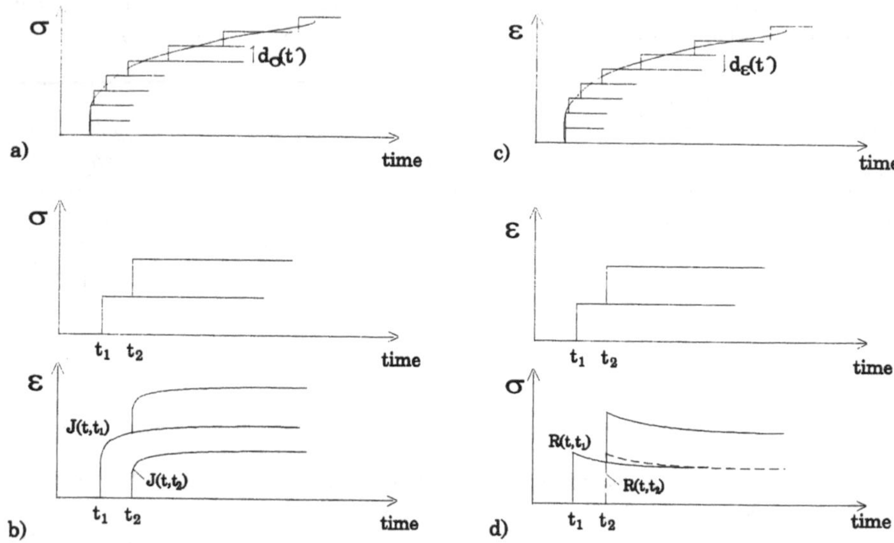

Fig 7.4 Use of principle of superposition at the modelling of viscoelastic response: **a)** Representation of an arbitrary stress history σ(t) as a sum of infinitesimal stress increments dσ. **b)** Representation of arbitrary strain histories ε(t) by superposition of response from different stress histories $\sigma_a(t)$ and $\sigma_b(t)$. (J(t, t'_1) denotes the compliance function at time t for a loading at time t_1 etc.). **c)** Representation of an arbitrary strain history ε(t) as a sum of infinitesimal strain increments dε. **d)** Representation of arbitrary stress history by superposition of response from different strain histories. (R(t,t'_1) denotes the relaxation function at time t for a loading at t'_1 etc).

in which R(t,t') is the relaxation function of time for a deformation at time t' (Fig 7.4)

dε(t') is the strain increment introduced at time t'

A discretization of Eq (7.9) for a time step $\Delta t = t_{r-1} - t_r$ leads to

$$\Delta\varepsilon_r = \sum_{s=1}^{r} J_{r,s-1/2}\Delta\sigma_s + \Delta\varepsilon_r^o \tag{7.11a}$$

or

$$\Delta\varepsilon_r = \sum_{s=1}^{r} \frac{(1+\varphi_{r,s-1/2})}{E_s}\Delta\sigma_s + \Delta\varepsilon_r^o \tag{7.11b}$$

if a creep function φ(t, t') is used instead of compliance function J(t, t') (notations see below).

Equation (7.11a) may be reformulated for change of stress $\Delta\sigma_r$ at a given strain history

$$\Delta\sigma_r = \frac{\Delta\varepsilon_r - \Delta\varepsilon_r^o - \sum_{s=1}^{r-1} \Delta J_{r,s-1/2}\,\Delta\sigma_s}{J_{r,r-1/2}} \tag{7.12}$$

or

$$\Delta\sigma_r = E_r^{''}\left(\Delta\varepsilon_r - \Delta\varepsilon_r^{''}\right) \tag{7.13}$$

where $\Delta\sigma_s$ is the stress increment during time step s

$\Delta\varepsilon_r$ is the change of (external) total deformation (e g flexibility of adjoining structures)

$\Delta\varepsilon_r^{''} = \sum_{s=1}^{r-1} \Delta J_{r,s-1/2}\Delta\sigma_s + \Delta\varepsilon_r^o$ i e strains due to relaxation, thermal strain and shrinkage

$\Delta\varepsilon_r^o$ is the change of the thermal strain $\Delta\varepsilon_r^o = \alpha\,\Delta T_r$ where ΔT_r is the increment of temperature during time step

$E_r^{''} = 1/J_{r,r-1/2}$ is an effective E-modulus for a load during the time step

An important feature of Eqs (7.12) and (7.13) is that the decrease of stress due to relaxation is modelled by a sum term. Effective E-modulus is related to the compliance function according to Eq (5.14).

The case of 100% restraint implies an external deformation $\Delta\varepsilon_r = 0$. For an adjoining structure that is flexible, the external deformation is modeled in a conventional way by means of the stiffness of the support, as shown in chapter 6.

All of the equations above may easily be programmed for evaluation by computer. However, with the superposition method, the total history of stress must be used i e every previous stress increment ($\Delta\sigma_s$) must be stored which in the case of FEM calculation calls for a rather large capacity of the computer.

Further, a use of the principle of superposition is permitted only if each response is linear with respect to stress according to Fig 7.4.

In general, when modelling creep and relaxation, the assumption of linearity means that several restrictions must be fulfilled, (e.g. Bazant and Wittman, [15]):

- The stress is less than about 40 % of the strength or failure load. Otherwise high-stress nonlinearity (microcracking) may occur.
- The specimen undergoes no significant drying during creep. Otherwise drying creep (the Picket effect) occurs.
- There is no major increase in of the stress magnitude after initial loading. Otherwise low-stress nonlinearity (plastic flow) may occur.
- The rate of strains do not decrease markedly in magnitude (but the stresses may).

It is doubtful if these restrictions always hold for thermal stress computation in young concrete and the superposition method may therefore give uncertain results.

Several methods for thermal stress analysis based on the superposition principle are reported in literature. Examples of use of such a methods at the Munich Rilem Conference were presented by Maatjes *et al.* [16], Matsui *et al.* [17], Onken and Rostásy [18], Pedersen [19], Torrenti *et al.* [20], Paulini and Bilewicz [21], Nagy and Thelandersson [22].

However, many of models based on the superposition principle do not incorporate relaxation of accumulated stresses (expressed by the sum term of strain ε'' in Eq (7.13)). Other models in computer programs do not reflect creep during the time step (expressed by the effective E-modulus). From Fig 7.5 it is obvious that these simplifications of the real behavior of the young concrete can involve errors in stress computations.

7.4 CONSTITUTIVE EQUATIONS BASED ON DIFFERENTIAL FORMULATIONS

With an expression of differential form we do not need to store all previous stress increments that were needed with use of superposition principle. Further, nonlinear effects due to high stress levels and effects of temperature on creep behavior may be incorporated in a differential form of equation. This can not, as been mentioned earlier, be done accurately in the method based on the principle of superposition.

Various differential forms of constitutive equations may be found in literature see e.g. Neville, Dilger and Brooks [23] and Dilger [24], see chapter 5.

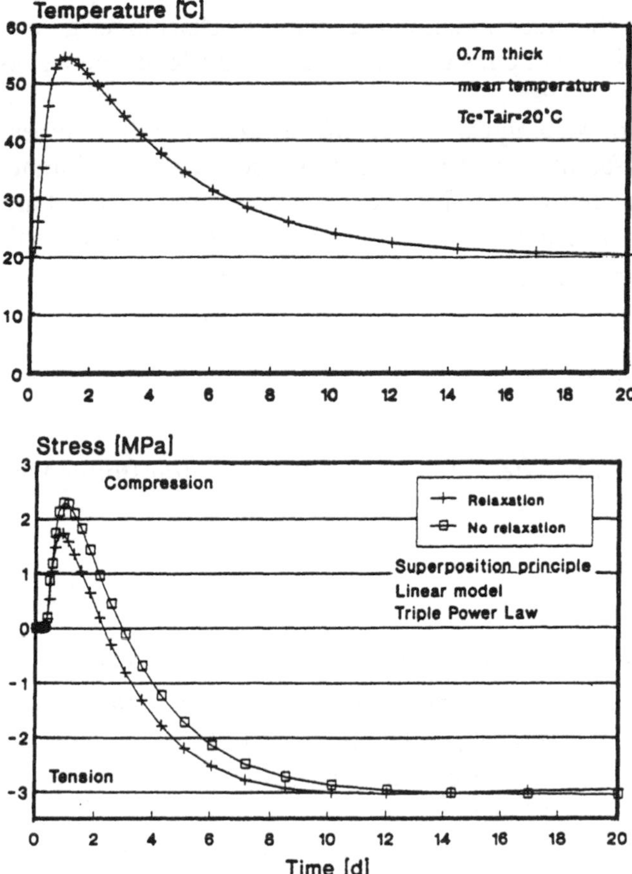

Fig 7.5 Examples of thermal stress evaluations with the principle of superposition according to Eq (7.13) where relaxation of accumulated stress respectively is treated or not treated.

Bernander [25] used the Rate of Creep Method for computation of thermal stresses which according to section 5.4.2 leads to the following relation for the thermally induced strain rate.

$$\frac{d\varepsilon(t)}{dt} = \frac{\sigma(t)}{E_c} \cdot \frac{d\varphi(t)}{dt} + \frac{1}{E_c}\frac{d\sigma(t)}{dt} \tag{7.14}$$

which can be discretized for a time step Δt and reformulated for evaluation of stress increment $\Delta\sigma$, see e.g. Emborg [26].

The Rate of Creep Method is based on a creep function according to Dischinger [27] where creep at very early ages which was by Bernander modeled with additional terms of Dischinger type. Eq (7.14) is based on the fact that the creep function according to Dischinger give parallel creep curves for different ages at application of the load, see e.g. Neville *et al.* [23].

By use of rheological models of spring and dashpot types, other equations of differential type may be expressed. Kelvin -Voight Models and Maxwell Models etc, see Fig 7.6, leads to expressions that are suitable for computer programming. As an example, the following formula is valid for the μth element of the Maxwell Chain Model

$$\dot{\sigma} + \frac{E_\mu(t)}{\eta_\mu(t)}\sigma_\mu = E_\mu(t)(\dot{\varepsilon} - \dot{\varepsilon}_o) \tag{7.15}$$

which for a time step, $\Delta t_r = t_{r+1} - t_r$, may be transformed into (compare with Eq (7.13))

$$\Delta\sigma_r = E_r^{''}(\Delta\varepsilon_r - \Delta\varepsilon_r^{''}) \tag{7.16}$$

where $\Delta\varepsilon_r$ is increment of total strain ($= \varepsilon_{r+1} - \varepsilon_r$)
$\Delta\varepsilon_r^{''}$ is increment of load-independent strains including relaxation strain and thermal strain ($= \Delta\varepsilon_r^{0} + \sigma''_r/E_r''$)
$E_r^{'}$ is an effective E-modulus also denoted the relaxation modulus
$\sigma_r^{''}$ and $\Delta\sigma_r$ is the stress level and increment of stress respectively

By taking into account nonlinear effects at high tensile stress levels the equation may be reformulated to (Bazant and Chern, [28])

$$\Delta\sigma_r = B(\Delta\varepsilon_r - \Delta\varepsilon_r^{''} - \Delta\zeta^{''})\,; \quad \frac{1}{B} = \frac{1}{E^{''}} + \frac{1}{D} \tag{7.17}$$

where D is a modulus describing the non-linear cracking response of the concrete (strain softening modulus) and $\Delta\zeta$ is increment of nonlinear strain due to cracking, see Fig 5.33

This expression has been used in several analyses of thermal stresses and may be visualized by the rheologic model of Fig 5.32. Format of input data, evaluations of response of the individual elements of the Maxwell Chain Model etc. are fully described in Emborg [29].

Examples of utilizing an aging Maxwell Chain Model in a finite element program for thermal stress analysis are reported by Roelfstra, Salet and Kuiks [30], Bernander and Emborg [31], Emborg and Bernander [32].

Further, Dahlblom [33] uses in a 2D FEM-program a differential model with a numerical procedure similar to the one of the Kelvin-Voigt Chain Model for the evaluation of thermal stresses. Thus, the so called strain rate tensor $\dot{\varepsilon}$ is in the finite element formulation assumed to consists of the sum

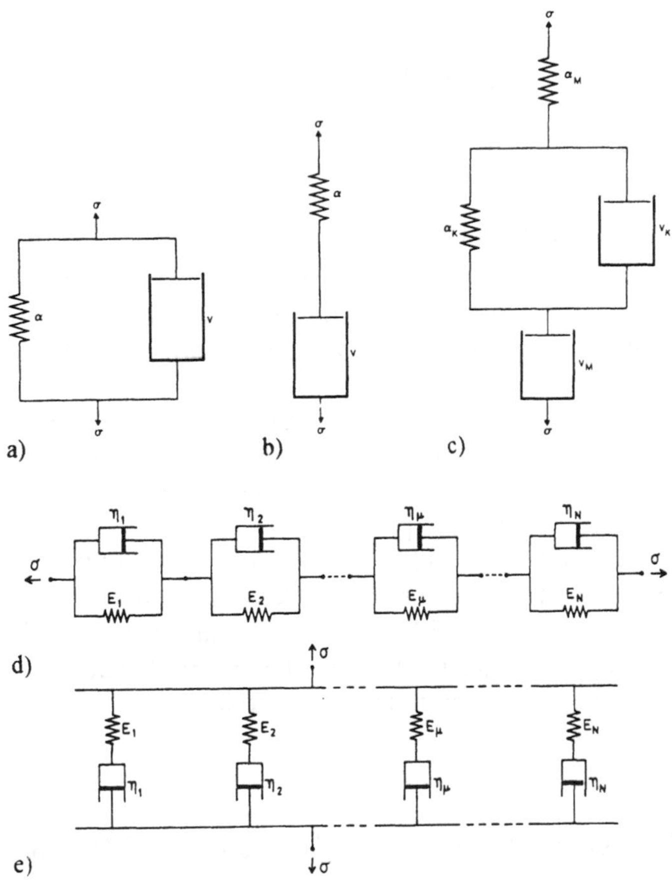

Fig 7.6 Rheological models: Kelvin Voigt element (a), Maxwell element (b), Burger element (c), Kelvin-Voigt Chain Model (d) and Maxwell Chain Model (e)

$$\dot{\varepsilon}_{ij} = \dot{\varepsilon}_{ij}^{e} + \dot{\varepsilon}_{ij}^{f} + \dot{\varepsilon}_{ij}^{T} + \dot{\varepsilon}_{ij}^{T\sigma} + \dot{\varepsilon}_{ij}^{C} \qquad (7.18)$$

where the strains rates $\dot{\varepsilon}_{ij}^{e}, \dot{\varepsilon}_{ij}^{f}, \dot{\varepsilon}_{ij}^{T}, \dot{\varepsilon}_{ij}^{T\sigma}$ and $\dot{\varepsilon}_{ij}^{C}$ are the elastic, fracturing, thermal, stress-induced thermal and creep strain respectively, see sections 5.4 - 5.6. Eq (7.18) may in the finite element model be reformulated into

$$\dot{\varepsilon}_{ij} = a_{pi}a_{qj}\overline{C}_{pqrs}a_{rk}a_{sm}\dot{\sigma}_{km} + \dot{\varepsilon}_{ij}^{o} \tag{7.19}$$

where

$$\dot{\varepsilon}_{ij}^{o} = \dot{\varepsilon}_{ij}^{T} + \dot{\varepsilon}_{ij}^{T\sigma} + \dot{\varepsilon}_{ij}^{C} \tag{7.20}$$

from which the stress rate may be evaluated

$$\dot{\sigma}_{km} = D_{kmij}\dot{\varepsilon}_{ij} - \dot{\sigma}_{km}^{o} \tag{7.21}$$

where is D_{kmij} the material stiffness and $\dot{\sigma}_{km}^{o}$ is the so called pseudo stress rate, see the reference. With this formulation, thermal stresses are computed with due consideration to important characteristics of the young concrete such as nonlinear effects from to cracking, viscoelastic strain and stress induced thermal strain.

Other combinations of spring and dahpots elements may also been used. For example, in a FEM program for thermal stress analysis developed in Denmark, two Kelvin-Voight elements in series are implement, Haugaard *et al.* [34], Pedersen [19].

Another example is Huckfeldt [35] and Huckfeldt, Duddeck and Ahrens [36] who utlize a nonlinear Maxwell-model in stress anlysis. Thermal cracks are modelled by local crack band approach. With a rate-type model, van den Bogert *et al.* [37] have performed thermal stress analysis of breakwater structure. The model was also verified with experimental data.

7.5 OTHER CONSTITUTIVE RELATIONS OR METHODS

Of course, there exists other constitutive relations than the above mentioned for use in thermal stress analysis. For example, in work by Mazars *et al.* [39], damage theory (see e.g. Mazars, [39]) is applied to analyses of effects of hydration in large concrete dams. Two main state variables are treated: maturity (acting on shrinkage, creep and evolution of elastic properties) and damage (affecting the stiffness of the concrete).

The thermal stress can according to Mazars et al be expressed as

$$\bar{\sigma} = \Lambda(M,D):\bar{\varepsilon}^{e} \quad ; \quad \bar{\varepsilon}^{e} = \bar{\varepsilon} - (\bar{\varepsilon}^{d} + \bar{\varepsilon}^{c}) \tag{7.22}$$

where ε is total strain ε^{d} is thermal strain and shrinkage and ε^{c} is creep strain. Further, Λ is the elastic tensor, D is damage and M is maturity. Damage is assumed to be isotropic. The elastic tensor for the damaged material at a given maturity is expressed, see Bournazel and Moranville-Regourd [40].

An other method for stress analysis has been reported by Mangold [41]. In laboratory tests carried out by the cracking frame (see Chapter 2), Mangold examines the zero-stress temperature, T_z, which is the temperature at which restrained concrete element is stress free under a given degree of restraint. If T_z is known, the stress at a certain temperature T may be expressed as

$$\sigma = E\alpha_T(T_z - T) \tag{7.23}$$

where α_T is the thermal expansion coefficient. Temperatures of the cross section are computed and for different laminars, cracking frame tests produce corresponding calues of T_z whose distribution gives thermal stesses.

7.6 METHODS FOR STRUCTURAL ANALYSIS

7.6.1 Compensation Line Method (Plane Bending Method - 1D)

Although many situations of casting may be analyzed for the case of uniaxial stress state with 100% end restraint or with flexible according to earlier sections, more complicated situations often are present where a method for structural analysis is needed.

One way of carrying out such a analysis without making the calculations difficult to survey is to approximate the structure to a beam subdivided into discrete laminar layers, Fig 7.7. Different temperature evolution rates can be assigned to each of the individual layers thus leading to individualized maturity levels and mechanical properties. By varying the layer width b, just about any type of section can be modelled.

Stress development is modelled in the time domain with the principle of superposition or even better with the differential method, described earlier. Thus, the viscoelastic behaviour and, in case of the differential method; also the non-linear behaviour of the young concrete is modelled as well as the geometry and restraint conditions of different parts of the structure. Computed tensile stress of each layer is compared with the tensile strength thus establishing a critical stress level as the criterion of the risk of cracking.

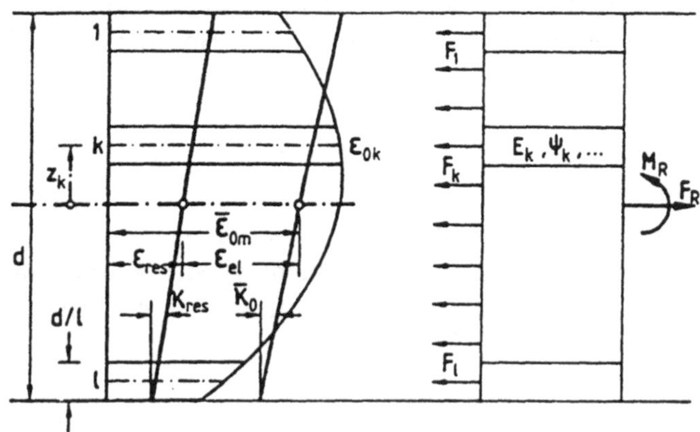

Fig 7.7 Division of structure in *l* discrete laminar layers with individual thicknesses and material properties (here modeled with E_k and ψ_k- the early age E-modulus and relaxation function of layer *k*). ε_{res} is the strain due translation of the structure, κ_{res} models the influence of bending (see Fig 7.8). z_k is the distance from neutral layer. M_R and F_R is the moment and normal force of the cross section. From Onken and Rostásy [18].

A presumption of the method is that the cross section is plane during 1D bending similar to beam theory (according to the theorem of Bernoulli), as indicated in Fig 7.7. The freedom of movement is taken to be of two kinds, see Fig 7.8: translation due to mean temperature change in the structure ((b) in the figure), 1D bending due to assymetric temperature change in the structure (c). A slight exception from the beam theory is to allow a third freedom of movement; the elastic resilience of individual layer due to difference distances from a possible stiff base (d).

Influences of the translation and bending are be modelled with restriction of equilibrium of normal force and moment (in one direction) which leads to

$$\begin{aligned} &\int \sigma \, dA = 0 \quad ; \text{(normal force)} \\ &\int \sigma \, y \, dA = 0 \quad ; \text{(moment in one direction)} \end{aligned} \tag{7.24}$$

where σ is thermal stress and y is distance from the centre of gravity. Adjoining structures or foundation are modelled as own layers i e full bond in the joint is expected.

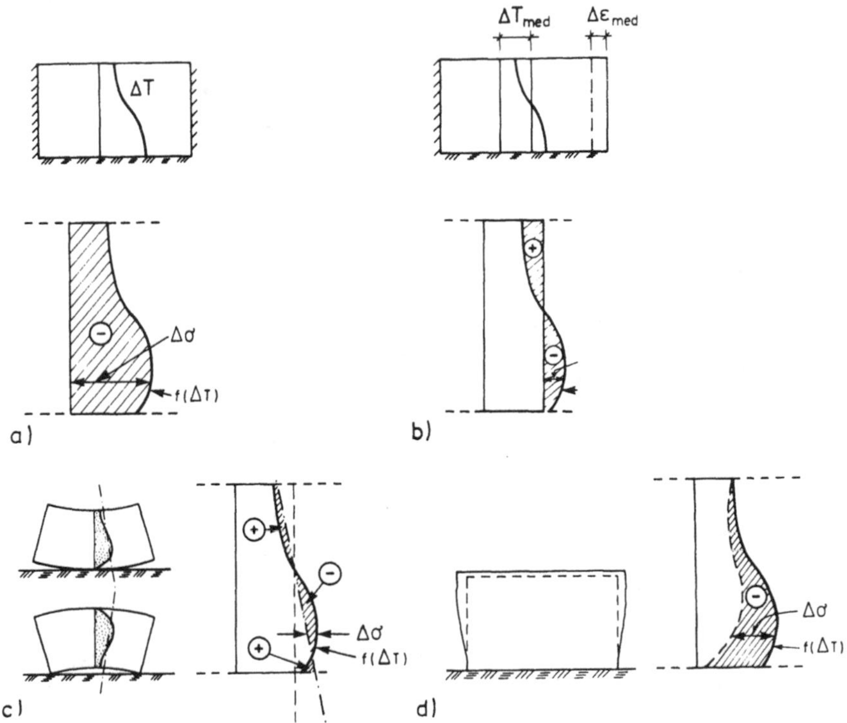

Fig 7.8 Freedom of movement with method according to Compensation Line Theory (a) inflexible supports (100% end restraint) (b) translation (c) bending and (d) elastic resilience, whereas influences are modelled with restriction of equilibrium of normal force and moment (Emborg and Bernander, [6]).

In fact, the assumption of plane bending leads to so simple equations that they could be solved in manual calculations. Such formulas for e.g. centre of gravity and the integrals of Eq. (7.24) for a section of an ageing concrete may be found in Thurston *et al.* [42], Bernander [43] and Tanabe [44, 45].

This analysis considers stresses in one direction and is therefore strictly applicable only when the main principal stress is very dominant. However, even in situations where the stress field is typically two- or even three-dimensional the method may be applied successfully. This is achieved by first establishing the principal stresses and restraint factor in the direction of this principal stress. The principal stress field may be determined on the basis of a 2D (or 3D) FEM study by means of the linear theory for the specific temperature field under investigation. Once the restraint factor is known in the direction, see Fig 7.9, the Compensation Line Method may be used to calculate the principal thermal stresses.

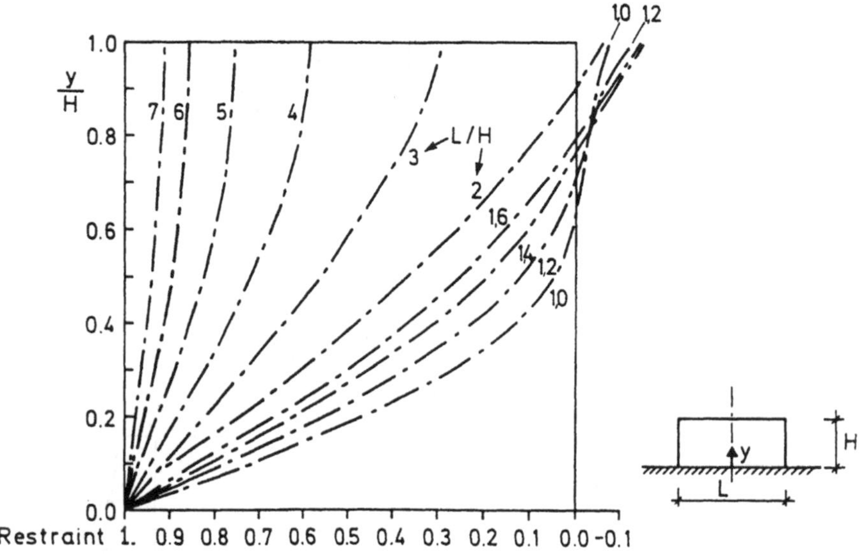

Fig 7.9 Example of restraint factor evaluated with 2D FEM program for a wall cast on inflexible support (Emborg, [29]).

There exist many programs for thermal stress evaluations that is based on the one-dimensional Compensation Line/Plane Bending Method, see Appendix A and several papers of the Munich Rilem Conference report on use of this method e.g. Sato *et al.* [46], Onken and Rostásy [18] and Matsui *et al.* [17].

7.6.2 Compensation Plane Method (Plane Bending Method - 2D)

The Compensation Plane Method (the two dimensional Plane Bending Method) has the same feature as the one dimensional method above with the only difference that bending also is allowed in the orthogonal direction, see Fig 7.10. Hence, three equations of equilibrium are set up for the plane surface by the location of the centre of gravity: one for the normal force and two for the moments of two ortogonal directions

$$\int \sigma \, dA = 0 \quad \text{; (normal force)}$$
$$\int \sigma \, y \, dA = 0 \quad \text{; (moment in one direction)} \qquad (7.25)$$
$$\int \sigma \, x \, dA = 0 \quad \text{; (moment in orthogonal direction)}$$

The cross section is divided into elements with individual temperature developments and maturity levels. The internal stresses of the elements are obtained by locating a compensation plane in such way that the integrals above become zero. Fig 7.10b shows the compensation plane for the temperature and strain distribution of Fig 7.10a. Different age dependent stiffnesses of various parts of the crossection may be considered, see Fig 7.11 where strain distribution and compensation plane are shown for a wall cast on a concrete supported by a rock bed. Input temperatures may be obtained by 2D FEM.

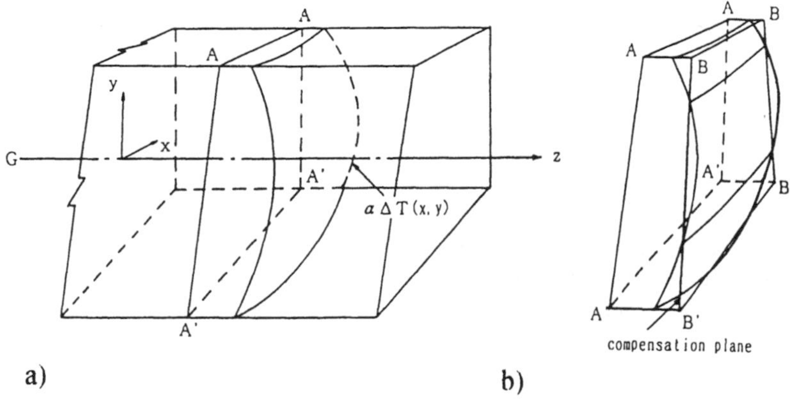

Fig 7.10 Compensation Plane Method (Plane Bending Method - 2D) Example of strain distribution due to 2D temperature changes of a cross section (a) and corresponding compensation plane (b) established by means of equations of equilibrium of normal force and moment in two directions (Tanabe, [45]).

In general, for 3 D analysis, the restraint conditions can be determined realistically and directly by a correct assumption of boundary conditions. For 2 D (and 1 D) analysis the restraint is obtained by the presumptions according to Eq (7.25). Thus, general kinematic boundary conditions for the direction perpendicular to the section being analyzed (the z direction - the out of plane direction) are set up by the integrals above. This technique of establishing boundary conditions and restraint in 2D analyses has been adopted in several FEM codes.

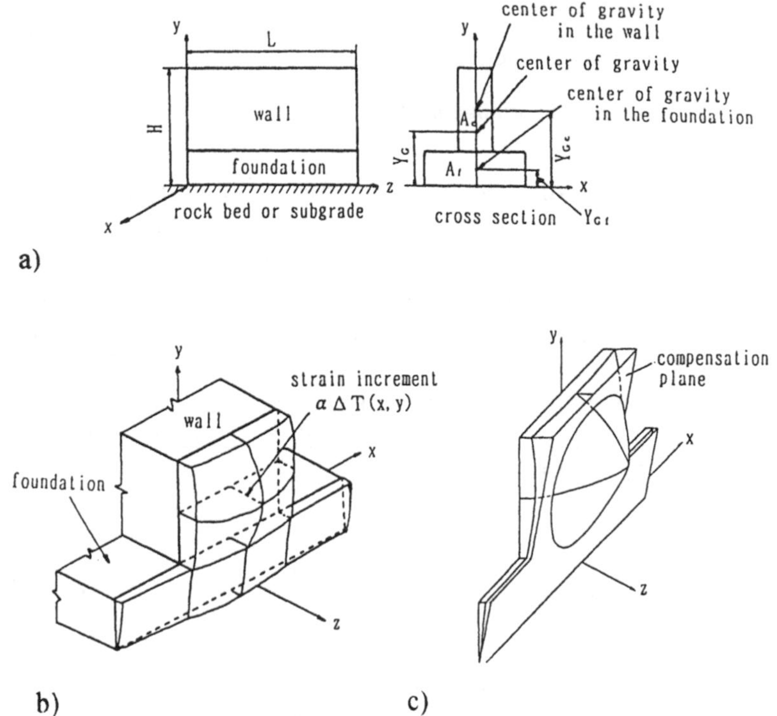

Fig 7.11 Compensation Plane Method for a wall cast on a foundation supported by a rock bed. Different maturity of the structures is considered. Geometry (a), strain distribution (b) and application of compensation plane (c) (Tanabe, [45]).

7.6.3 Finite Element Method

The finite element method is often used in conventional structural analysis both for the modelling of pure linear elastic behaviour but also for nonlinear behaviour such as plasticity, cracking, buckling and simulation of hydration heat. It is thus possible to use FEM-program for the analysis of thermal stresses in young concrete. The main advantage compared to the methods described earlier is that much more complicated geometries and boundary conditions may be studied with the finite element method. Further, consultant engineers, researchers etc have already experience of the use of FEM to establish the 2D and 3D temperature distribution of the newly cast concrete structure.

It must however be stated that often in commercial FEM-programs of standard types, there is a lack of models for the behaviour of young concrete. Therefore, it is important to check whether the program to be used on the whole can model young concrete as well as to check which constitutive model that is utilized. It is also essential to control that input material data is representative for the actual concrete. When restricting a FEM analysis to two dimensions, an important task is to define the boundary conditions of the third direction - the "out of plane direction". Further, it must be kept in mind that, for many applications, results from FEM-computations may - especially if 3D calculations are performed - be difficult to grasp and every prediction may be time consuming.

There exist today several commercial FEM programs for application on thermal stress analysis of which establishment of boundary conditions and restraint, models of mechanical and thermal properties as well as crack criterion etc differ substantially, see Appendix A.

Norman and Anderson [47] applied smeared crack approach and blunt crack band approach of an existing finite element code for the analysis of cracking in large concrete dams. Zhu [48] introduce an "implicit method" for the numerical analysis of thermal stresses in mass concrete structures. New formulas for basic mechanical parameters of young concrete were evaluated and accuracy and speed of computations were according to Zhu increased markedly. An application of a nonlinear incemental structural code (ABAQUS code) for stress analysis has beens reported by Truman *et al.* [49].

The program HACON-S developed by Dahlblom [33] has been used in some rather complicated cases of concreting e g when evaluation of effects of cooling with embedded pipes. We may also mention the program CIMS - 2D, Haugaard *et al.* [34], Pederson [19] and HEAT developed by Roelfstra, Salet and Kuiks [30]. Further, the TEXO module modelling temperature field and degree of hydration and the MEXO module modelling mechanical effects are used in the CESAR-LCPC calculation code, Torrenti *et al.* [20], Acker *et al.* [50]. In the references, application of the modules to different parts of French bridges are presented.

As an example of the use of the finite element method, Huckfeldt, Duddeck and Ahrens [36] demonstrates how FEM investigations applied during planning stages enable designing and construction of massive concrete structures with low cracking risk, see Fig 7.12.

Finally, it has been observed that an important use of finite element method is also *prior* to the thermal stress analysis. In this stage, linear elastic models for hardened concrete in standard FEM programs may successfully be utilized to study 2D or 3D dimensional boundary and restraint conditions of the structural member of interest. Fig 7.13 shows such an analysis where translation restraint for a wall was evaluated for different widths of ground slab.

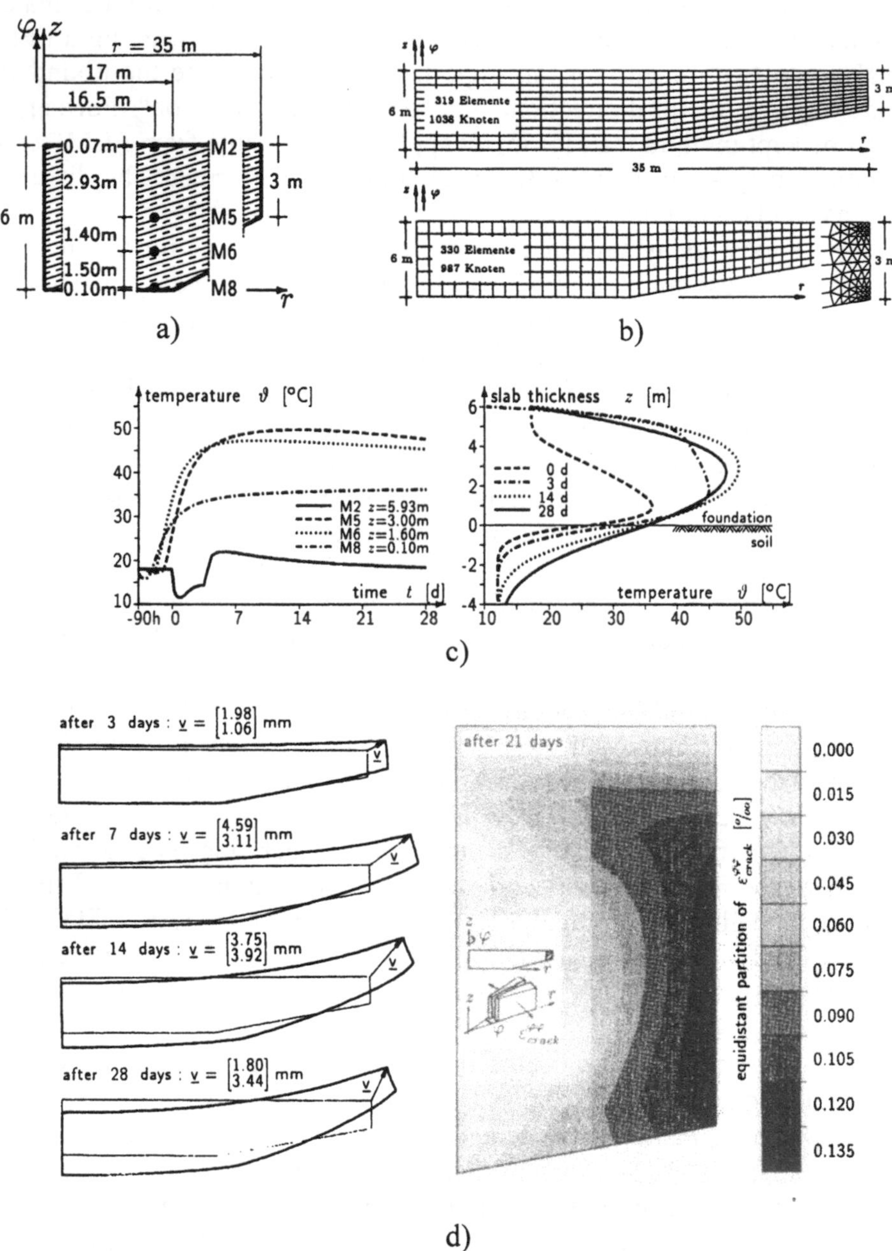

Fig 7.12 Example of application of FEM on massive foundation slab: (a) geometry (b) element mesh (c) temperature distribution (d) deformed shapes (left) and cracks at outer region (r =33 to 35 m) of slab (right). Huckfeldt *et al.* [36] and Huckfeldt [35].

7.7 CRACKING RISKS

The cracking risk may preferably be defined as the maximum value of the stress level

$$\eta^{max} = \left[\frac{\sigma_t(t)}{f_{ct}^*(t)}\right]^{max} \tag{7.26}$$

where $\sigma_t(t)$ is the tensile stress at a certain time t and $f_{ct}^*(t)$ is the tensile failure stress at the same time. The tensile failure stress is a strength value reduced due to the slow loading rates which are present in early age thermal loading, see also section 5.3.4. Alternatively, the restrained tensile strain $\varepsilon_t(t)$ may be compared with the ultimate tensile strain $\varepsilon_u(t)$ at the same time

$$\chi^{max} = \left[\frac{\varepsilon_t(t)}{\varepsilon_u(t)}\right]^{max} \tag{7.27}$$

The safety factor against cracking may be expressed as

$$\Gamma = \frac{1}{\eta^{max}} \quad \text{or} \quad \Gamma = \frac{1}{\chi^{max}} \tag{7.28}$$

It is difficult to define which stress level that is acceptable regarding cracking risks. Conside-ring general uncertainty in early age concrete behaviour and material modelling as well as simplifications and approximations done in application to structural analysis a maximum value of 0.7 is suggested for stress level η^{max}. This corresponds to a safety factor $\Gamma \approx 1.4$.

In Eberhardt, Lokhorst and van Breugel [51] this uncertainty in material behaviour and theoretical modelling is treated. As a consequence of scatter of material properties the respec-tive standard deviations of tensile strength and thermal stresses is considered and the propa-bility of cracking for 5% confidence interval was evaluated, see Fig 7.14, where this procedure is described schematically. Thus, according to the authors, the standard deviation of the tensile strength was estimated at 0.5 MPa and the standard deviation of stresses at 0.6 MPa.

7.8 APPLICATIONS

Modern simulations methods of thermal stresses and cracking can be very useful in the predesign stage, the design stage and at the construction of structural members in which hydration stresses may become critical.

However, as stated earlier, it is important that the methods used permit consideration of - besides temperature - restraint, transient mechanical properties, non-uniform maturity develop-ment, construction sequences and various environment aspects. In that case, by means of precalculations in the predesign stage, the cracking risks can be checked preliminary. Thus, different acceptabel levels of cracking risks are defined with due to the consequence of cracking. If the risk is too high, the relative efficiency with regard to cracking of various constructional measures can be quantified and optimized as demonstrated in several papers of the Rilem Munich Conference [52].

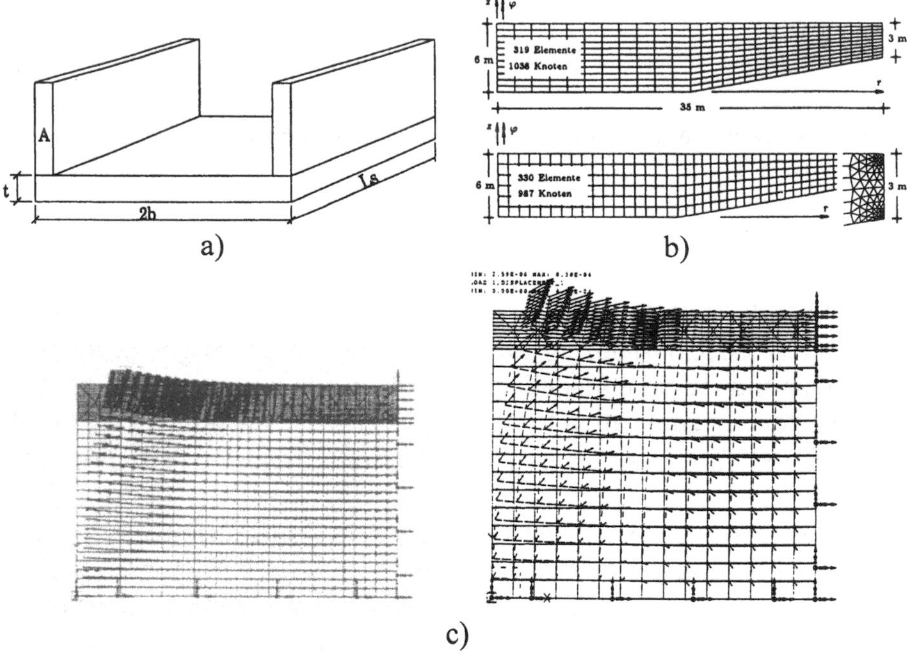

Fig 7.13 Evaluation of restraint from slab of a railway box tunnel with linear elastic FEM. **a)** geometry of slab and walls. **b)** loading condition on slab. A horisontal triangular load is applied on the slab on order to similate the forces on the joint from restrained thermal dilatation of the newly cast wall (due to symmetry, only a quarter of slab is analyzed). **c)** deformations of slab providing values of translation restraint for a thermal stress analysis. Here, restraints were evaluated for slab widths 4.5 m (left) and 6 m (right) respectively, Lindmark and Persson [53].

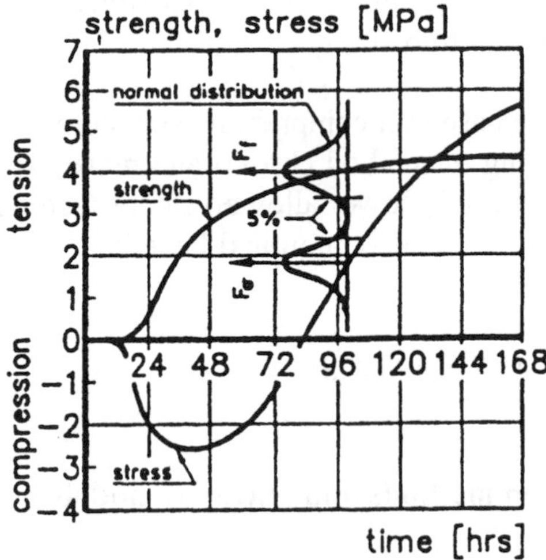

Fig 7.14 Propability of cracking - schematic representation of the calculation procedure, Eberhardt, Lokhorst and van Breugel [51].

Then, in the design phase, which is carried out closer to pouring, more information about material and casting conditions are available thus providing more exact analysis. A reanalysis may be carried out some day before pouring. It is important that during and after casting the behaviour of the young concrete is followed up, thus offering an input to predictions for next pouring sequences.

An important task in future activities is to implement new technology in codes of practice. Still, in many codes, restrictions on cracking risks are dealt with in terms of compute and control temperatures during hardening. More rational criteria shall be defined in terms of strain of stress levels and it is essential to harmonize different national technical practices (e.g. in future Eurocode) in respect to early age crack prevention. One example of tentative code of this kind has been prepared by the Swedish Board for Road Administration. According to this code (the Swedish Code for Bridge Construction - Bro 94) three possibilities of crack control are provided:

- Method 1: Simple temperature related criteria for surface cracking and through cracking respectively.
- Method 2: Design diagrams of some typical cases where cracking risks are related to air and casting temperature, member thickness, concrete mix, *etc*.

- Method 3: Own thermal stress and cracking analysis with well documented computer program and with experimentally obtained material data of concrete mix to be used.

Method 1 and 2 are based on comprehensive theoretical and experimental studies. Higher rating of method chosen and more experimental information of the concrete imply lower allowed safety factor and larger flexibility - i e several practical and economical benefits.

7.9 CONCLUDING REMARKS

7.9.1 Computation methods and material models

Which one of the described methods for the calculation of thermal stresses that should be used depends on the kind of application and requirement on accuracy.

The differential methods based on e g Maxwell Chain and Kelvin Chain models are, as far as the author concerns, the most exact way of evaluation thermal stresses. However, the Maxwell Chain model calls for data on the relaxation properties of the concrete. Although these properties may be obtained by separate computations (ex. programs by Bazant, [54], and Jonasson, [55]) the method may be considered as somewhat too complex and time-demanding, see also section 5.4.7. For the Kelvin Chain model, creep properties (in the form of a "creep spectra") can directly be used as input data to the stress evaluations.

The Rate of Creep Method leads to much simpler expressions than with the Chain Models above. The method has, however not been compared thoroughly with tests on modern concrete qualities. Some experiences may be found in Bernander and Gustafsson [56] and Emborg [26]. The method deserve a more broad use in future thermal stress analysis.

Methods based on the superposition principle may easily be implemented in computer programs and are frequently used for actual applications. The principle implies that restrictions of linearity with stress should be fulfilled. In thermal cracking analysis, however, nonlinear phenomenons always are present, see Fig 7.15. Further, difficulties occur when modelling temperature effects on creep behaviour. This means that the use of superposition principle may give uncertain results why evaluated stresses should be controlled against laboratory tests or more accurate methods.

Manual calculations with stress or strain related methods are with its simplicity and efficiency very suitable for practical use e.g. in rough estimations. It is however very important to check the accuracy of the methods against more exact methods or against laboratory tests.

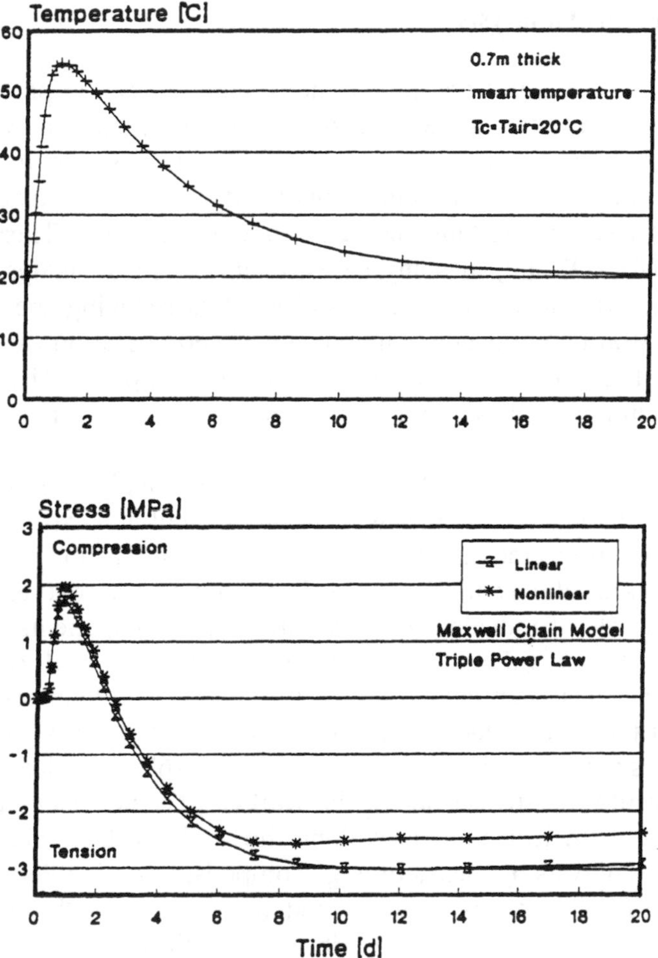

Fig 7.15 Example of thermal stress computations where nonlinear effects at high tensile stress levels respectively are treated and not treated. a) Mean temperature rise for a 0.7 m thick member. b) Thermal stresses computed for the case of 100 % end restraint. (Often, the deviation between linear and nonlinear evaluation is larger).

Temperature related methods should with its uncertainty only be utilized if a broad base of experience is present. However, if temperature based cracking criteria has been converted from thermal stress analyses for certain typical cases the accuracy is increased markedly. Example of such a conversion is the diagram of Fig 7.2 where restriction on initial concrete temperature can be found for a certain thickness and air temperature.

7.9.2 Structural analysis

By nature, all structures are three-dimensional, but can in practical situations be reduced to two dimensions or even to one dimension without any important loss of accuracy. This is specially the case for infrastructure buildings where the structural elements (bridge columns, tunnel elements etc.) often have a constant geometry in one direction. Therefore, two dimensional (2D) and, in many cases, even 1D stress analyses may, as described earlier in the text, be justified, thus reducing the number of stress and strain components. Of course a 3D computation can always be performed but calls for more capacity of the computer. They are more time-consuming and leads to a pronounced increase of output data to be analyzed.

One approach of simplifying thermal stress analysis is reported by König and Puche [57]. By means of finite element analysis, simple models for computing cracking risks were derived, describing the effects of internal stresses alone and in combined action with external restraint.

An example of reducing the stress and strain components and simplifying the analysis is the well-known case of pouring a wall on an earlier cast slab that has been cooled down, Fig 7.16a. The temperature development of the wall induces stresses that are oriented in all three directions. The stresses of the cross section (Fig 7.16b) may lead to surface cracking while the stresses along the wall (Fig 7.16c) may cause cracks through the wall. Cracking risks along the wall can be evaluated with a 2D or a 1D analysis by means of e. g. Compensation Line/Plane Methods. Also a 2D FEM analysis may be used for the wall if the equations of equilibrium are atttached as described in section 7.6.2. Often, due to symmetry reasons, no second bending is present. At the same time the stresses of the cross section and risks of surface cracking may be computed by conventional 2D FEM analysis, or in many cases, also her by a 1D analysis that can be coupled to the longitudonal direction by the Poissons Ratio.

The example shows that thermal stress analyses may be performed without any 3D program. This technique may in fact be more accurate than conventional 2D FEM stress analyses where common restrictions of plane stress or plane strain always are attached.

7.9.3 Reliability of thermal cracking analysis

When discussing computation methods and material models it cannot be avoided to call attention on having reliable data and models for the temperature evaluation, determination of restraint and evaluation of mechanical behaviour.

The importance of accurate *temperature computation* may be demonstrated for a massive monolith of a concrete dam. Fig 7.17a shows an example of differences between temperatures evaluated with a 2D analysis and obtained at in-situ measurements. Fig 7.17b shows the corresponding influence on horisontal thermal stresses evaluated with 1D analysis near the joint to earlier cast section. It is concluded that the scatter is unacceptabel. Similar scatter is expected for medium size structures and slender structures. Reliability of temperature evaluations are further treated in Chapter 4.

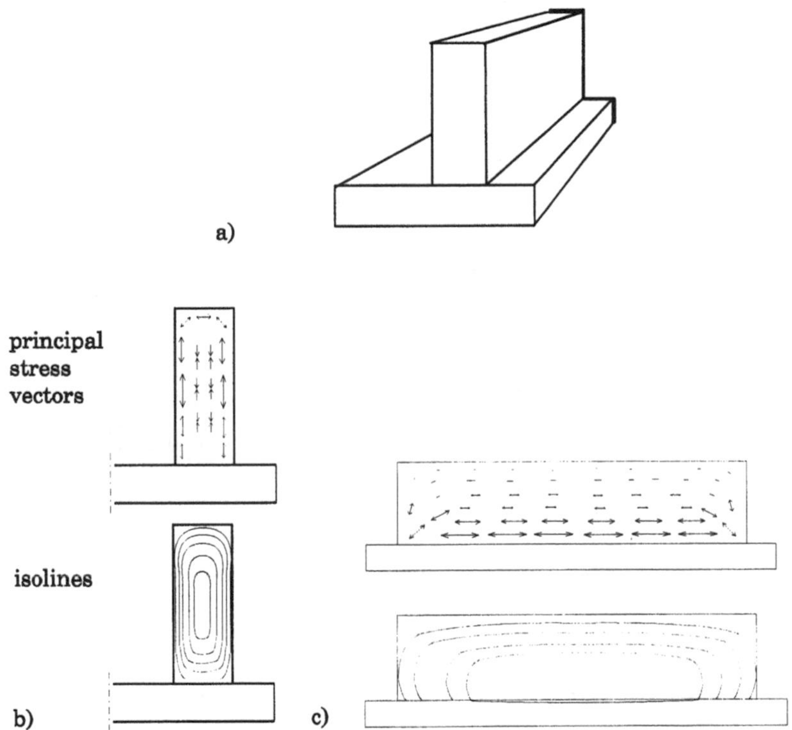

Fig 7.16 Wall cast on a slab (a): stresses of cross section (b) and stresses along the wall (c) shown schematically.

The importance of an accurate evaluation of *restraint* of a structure has been discussed in e. g. chapter 6. This is examplified also in Fig 7.18 which shows that the degree of rotational restraint affects the thermal stress state significantly. Most often the restraint from a supporting structure (e g ground slab) can automatically be obatined by including also the supporting member in the thermal stress analsysis. It is here essential to model possible heating and thermal movement of adjoing structure thus reducing the restraint. Also, friction to subbase should be considered accordingly. Modelling of possible bending is done by treating dead weight, lever of moment and stiffness of subbase, see Fig 7.19. Diagrams for evaluation of bending restraint for different types of subbase has been developed by Emborg and Bernander [32].

Study of the the crucial influence of restraint is to be a major part in the assignement of future research. Theoretical analysis and laboratory tests should be carried out to assess the effect of bonding and possible slip in joints between young and old concrete. The effects of subgrade on restraint should also be addressed in future research.

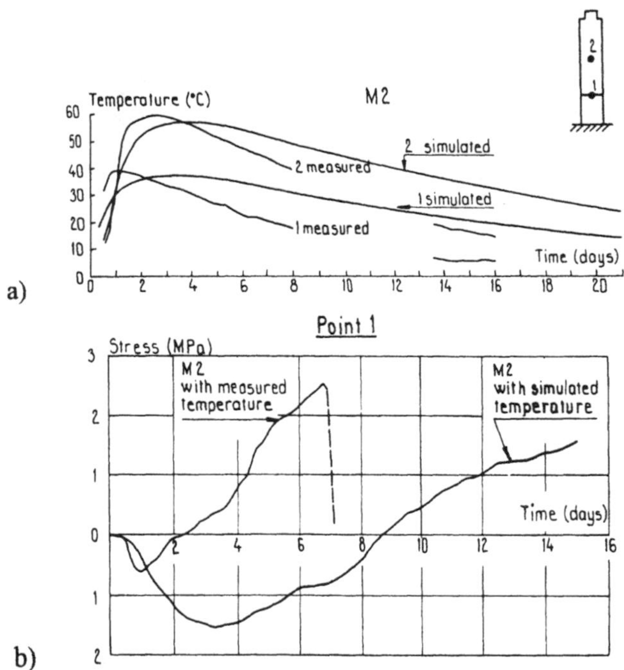

Fig 7.17 **a)** Comparison of measured and simulated temperatures in points 1 and 2 of massive monolith of a concrete dam. Temperatures are evaluated with a 2D FEM program. **b)** Corresponding horisontal thermal stresses of point 1. Simulations with a 1D method based on the principle of superposition, Nagy [58, 59]

The reliability of a certain theoretical model for *mechanical behaviour* may be checked as follows:

- Does the model describe the growth of mechanical behaviour at early ages properly (maturity, strength growth, growth of stiffness etc)?
- Is the viscoelastic behaviour treated in an accurate way (concerning e.g. creep during time step, relaxation of accumulated stresses)?
- Is nonlinear behaviour at high tensile stresses (fracture mechanics behaviour) included?
- Is nonlinear behaviour at unloading of compressive stresses (creep recovery etc) treated?
- Is input material characteristics representative for the concrete quality chosen? If not - how easy can the model be calibrated to the behaviour of a certain concrete mixture?
- Is the model verified by laboratory investigations of the total response?

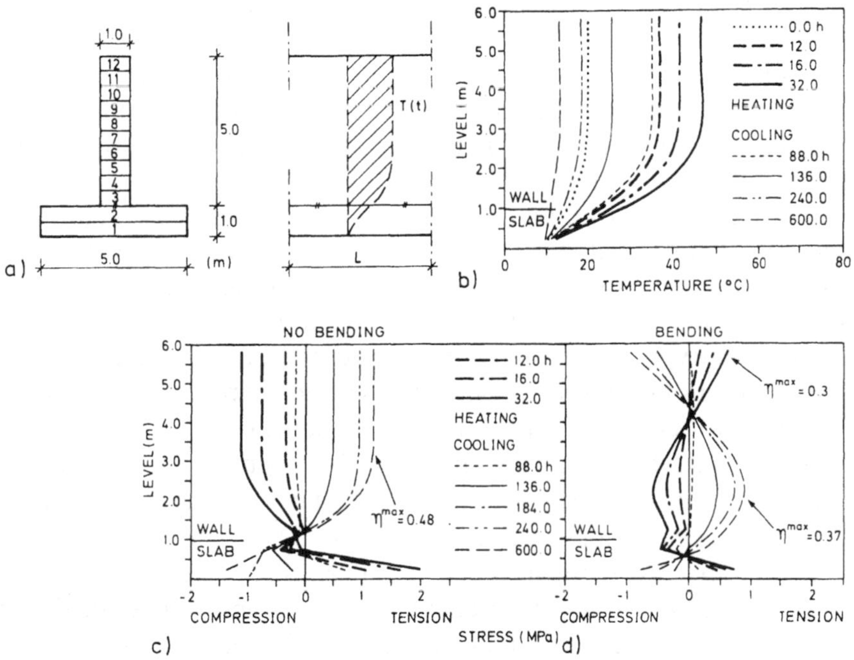

Fig 7.18 The influence of bending restraint of a wall cast on a concrete slab. (a) geometry and divison into elements in the stress analysis, (b) theoretically obtained mean temperatures during heating an d cooling process, (c) and (d) computed thermal stresses at full rotational restraint and free rotation respectively. The wall and slab are free to translate, Emborg and Bernander [6].

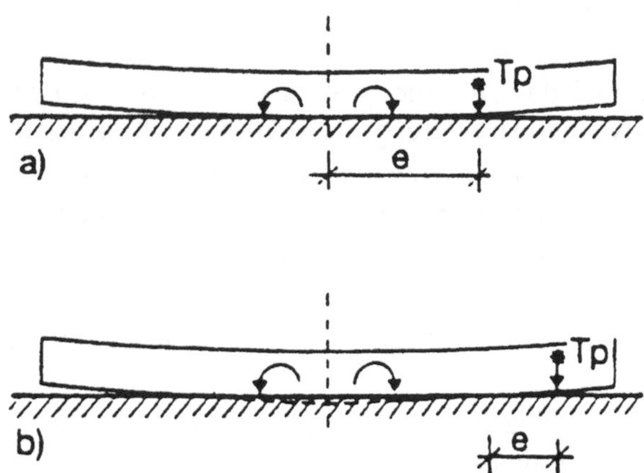

Fig 7.19 Influence of moment of dead weight counteracting bending due unsymmetric temperatures. (a) Inflexible sub-base (b) flexible sub-base. T_p is the centre of gravity of the "cantilever slab concrete", Jonasson, Emborg and Bernander, [14].

Concerning structural analysis it is important to check how every approximations and simplifications that are done influence on the results (reduction of 3 D stresses to 2 D plane stress/strain or even to 1 D, influence element lengths etc).

An essential area of future research is to, for some certain typical cases, compare thermal stress evaluations with the various methods mentioned earlier. One such interesting comparison has been performed by Sato, Dilger and Ujike [46]. Different constitutive models (superposition methods, rate-type formulations, linear and nonlinear methods) may also be compared in the future. In Emborg [26, 29] thermal stresses obtained with Eqs (7.13) and (7.16) are compared for some constitutive models.

In this context, we may also mention the essential future area of sensistivity analysis. One such a sensistivity analysis of cracking risk computations with due to input material characteristic is reported by Matsui *et al.* [17]. Further, At the University of Luleå, the Monte Carlo simulation method has been used for comprehensive evaluations of individual importances of separate material parameters in calculations of through and surface cracking risks, Westman and Hedlund [60].

Finally, the decisive question will be to which extend the results of the calculation method used correspond with the stresses *in situ*. Progress in this most essential question can only be gained by comparison of calculated and measured temperatures, stresses and displacements.

REFERENCES

[1] Roelfstra P. E., Salet T M: Modelling of heat and moisture transport in hardening concrete, see Rilem (1994), pp 37 - 44.

[2] Jonasson J-E: Modelling of temperature, moisture and stresses in young concrete. Divison of Structural Engineering. Luleå University of Technology, Division of Structural Engineering, Doctoral Thesis 1994:153D, 225 pp.

[3] Jonasson, J.-E.; Groth, P. and Hedlund, H.: Modelling of temperature and moisture field in concrete to study early age movements as a basis for stress analysis, see Rilem (1994), pp 45 - 52.

[4] Sellevold E, Bjöntegaard Ö, Justnes H, Dahl P. A: High performance concrete: early volume change and cracking tendency, see Rilem (1994), pp 229 - 236.

[5] Hedlund H: Stresses in high preformance concrete due to temperature and moisture variations at early ages. Luleå University of Technology. Division of Structural Engineering. Licentiate Thesis, 1996:38L, 238 pp.

[6] Emborg M, Bernander S (1994a): Assessment of the risk of thermal cracking in hardening concrete, *Journ. of Struc. Eng* (ASCE), **120**, No 10, October 1994.

[7] Mangold M and Springenschmid: Why are temperature related criteria so unreliable for predicting thermal cracking at early ages? see Rilem (1994), pp 361 - 368.

[8] Harrison T A: Early-age thermal crack control in concrete, Construction Industry Research and Information Association (CIRIA), Report No 91, 1981, 48 pp

[9] Harrison T A: A simple approach to the prediction of early age thermal cracking. Proceedings from "RILEM International Conference on Concrete at Early Ages", Paris, April 6 - 8, 182, pp 239-242

[10] Bernander S: Cooling of hardening concrete with cooling pipes, *Nordisk Betong*, no 2 1973 (in Swedish) , pp 21-30

[11] Byfors J: Plain concrete at early ages. Swedish Cement and Concrete Research Institute, Fo 3:80, Stockholm 1980, 350 pp

[12] Gutsch, A. and Rostásy, F.S.: Young concrete under high tensile stresses - creep, relaxation and cracking, see Rilem (1994) pp 111 - 118.

[13] Emborg M, Bernander (1994b): Thermal stresses computed with a method for manual calculations, see Rilem (1994) pp 321 - 328

[14] Jonasson J-E, Emborg M, Bernander S: Temperatur, mognadsutveckling och egenspänningar i ung betong (Temperatures, maturity and thermal stresses of young concrete, in Swedish), Chapter 16 in *Swedish Handbook for Concrete Design - Material*, Svensk Byggtjänst, Stockholm, 1994, pp 547 - 607.

[15] Bazant Z P and Wittman F H (editors): *Creep and shrinkage in concrete structures*, Wiley & Sons, New York 1982, 363 pp

[16] Maatjes E, Schillings, J J M, de Jong R: Experience in controlled concrete behaviour, see Rilem (1994), pp 247 - 254

[17] Matsui K, Nishida N, Dobashi Y, Ushioda K: Sensistivity analysis and reliability evaluation of thermal cracking in mass concrete, see Rilem (1994), pp 305 - 312

[18] Onken P, Rostásy F.S: A practical planning tool for the simulation fo thermal stesses and for the prediction of early thermal cracks in massive concrete structures, see Rilem (1994) pp 289 - 296.

[19] Pedersen: Prediction of temperature and stress development in concrete structures, see Rilem (1994), pp 297 304

[20] Torrenti J M, Larrard F, Guerrier F, Acker P, Grenier G: Numerical simulation of temperatures and stresses in concrete at early ages: The french experience. see Rilem (1994) pp 281 - 288

[21] Paulini P, Bilewicz D: Temperature field and concrete stresses in a foundation plate, see Rilem (1994), pp 337 - 344

[22] Nagy A and Thelandersson S: Material characterization of young concrete to predict thermal stresses, see Rilem (1994), pp 161 - 169.

[23] Neville A, Dilger W, Brooks J: *Creep of plain and structural concrete*, Construction Press, Longman Group Limited, New York, 1983, 361 pp.

[24] Dilger W: Methods fo structural creep analysis, chapter 9 in " Creep and shrinkage in concrete structures", see Bazant and Wittman [15], pp 305 - 340.

[25] Bernander S: Temperature stresses in early age concrete. Rilem International. Symposium on Early Age Concrete, Vol II, Paris 1981

[26] Emborg M: Temperature stresses in massive concrete structures. Viscoelastic models and laboratory tests. Division of Structural Eng, Luleå University of Technology, Licentiate thesis 1985:011L, 143 pp

[27] Dischinger: Elastische and plastische Verformung der Eisenbetontragwerke, *Der Bauingenieur*, **20**, No 5/6, 21/22, 31/32, 47/48, Berlin 1939, pp 55-63. 286-294, 426-437 and 563-572.

[28] Bazant, Z.P. and Chern J.C.: Strain softening with creep and exponential algorithm, *J. Eng. Mech. Div.* (ASCE), **111**, No. 3, 1985, pp 391-415.

[29] Emborg M: Thermal stresses in concrete structures at early ages, Div of Struct Eng, Luleå Univ. of Technology, Doctoral Thesis 1989:73 D, revised edition, Luleå 1990, 286 pp

[30] Roelfstra P. E., Salet T M, Kuiks J. E (1994).: Defining and application of stress-analyis-based temperature difference limits to prevent early-age cracking in concrete structures, see Rilem (1994), pp 273 - 280

[31] Bernander S, Emborg M (1994) Risk of cracking in massive concrete structures - New developments and experience. see Rilem (1994), 8 pp.

[32] Emborg M, Bernander S (1994c): Avoidance of early age thermal cracking in concrete structures - predesign, measures, follow-up. see Rilem (1994) pp 409 - 416.

[33] Dahlblom O: HACON-S A program for simulation of stress in hardening concrete. Vattenfall Hydro Power Generation, Vällingby, 1992, 77 pp

[34] Haugaard M, Berrig A, Fredriksen J O: Curing technology - a 2-dimensional simulation program, Proceedings from the Nordic Concrete research Meeting, Gothenburg, August 1993. The Nordic Concrete Federation, Oslo, 1993, pp 222 -224

[35] Huckfeldt J: Thermomechanic hydratisierenden Betons, Technical University of Braunschweig, Institut für Statik, Dissertation, No 93-77, Braunschweig, 1993

[36] Huckfeldt J, Duddeck H and Ahrens H : Numerical simulation of crack-avoiding measures, see Rilem (1994), pp 255 - 264

[37] van den Bogert P, de Borst R, Nauta P: Simulation of the mechaincal behaviour of young concrete. IABSE-Colloqium on "Computional Mechanics of Concrete Structures: Advances and Applications", Delft August 1987, Delft University Pressm Delft, pp 339-347

[38] Mazars J, Bournazel J P, Moranville - Regourd M: Thermomechanical damage due to hydration in large concrete dams. Proc of the First Material Engineering Concress, ASCE, August 1990, Denver, Colorado, Vol 2, 1061-1070

[39] Mazars J: A description of micro and macroscale damage of concrete structures. *Engineering Fracture Mechanics*, **25**, No 5/6, pp 729 - 737

[40] Bournazel J P, Moranville - Regourd M: Thermal effects, cracking and damage in young massive concrete, see Rilem (1994), pp 329 - 336

[41] Mangold M: Thermal prestress of concrete by surface cooling, see Rilem (1994), pp 265 - 272

[42] Thurston S J, Priestley M J N and Cooke N: Thermal analysis of thick concrete sections, *ACI Journal*, Sept-Oct 1980, Technical paper 77-38, pp 347-357

[43] Bernander S: Modell för beräkning av egenspänningar vid enaxliga spänningstillstånd. (Model for calculation of uniaxial thermal stresses). in: Daerga, Emborg and Bernander: Egenspänningar i undervattensgjutna bottenplattor på grund av temperaturförloppet under hydrationen (Thermal stresses due to hydration in bridge foundations placed under water, in Swedish with english summary), Division of Structural Engineering, Luleå University of Technology, Technical Report 1986:027T, 100 pp

[44] Tanabe T: Report on review of studies of thermal and stresses and proposal for calculation method of thermal stresses using the Compensation Line and Compensation Plane Method, Research Committee in Thermal Stesses in Massive Concrete Structures, JCI, Tokyo, 1985, pp 47 - 61 (In Japanese, see also Tanabe [45]).

[45] Tanabe T: Compensation plane and compensation line method, Nagoya University of Technology, Internal report, 8 pp

[46] Sato R, Dilger W, Ujike I: Deformations and thermal stresses of concrete beams constructed in two stages, see Rilem (1994), pp 313 - 320

[47] Norman D, Anderson F: Reanalysis of cracking in large concrete dams in the US Army Corps of Engineers. Proceeding from the International Conference of Large Dams, Commission Internatioal des Grands Barrages, Lausanne, 1985, pp 157-169

[48] Zhu B: Computation of thermal stesses in mass concrete with consideration of creep effect, Proceeding from the International Conference of Large Dams, Commission Internationale des Grands Barrages, Lausanne, 1985, pp 527-545

[49] Truman K, Petruska D, Norman D: Creep, shrinkage and thermal effects on mass concrete structure, *Journal of Engineering Mechanics*, (ASCE), **117**, No 6, June 1991 pp 1274-1288

[50] Acker P, Foucrier C, Malier Y: Temperature related mechanical effects in concrete elements and optimization of the manufacture progress, ACI Symposium on Properties of Concrete at Early Ages, Chicago, pp 33 - 47.

[51] Eberhardt M, Lokhorst S, van Breugel K: On the reliability of temperature differentials as a criterion for the risk of early-age thermal cracking, see Rilem (1994), pp 353 - 360

[52] Rilem: Proc. from Rilem Int. Symp. "Avoidance of Thermal Cracking in Concrete", Munich Oct. 1994 (Edited by Springenschmid R), E & FN Spon, London, 464 pp.

[53] Lindmark M, Persson J: Risker for temperatursprickor vid gjutning av betongtunnlar, Hallandsås. Sprickriskanalyser och förslag till åtgärder (Risks of thermal cracking at the casting of concrete tunnels, Hallandsås. Crack risk analyses and proposals of measures, in Swedish), Luleå University of Technology, Division of Structural Eng., Diploma Work 1994:189E.

[54] Bazant Z P: Input of creep and shrinkage characteristics for a structural anlaysis program, *Materials and structures*, **15**, No 88, pp 283-290

[55] Jonasson J-E: Datorprogram för icke-linjära beräkningar i betong med hänsyn till svinn, krypning och temperatur (Computer program for non-linear analyses of concrete in view of shrinkage, creep and temperature, in Swedish), Swedish Cement and Concrete Research Institute, Fo 7:77, Stockholm, 1977, 161 pp

[56] Bernander S, Gustafsson S: Temperature induced stresses in concrete due to hydration. *Nordisk Betong* , No 2, 1981 (in Swedish)

[57] König G, Puche M: Zur Rissbreitenbeschränkung und Mindestbewehrung bei Eigenspannungen und Zwang. *Bauingenieur*, No 66, pp 149 - 156

[58] Nagy A: Cracking of concrete structures due to early thermal deformations. Lund Institute of Technology, Department of Structural Engineering, Licentiate thesis, Report TVBK-1009, Lund 1994, 110 pp

[59] Nagy A: Cracking in reinforced concrete structures due to imposed deformation, Dissertation, Report TVBK - 1012, Lund 1997, 90 pp

[60] Westman G, Hedlund H: Monte Carlo simulering av temperaturspänningar och sprickrisker (Monte Carlo simulations of thermal stresses and cracking risks, in Swedish), Division of Structural Engineering, Luleå Univ. of Technology, Internal Report, 30 pp.

APPENDIX A: List of Computer Programmes for Calculating Thermal Stresses and/or Cracking Risk

Name	*Contact*	*Description (modelling, etc...)*		*Reference*
		Material	*Structural*	
1 HYDCON	TU Braunschweig, Institut für Statik Beethovenstrasse 51 D-38106 Braunschweig Phone 49 531 391 3667 Fax 49 531 391 8116 Dr.-Ing. Huckfeldt J, Prof. Dr.-Ing. Ahrens H., Prof. Dr.-Ing. Duddeck H.	Thermomech. planning tool for efficiency of constructive, techno-logical and executive methods to avoid cracking Temperature comput. Nonlinear elastic, creep and relaxation, post failure, damage, degr of hydration total strain divided into elastic, thermal, creep and crack parts	Temp: 2D and rotational Stress: FEM 2D plain stress, plain strain, rotational Restraint: bending, friction etc	Foundation Bank, Frankfurt/Main Immersed Tunnel, Öresund (tender phase) Tie-in Chamber Europipe, Statoil
2 ASTEA, MACS	Research Center of Computional Mech. Inc L-Kauei Togoshi Bldr 1-12-9 Hiratsuka Shinagawa-ku Tokyo, Japan 142 Phone (03)3785-3033 Fax (03)3785-6066 Mr S Yoshikawa	Transient thermal conduction and stress analysis for concrete Nonlinearity heat and mechanical properties due to hydration Compressive yielding, creep, tensile collapse, Material database according JSCE (Japans Society of Civil Engineers) - could be modified with 19 items	2D FEM, changes of boundary and constraint Conditions and addition of elements due to concreting and cracking. Mesh generation. gravity loads. Steel bars in 4 directions Windows system	(no given) program sold in a no of 25 since 1994

3 TEMP TSAPE	Instituto Superior Técnico Departemento de Engenharia Civil Av Rovisco Pais 1096 Lisboa Codex Portugal Phone 351-1-847-3457 Fax 351-1-848 8481 Prof Branco	TEMP: temperature and stress analysis due to environmenatal and heat of hydration. TSAPE: nonlinear structural analysis including prestress, creep and shrinkage effects (based on TEMP)	TEMP: 2D FEM stress analysis based on self equilibrium stresses TSAPE: beam type elements. Loads due to prestressing	several studies for bidges and dams see e.g. Branco, Cachadinha "Thermal Actiohns for Concrete Bridge Design", Journal Struct Eng Vol 119 no 8, 1993
4 CPTC	Dr R Sato University of Utsunomiya 2753 Ishii Utsunomiya Tochigi 321 JAPAN	Effective Youngs modulus, Principle of superposition Combined actions of thermal stress and autogenious stress	2D FEM analysis	
5 TEEMU	Tauernkraftwerke AG Rainerstrasse 29 Postfach 161 A-5021 Salzburg Austria Phone 43 (0)66 3/86 82-0 Fax 43 (0)66 2/88 9 50-63 Dipl Ing Proper R Dipl Ing Leobacher A	Nonlinear temp calcultaion of layers Thermal stresses with a nonlinear material model	Finite difference model Incremental con-struction by layers Specially designed for dams. For contact stresses near foundation FEM is used	Arch dam, Shahid Rajee, IRAN Concrete structure earthfill dam, Birecik, Turkey Arch dam, Zillergründl, Austria Arch dam, Kölnbrein, Austria

6 HYTEMP	Baustoffprüfstelle der Tiroler Wasser- kraftwerke AG (TIWAG) A-6430 Ötztal- Bahnhof Austria Phone 05266-88545 Fax 05266/87381 Dr Werthmann E	Predictions of hydration heat temperatures Thermal stress computations not included	Numeric calculation method, 1D	Turbine foundations power station, Sellrain Silz, Kuhtai High Way Bridge, Scott-Wien, Vienna
7 THYDWL CATOLD	National Building Research Institute Techion City, Haifa 32000 Israel Phone 972-4-8292242-3 Fax 972-4-8324534 Ass Rachel Becker	Predictions of hydration heat temperatures Thermal stress computations not included	cross sections of walls and slabs, 1D	
8 HEAT/ FEM- MASSE	Intron Structural and Material Engineering P O Box 26 3990 GA Houten Netherlands Phone 31 3403 79580 Fax 31 3403 79680 Dr ir Salet T Dr ir Roelfstra	Temperatures, maturity, degree of hydration, strength, stresses, displacements, strength due to loading rate, cracking risk 4 different temperature models 4 different mechanical models (including ageing concrete by means of Maxwell chain, Burger or Kelvin models), moisture module	Temp: 2D cross section (also 3D), Build phases, solar radiation, wind, heat. wires, cool. pipes Stress: 2.5 D restraining in out-of-plane direction Restraint: friction, springs, settlements etc pre- post processor	West/Eastern Bridge, Storebaelt, Denm. Storm barrier, Neuwe waterweg, Hartel Channel Netherlands, Laffan harbour, Qatar Railway bridge Dordrechtm, Netherlands Dry docks, Dubai Harbour, Rotterddam Öresund bridge/tunnel N Humberland Bridge, Canada

9 Computer Program for Calculation of Temperature, Stress, and Crack Width of Massive Concrete	Japan Concr Inst. (JCI) TBR-Building, room 708 5-7, Kohjimachi, Chiyoda-ku Tokyo Japan Phone: Fax: Dr Ishikawa Prof Tanabe	Temperatures, maturity, strength, Young´s modulus, creep, Cracking risks, crack index related to probability for cracking Reinforcement, crack width,	Temperatures: 2D FEM Stresses: Compensation Plane Method (extended for reinforcement), bending 2 directions, translation (axial restraint) modelled with restraint factors Windows	several (walls, culvert box, slab etc) e g found in proceedings of Munich conference and in JSCE (Japans Society of Civil Engineers) Standard Specifications
10 TN-Term	Nishimatsu Construct Co Technical Res Inst, Dep of Civil Eng 2570-4, Shimotsuruma, Yamato, Kanagawa 242, Japan Phone 81-462-75-1135 Fax 81-462-75-6796 Nishida N, Ushioda K	Temperatures and stresses Effective Young´s modulus, compressive and tensile strengths Sensitivity analysis with respect to errors in input parameters	1D thermal and stress analysis FEM for thermal and stress analysis by Compensation Plane method, Axial and bending restraint, Internal restraint	(no given)
11 CARC	Information Processing Center, KAJIMA Corporation 1-2-7, Motoakasaka, Minato-ku, Tokyo 107 Japan Morikawa H	Temperatures, heat transfer stationary and non-stationary analyses Stress and deformation non-linear analysis (crack in tension and plasticity in compression, time and stress dependent viscosity)	2D and 3D temperature and stress analysis for super computer (of 1996) use	Yukio H et al "Nonlinear analysis of shear walls subjected to cyclic in-plane shear loading", Proc of 4th international conference, Nonlinear Engineering Computations (NEC-91), 1991

12 HACON-T HACON-S HACON-H	Vattenfall Utveckling AB S- 814 26 Älvkarleby Sweden Phone 46 26 83500 Fax 46 26 83670 M Sc Nordström E Prof Alemo J	Temp and maturity (HACON-T). Temp, maturity, strength and stresses (HACON-S). Material params (HACON-H). degr of hydr, compr and tensile strengths, Young´s modulus, creep, Kelvin Chain model, ageing therm dilat coeff., crack dev. autogen. shrinkage, stress induced therm strain	2D and axisymm by HACON-T HACON-S. Quasi 3D by combin. of programs. Heat exchange with unbounded region by infinite elements, mesh extended simul. casting stages heating cables, cooling pipes	HACON-T: Öland, Höga Kusten, Store Baelt bridges, Sweden, Denmark Nuclear power plant, Ringhals, Sweden Foundation, Fort Georg Power Station., Mauritius HACON-S: Hydro power station, Swed Lee River Tunnel, Ireland Drogden Tunnel, Öresund Link (tender phase)
13 FIESTA	Dansk Konstruktions Analyse A/S Sortemosevej 2 DK-3450 Alleröd Denmark Phone 45 48140764 Fax 45 48140033 M Sc Grodkjaer E	Transient therm and stress analys. All therm prop. (heat conductivity, heat generation etc) dependent on maturity. Arbitrary curves of heat generation possible All mechan prop (E-modulus, Poiss. ratio, creep, strengths, thermal expansion coeff.) depend on maturity, stress-strength ratio	3D thermal and stress casting in several stages Influence of adjacent structures, Formwork insulation etc Cooling pipes and heating cables	

14 CIMS-2D	Danish Technological Inst Gregerensvej P O Box 141 DK 2630 Taastrup Phone 45 43 50 40 65 Fax 45 43 50 40 69 B Sc Pedersen Erik Steen Mr Pedersen Erik Jörgen	All material properties for thermal and stress analyses are dependent on maturity Viscoelastic properties with two Kelvin elements in series, strength, autogenious shrinkage,	2D FEM therm analysis 2.5 D FEM stress anal. Compensation Plane th. in combination with disc stresses of cross section, bending, translation. Casting sequences, cooling pipes heating wires, insulation etc	Great Baelt Bridge, Denmark Öresund Link Bridge and tunnel, Denmark-Sweden and others
15 TEMP-SPANN	TU Braunschweig Institut für Baustoffe, Massivbau und Brandschutz Beethovenstrasse 52 D-38106 Braunschweig Phone - Fax 49 531 391 4573 Prof Rostásy F Dr Ing Laube M	Thermal and mechanical properties dependent degree of hydration Compressive strength, E-modulus, Creep function transformed into relaxation function Tensile strength, cracking risk,	Compensation Line Method (Finite laminae method, 1.5 D), Bending, translation, Friction Internal and/or external restraint	several, see references of Proceedings Munich conference

16 TEMPSPAN HYMO- STRUC, CRACK	Delft Univ of Techn, Fac. of Civil Eng. Concr Struct P O Box 5048 2600 GA Delft The Netherlands Phone 31 (0)15 278 4954 Fax 31 (0)15 278 7438 Dr Ir van Breugel K M Sc Lokhorst S	Hydration heat *computed* with HYMOSTRUC,(effects of concr mix, cement type etc) or modelled from calorimetric tests. Comput. of heat flow boundary conditions . Thermal stress: strength, creep and relax and nonlinear behav at cracking, autogenious shrinkage Normal and high strength concrete	1- 3D heat flow comp. in hardening concrete structures, insulation, subbase of e g soil. Stress calculations due to completely or partially restrained thermal movement and shrinkage.	several types of structures e g walls on slabs, T-shaped beams, walls and plates,
17 HYDRA1	Mangel Lab for Concr Res Deparment of Struct Eng Univ of Ghent B-9052 Gent Belgium Phone 32 9 264 55 21 Fax 32 9 264 58 45 de Shutter G	Heat generation and temperature development function of degree of hydration Mechanical behaviour (strength and deformability) function of degree of hydration. Nonlinear creep and shrinkage	2D, 3D versions for thermal stress computations	several, see references of Proceedings Munich conference
18 HIPERPAV HIPER- BOND	Transtec Inc 1012 Est 38 1/2 Street Austin Texas USA Phone Fax Mr Rasmussen R O Dr Zollinger D Dr McCullogh F	Early-age stresses due to temperature movements and shrinkage. Dynamic values of thermal exansion coefficient. E-modulus, specific heat, creep and relaxation, shrinkage, strength. Potential for uncontrolled cracking. Degree of hydration is key parameter	HIPERPAV - stesses in jointed concrete pavements restraint due to slab-base friction, curling and warping HIPERBOND - bond stress of concrete overlay	(no given)

19 HETT5, HETT 2D, HETT 97 TEMPSTRE CONSTRE	Luleå Univ of Technology Division of Structural Eng S-97187 LULEÅ SWEDEN Phone 46 920 91000 Fax 46 920 91913 Dr Emborg M Dr Jonasson J-E	Heat generation dependent on degr. of hydrat: Temp. developm Mech. prop. (compr. and tensile strengths, viscoelastic beh., plasticity in compr., strain softering in tension, failure strength due to loading rate) depend. on maturity: Early-age thermal and contraction coefficients, cracking risks	HETT : 1-3D, insul., cast sequence heat cables, coolign etc TEMPSTRE: Comp Line Meth, friction, bend transl, el resilience. CONSTRE: 2D plane stress or plane strain, axisymmetric,	Carhuacuero power plant, Peru., Rock storage cavern. Middle East, Massive walls of shelters, Bagdad, Iraq. Railway tunnels, Helsingborg The Öland, Höga Kusten, Igelsta Bridge, The Great Baelt Bridge, Denmark and the Drogden Tunnel, Öresund link Purifications plants, Stockholm Diagrams Code of Bridge buildiing
20 TEXO MEXO	CESAR LCPC Laboratorie Central des Ponts et Chaussées 58 Boulevard Lefèbvre 75732 Paris cedex 15 Phone: 33 1 40 43 50 00 Fax: 33 1 40 43 54 98 Dr. Humbert P.	TEXO numerical simulation evolution temperature fields, degree hydration, boundary conditions etc MEXO mechanical effects dependent on degree of hydration,linear behaviour		Elorn and Normandie Bridges, France etc

21 FeC_3S (ANSYS)	HBW, Hollandsche Beton- en Waterbouw bv, P O Box 63 2800 AB Gouda, The Netherlands Phone Fax Maatjes E Schillings de Jong	Heat development modules and modules for modelling mechanical properties at early ages (strength, creep etc) compiled in the ANSYS package Calculation of linear viscoelastic response (thermal stresses) - a standard feature of ANSYS.	2 D temperature boundary conditions (form, insulation etc), cooling pipes, 3 D stress considering restraint	see reference of Proceedings Munich conference
22 FETAB- HYDRA	Dept of Civil Engineering University of Calgary 2500 University Dr. N. W. Calgary, Alberta T2N 1N4 CANADA Phone: 1 403 220-5832 Fax: 1 403 282-7026 Dr Wang C. M. Prof Dr Dilger W. H.	see the reference: Wang C Dilger W H (1995): Modelling of development if heat of hydration in high-performance concrete:, ACI Spec Publ 154, Advances in Concr. Techn, 473-488	2D FEM thermal analysis Eigen and fixed-end thermal stresses in hardening concrete during construction in field conditions	see Wang and Dilger (1994): Rilem Conf Munich and: Elbadry M M and Ghali A (1984, 1982): User Manual for Computer Program FETAB, Research. Report, Dept of Civ Eng, Univ of Calgary, Canada, Oct 1982 revised 1984.
23 LUSAS	FEA Ltd. Kingston-upon-Thames Surrey KT1 IHN United Kingdom	(no information)	(no information)	(no information)

8

Measurement of Thermal Stresses In Situ

T. Tanabe

8.1 INTRODUCTION

It is a typical method in most countries that c-c thermocouples and strain gauges are embedded in the concrete for massive concrete structures in order to measure temperature change and strain. In this case, the purpose of field measurements is to determine the generated strain which corresponds to the stress and to predict whether thermal cracks occur or not. However, it has been the long standing problem in this case how to identify non-stress components of measured strain and the Young's modulus to be used for stress conversion. In response to these situations, a stress gauge with the specific purpose to be used for young concrete has been developed in Japan and has been used there extensively, whereas in France a different type of stress gauge has been developed and is in use. In this section, these direct measuring methods will be described first and afterwards the indirect measuring method using strain gauges.

8.2 STRESS GAUGES DEVELOPED IN JAPAN

The Young's modulus and creep properties of concrete at very early ages after placing, change extremely with time, so there is no reliability on the thermal stress estimated by strain if non-stress components and varying Young's modulus be not accurately evaluated. In the laboratory, an apparatus has been designed to measure the thermal stress of concrete at an early age. An apparatus for in situ measurement was not available.

Prevention of Thermal Cracking in Concrete at Early Ages. Edited by R. Springenschmid. RILEM Report 15. Published in 1998 by E & FN Spon, 11 New Fetter Lane, London EC4P 4EE, UK. ISBN 0 419 22310 X

An instrument to measure thermal stress independently of the change of concrete properties at early ages has been desired for long time and a stress gauge was developed about fifteen years ago in Japan as well as in France. The basic mechanism of the stress gauge developed in Japan for measuring thermal stress in massive concrete structures is shown in Fig. 8.1 and the actual gauge is shown in Photo 8.1.

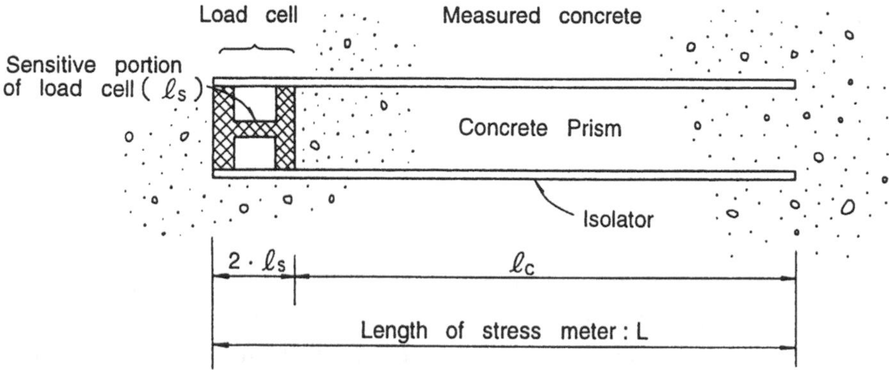

Fig. 8.1 Basic mechanism of stress meter.

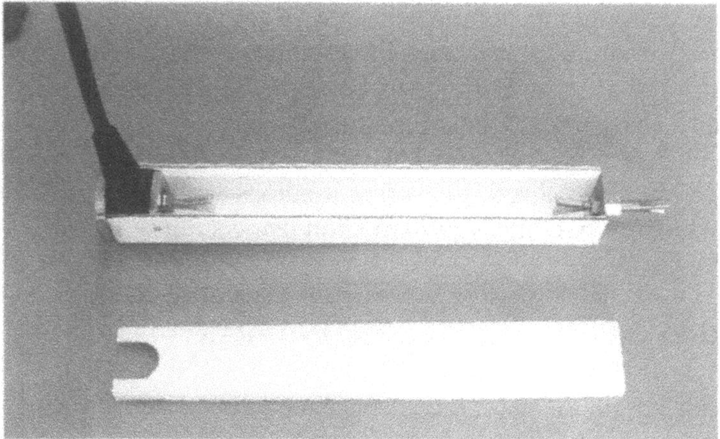

Photo 8.1 Stress meter for practical use.

A load cell and a concrete prism are connected in series each other. The concrete in the prism of the stress gauge is poured at the same time as the surrounding concrete so that the concrete in the prism should have the same quality as the surrounding concrete. Two planes at right angles to the longitudinal axis of the gauge touch the surrounding concrete. Flange of the load cell at one end and special anchor bolts at the other end tightly connected the surrounding concrete to the stress meter. These bolts and flange are able to transmit not only the compressive stress but also the tensile stress. Other four planes parallel to the longitudinal axis are covered by blotting paper, porous sheet and felt and separate the concrete prism of a stress gauge from surrounding concrete, but moisture migrations between the concrete is possible through these materials. Due to such mechanisms of the stress gauge, the rigidity of stress gauge is able to follow that of the measured concrete, which varies much with the time after placing concrete. According to moisture migration of concrete, it is assumed that the property of concrete prism in the stress gauge is equal to that of the surrounding concrete from the beginning of age of fresh concrete to the latter age of hardened concrete.

The stress gauge can measure thermal stress in concrete even if Young's modulus of concrete is unknown, only if the relationship between stress and deformation of the load cell is known. Theoretical accuracy of the stress gauge is shown in the reference [l] in three assumed ideal environmental states, i.e., a) the state that concrete is subjected to sustained constant external load, with assumption that surrounding concrete of the gauge and concrete in the gauge having the same creep strain rate of m, b) the state that concrete is subjected to uniform temperature rise of ΔT with assumption of no creep and no shrinkage, c) the state that concrete is subjected to uniform shrinkage strain alone with no creep. Basing on the equilibrium equation of forces and the compatibility condition of deformation together with the following hypotheses that the stress gauge is perfectly connected with the surrounding concrete, and the load cell is also perfectly connected with the concrete prism in the stress gauge, the relationships between the stress acting on the surrounding concrete, σ, and the stress measured by the load cell, σ', is derived in the following equations. Notations in Table 8.1 are used in the formulation.

Table 8.1 Notations

Length of concrete prism in the stress gauge: l_c,
Length of sensitive portion in the load cell: l_s
Thickness of supporting plates of the load cell: l_s
Whole length of the stress gauge: $L = l_c + 2\,l_s$
Cross sectional area of the concrete prism and the supporting plates: A_c
Cross sectional area of the sensitive portion of load cell: A_s
Young's modulus of concrete at time t_0: E_c
Young's modulus of load cell: E_s
Creep coefficient of concrete after time t from time t_0: m

Coefficient of the thermal expansion of concrete: α_c
Coefficient of thermal expansion of load cell: α_s
Shrinkage strain of concrete after time t' from time t_1: ε_{cs}

a) For a state when concrete is subjected to sustained constant external load, theoretical stress ratio, σ'/σ, is written as

$$\left(\frac{\sigma'}{\sigma}\right)_a = \frac{1+2n}{1+\frac{n}{k}\left(1+\frac{1}{\beta}\right)\frac{1}{1+m}} \tag{8.1}$$

where σ' is the stress measured by the stress gauge and σ is the stress generated in the surrounding measured concrete after time t and $l_s/l_c = n$, $A_s/A_c = \beta$, $E_s/E_c = \kappa$. Notations l_s, l_c, A_s, A_c, E_s and E_c are the length of load cell, the length of concrete prism portion, the sectional area of sensitive portion of a load cell, the sectional area of concrete prism, the Young's modulus of load cell metal and the Young's modulus of concrete respectively. No shrinkage strain is assumed and Young's modulus of concrete is assumed to remain constant after time to.

b) For a state when concrete is subjected to uniform temperature rise of ΔT, the ratio σ'/σ is written in the equation of

$$\left(\frac{\sigma'}{\sigma}\right)_b = \frac{1+2n\cdot\gamma}{1+\frac{n}{k}\left(1+\frac{1}{\beta}\right)} \tag{8.2}$$

where $\alpha_s/\alpha_c = \gamma$. Thermal deformation is absolutely restrained and no creep strain and no shrinkage strain is assumed to occur and Young's modulus of concrete is assumed to remain constant during temperature change.

c) For a state when concrete is subjected to uniform shrinkage strain ε_{cs} during time t' from time t_1, and the deformation due to shrinkage is assumed not to be restrained, σ' is expressed by the following equation, where σ' is the stress measured by the stress gauge,

$$(\sigma')_c = \frac{2n}{1+\frac{n}{k}\left(1+\frac{1}{\beta}\right)}\cdot E_c\cdot\varepsilon_{cs} \tag{8.3}$$

No creep strain is assumed to occur and Young's modulus of concrete is assumed to remain constant during time change.

By varying the values of parameters in Eq. (8.1) to Eq. (8.3) and performing numerical experiments, it is possible to examine the fluctuation of the values, σ'/σ, which is the indicator of theoretical accuracy of the gauge.

The dimensions of stress gauge of practical use are as such that $L = 50$ cm, $l_c = 45$ cm, $l_s = 2.5$ cm, $A_c = 25 cm^2$, $A_s = 1.32$ cm^2, shown in Photo. 8.1 corresponding to usual concrete of the maximum coarse Aggregate size 20 mm - 25 mm.

Numerical experiments are performed for the usual range of Young's modulus of concrete, 5.0×10^2 N/mm^2 - 4.0×10^4 N/mm^2 where $E = 2.1 \times 10^5$ N/mm^2 and $\alpha_s = 11 \times 10^{-6}/°C$.

The relationship between $(\sigma'/\sigma)_a$ of Eq. (8.1) and Young's modulus of concrete is shown in Fig. 8.2, where solid line is for the case of creep strain rate, $m = 0$, which corresponds to elastic deformation alone and dotted line is for the case of $m = 1$, which considers creep deformation.

The relationship between $(\sigma'/\sigma)_b$ of Eq. (8.2) and Young's modulus of concrete is shown in Fig. 8.3, where solid line is for the case of $\gamma = 1$, $\alpha_c = 8 \times 10^{-6}/°C$ and the dotted line is for the case of $\gamma = 1.375$, $\alpha_c = 8 \times 10^{-6}/°C$.

The relationship between $(\sigma')_c$ of Eq. (8.3) and Young's modulus of concrete is shown in Fig. 8.4, where the shrinkage strain of $\varepsilon_{cs} = 100 \times 10^{-6}$ is assumed. All those theoretical investigation assures practically allowable range of accuracy.

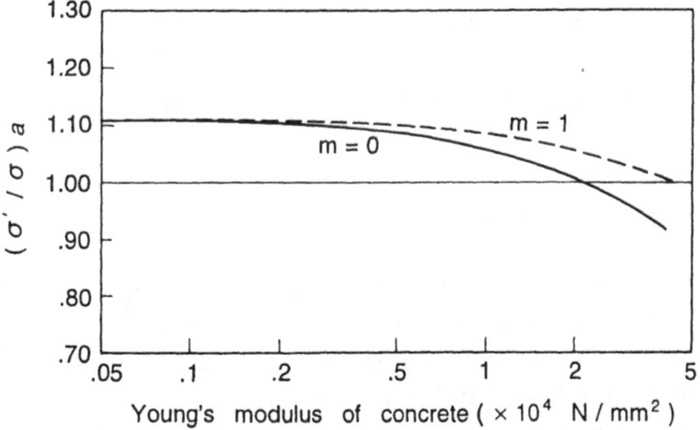

Fig. 8.2 The relationship between $(\sigma'/\sigma)_a$ and Young's modulus of concrete subjected to sustained constant external load.

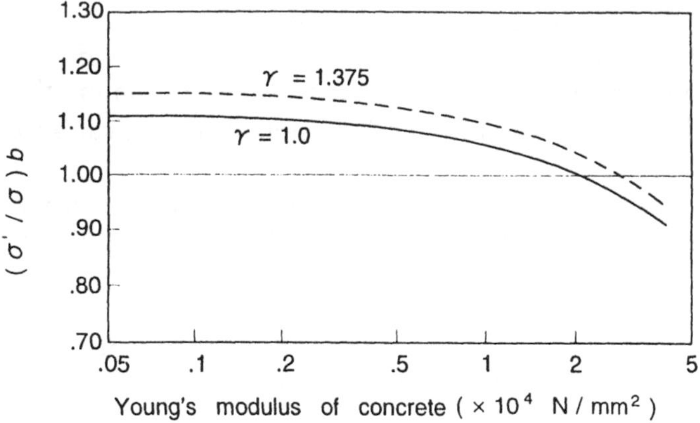

Fig. 8.3 The relationship between $(\sigma'/\sigma)_b$ and Young's modulus of concrete subjected to uniform temperature rise ΔT.

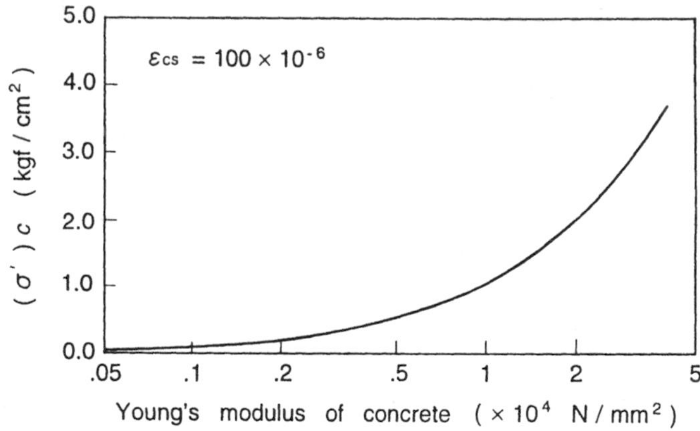

Fig. 8.4 The relationship between $(\sigma')_c$ and Young's modulus of concrete subjected to uniform shrinkage strain ε_{cs}.

Experimental Confirmation

In addition to the theoretical investigation of its accuracy, experimental confirmation has been performed in reference [2] for a) the instant loading state, for b) the state that sustained loading has been added, for c) the state drying environment has been made, for d) the state of rising temperature and for e) the state of constrained condition The results are summarised as follows.

a) For the instant loading state, the relationship between stress measured and stress given by the loading apparatus is shown in Fig. 8.5. The detected errors were practically small enough to be neglected. This indicates the rigidity of a stress gauge is consistent with that of concrete having different values of Young's modulus.

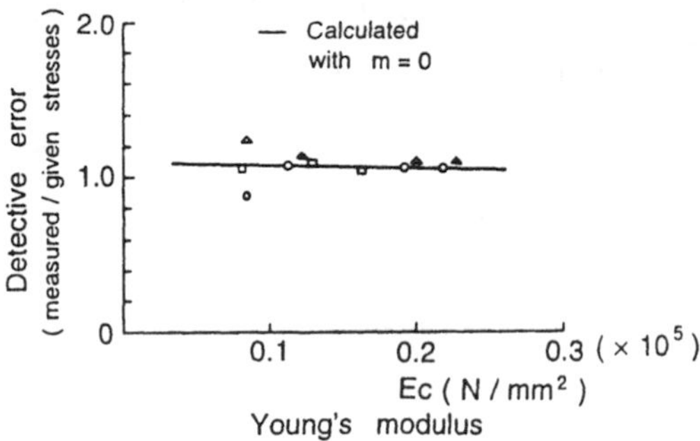

Fig. 8.5 Accuracy of the stress measurement at instant load.

b) As for creep effect, test set-up used is shown in Fig. 8.6. Fig. 8.7 shows the test results of measured stresses with time. The final creep coefficient was about 2. It is indicated that the measured stresses were consistent with the given stresses.

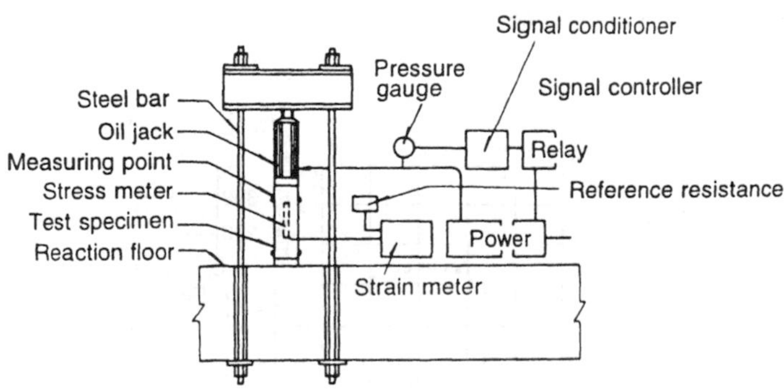

Fig. 8.6 Creep test setup.

c) As for the effect of drying shrinkage, the test set-up is shown in Fig. 8.8. Test specimens were cured under the condition of cyclic temperature change in the chamber. The test results are shown in Fig. 8.9. The measured stresses were in compression at the early stages of the test due to surface drying and internal constraint.

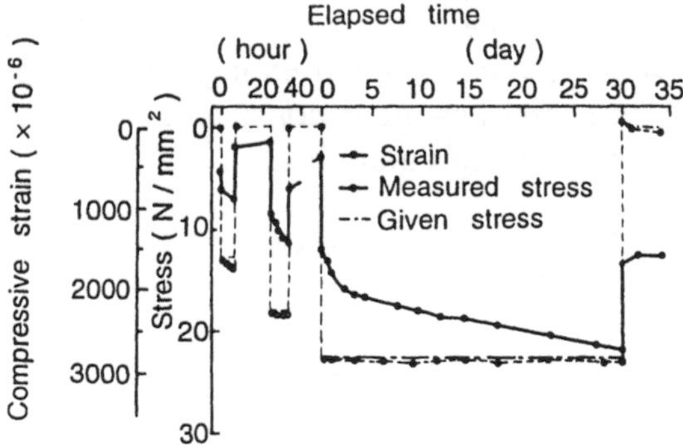

Fig. 8.7 Creep strains and measured stresses.

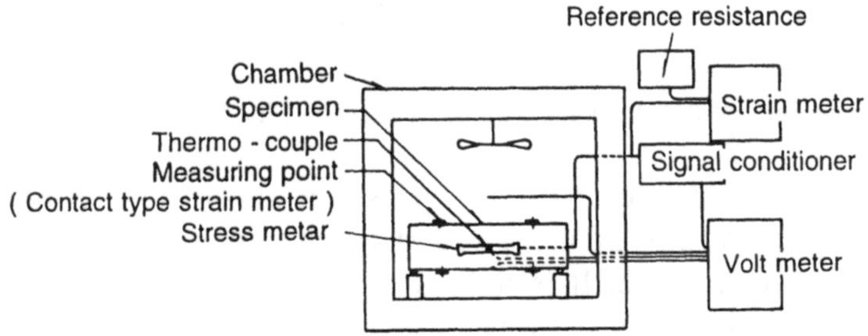

Fig. 8.8 Set-up for drying shrinkage test.

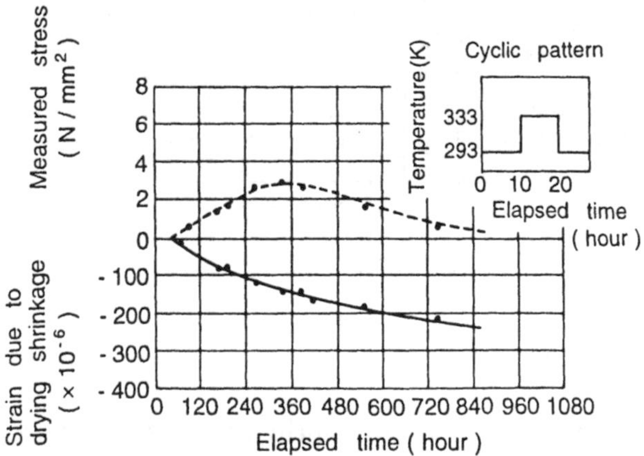

Fig. 8.9 Drying shrinkage strains and measured stresses.

However, the compressive stresses decreased and approached to zero after the completion of drying process at the age of 800 hours. Therefore, the effect of drying shrinkage is considered not to effect the measured values.

d) As for the effect of temperature change, the test set-up was same as shown in Fig. 8.8 and test specimens were coated by epoxy in order to prevent drying shrinkage of concrete. Fig. 8.10 represents the measured stresses with time. As shown in the figure, some stresses in both tension and compression due to internal constraint were observed at the processes of temperature change. However, since those thermal stresses diminished under the constant temperature stage, it was considered that correction against temperature change is not necessary.

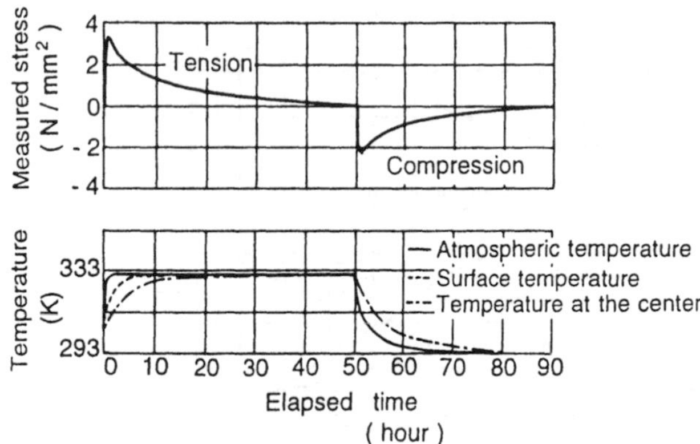

Fig. 8.10 Temperature changes and measured stresses.

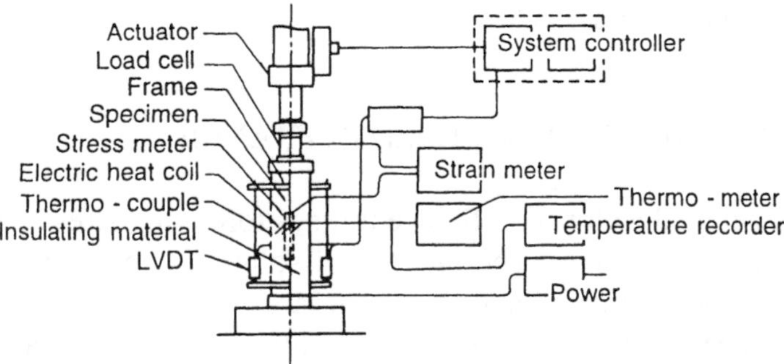

Fig. 8.11 Test set-up for constraint stresses.

e) As for the constraint condition, Fig. 8.11 shows the test set-up. Test specimen coated by epoxy were heated by the surrounding electric coils and the electric servo loading apparatus was used to constrain thermal strains. The constraining loads and the restraint loads did not agree with the stresses measured by the stress gauge in the process of temperature rise because of temperature difference in the interior. However, after the uniformly distributed temperature was attained, the value were also consistent with the calculated stress using the thermal coefficient of expansion α_c, Young's modulus E_c and the temperature rise ΔT as shown in Fig. 8.12. Therefore, the stress gauge used here is considered to be applicable to constraint stress measurement.

Through the above mentioned calibration tests together with theoretical investigations, it was indicated that the stress gauge has good performance as a measuring apparatus *in situ*.

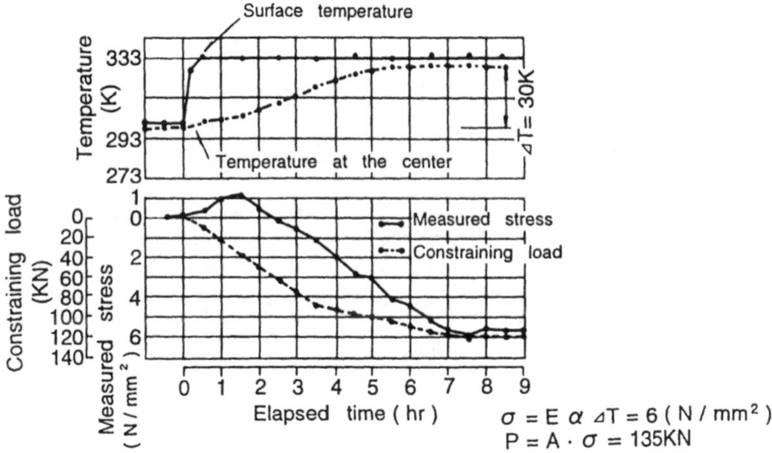

Fig. 8.12 Constraint stresses with time.

8.3 STRESS GAUGES DEVELOPED IN FRANCE

The stress gauge developed in France utilises following stress distributing mechanism in continuum. The theoretical solution shows that an elastic inclusion in elastic matrix gives different stress or strain condition in the inclusion depending on the relative rigidity of the two materials i.e., inclusion and matrix, E_c/E_b, and width over length ratio, R/L, of inclusion as shown in Eq. (8.4).

$$K_c = \frac{\dfrac{E_c}{E_b} - 1}{1 + \dfrac{\pi R}{L}\dfrac{E_c}{E_b}(1 - \nu^2)} \tag{8.4}$$

where K_c denotes stress concentration factor. Therefore by choosing the R/L ratio and the E_c/E_b ratio as the value which affect to K_c value to minimum degree, the inclusion is used as a stress gauge. Reference [3] shows that the dimensions of inclusion shown in Fig. 8.13 which is used now for a stress gauge is appropriate for measuring thermal stress in early age concrete.

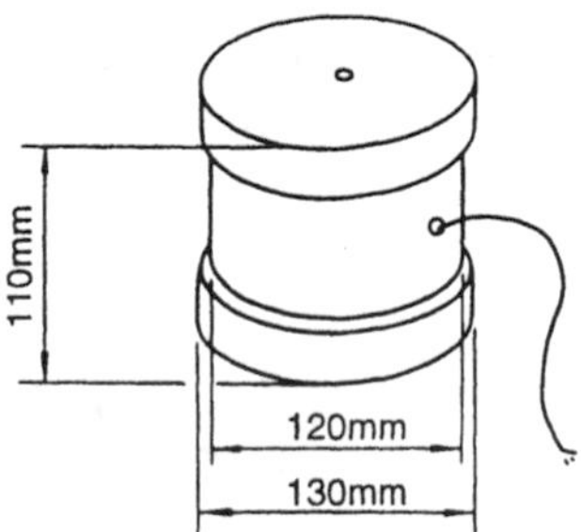

Fig. 8.13 Stress gauge of French type.

The experimental concrete specimen with stress gauge developed at the centre is shown in Fig. 8.14.

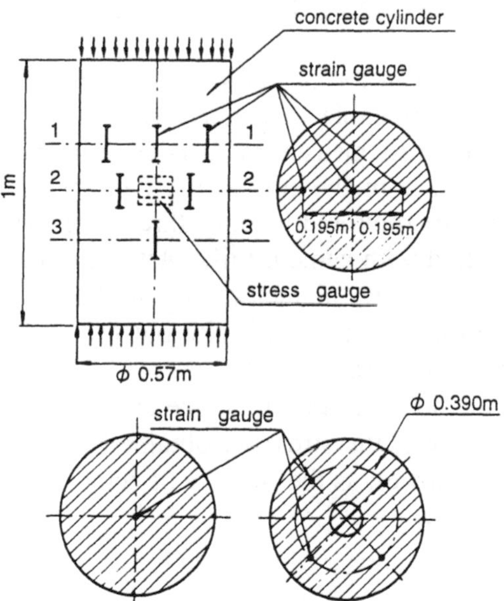

Fig. 8.14 Experimental specimen of stress measurement.

Stress variation measured during the period of 1000 days duration of constant load is shown together with strain variation in Fig. 8.15 which indicate good performance of the gauge as a stress gauge.

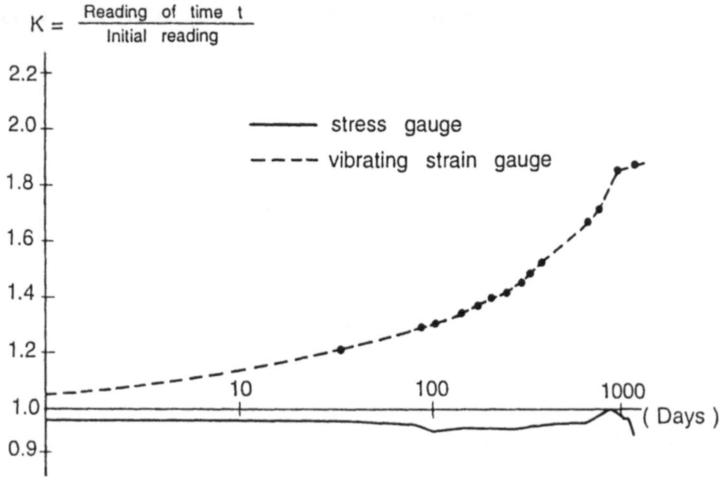

Fig. 8.15 Experimental result.

8.4 STRESS MEASUREMENT BY INCORE METHODS

As has been often used in the initial stress measurement of rock *in situ*, the same method is applicable to concrete in-situ. The most commonly used method in rock stress measurement is done in the following way. In the first place, it is to bore the small hole diameter of which is usually about 56 mm inside of rock and to bury three dimensional strain gauges and the hole is grouted with cement milk. After grouting material is hardened enough then, the larger diameter hole of about 218 mm diameter is bored outside of pre-bored hole and take out the specimen. As concrete cylinder now free from restraint undergoes free contraction or free expansion Then, the strain at that instance and the rigidity give the pre-existing stress. However, the stress so obtained is obviously affected by the core dimensions as well as the rigidity of cement milk. Therefore, pre-calibration is necessary.

8.5 INDIRECT MEASURING BY STRAIN GAUGES

Indication of the strain gauges, corrected by the temperature change measured by the thermocouples, can be converted to the stress by subtracting non stress component and multiplying the Young's modulus at the age. In this indirect method, it is most crucial to correctly estimate the Young's modulus and non-stress component of strain such as creep strain. However, since the stress gauges of Japan type must be filled with fresh concrete in placing concrete just before starting measurement and setting gauges where complicated placement of reinforcing bars are existing, is sometimes impossible since it is very difficult to approach the locations during concrete placement. Further, these stress gauges are relatively expensive. With these reasons, strain gauges are used to estimate thermal stress in-situ. To measure the strain of young age concrete, the following points should be at least considered carefully. Firstly, anchoring of strain gauges to young age concrete should be perfect. Reliability is low if anchoring is not sufficient. The experimental results that compare the effect is simply explained in the following example. Three kinds of anchoring is developed for a strain gauge the specific shapes of which are shown in Fig. 8.16.

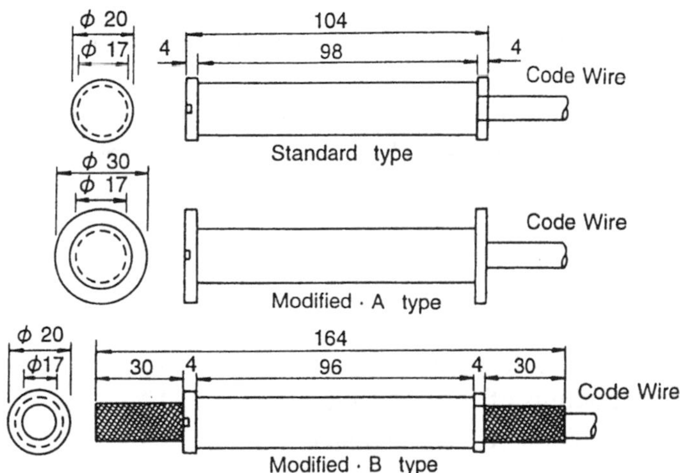

Fig. 8.16 Strain gauge with modified anchorage.

They are embedded in the same concrete wall specimen and strains are measured. The measured strain of a day, 3 days, 5 days and 10 days of concrete ages are shown in Fig. 8.17. It can be noted that there exist large differences in strain values measured though it should give same strain. Secondary, durability of strain gauge is important. In this regards, Carlson type strain gauge has been recognised as very stable for long duration of measurement period. It should be also noted that the low Young's modulus of concrete requires relatively softer stiffness of gauge itself In recent years, however, incased strain gauges stiffness of which is around 50.0 N/mm^2 and which has relatively good durability are often used due to the economical reasons. Besides these, a vibrating wire strain gauge is sometimes used for its accuracy of measurement of small strain.

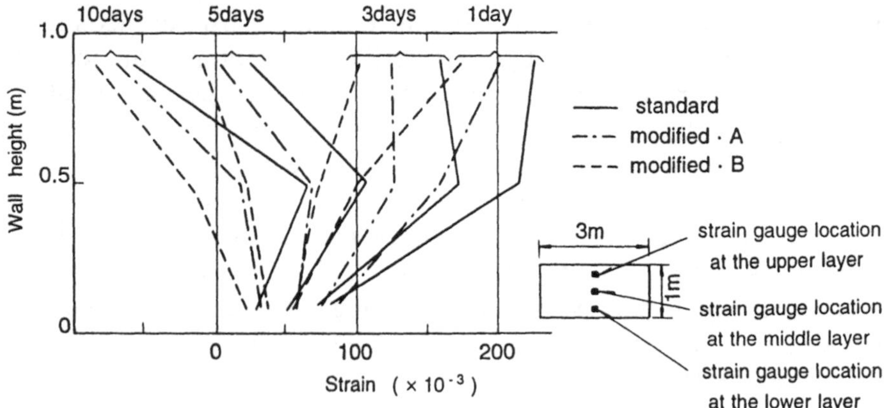

Fig. 8.17 Measured strain differences due to the anchorage of strain gauges.

In evaluating the thermal stress from the measured value of strain gauge, the following time-interval and superposed method is usually adopted.

$$\sigma_i = \sigma_{i-1} + k_i(\varepsilon_i - \varepsilon_{i-1})(E_i + E_{i-1})/2 \tag{8.5}$$

where

σ_i : thermal stress at the time e
E_i : Young's modulus of concrete at the time i
ε_i : total strain of concrete at the time e
ε_{ei} : elastic strain of concrete at the time i
ε_{cri} : creep strain of concrete at the time e
k_i : reduction factor due to concrete creep at the time $k_i = 1/(1+\phi_i)$
ϕ_i : creep coefficient at the time i, $\phi_i = \varepsilon_{cri} / \varepsilon_{ei}$

It should be pointed out that the reliability of estimated thermal stress depends exceedingly on the accuracy of adopted values of Young's modulus and creep property of concrete.

8.6 EXAMPLES OF MEASUREMENT OF THERMAL STRESSES IN SITU

Example-1 Footing- 1 [5]

A large scale footing of which dimensions are given in Fig. 8.18 has volume of some 1200 m³ of light weight concrete and supported by R.C. piles. Five stress gauges of Japan type are buried along a vertical line close to the centre of the footing as shown in Fig. 8.18 with five strain gauges and thermocouples in the same locations and measurements at each point were performed for several weeks starting immediately after the completion of placing concrete.

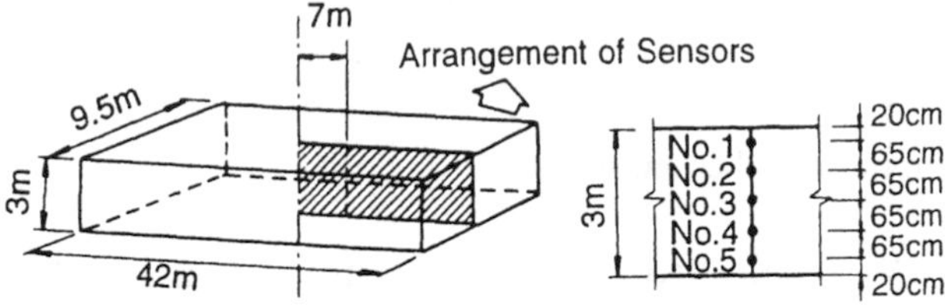

Fig. 8.18 Measured points and analytical model of the stricter.

Measured stresses are shown in Figs. 8.19 and 8.20. The main feature of the stress pattern in the structure was that a relatively larger compressive stress occurred in the middle of the depth and tensile stress occurred in the surface and in the bottom. The stresses increased with temperature rise and then levelled off after two weeks of casting.

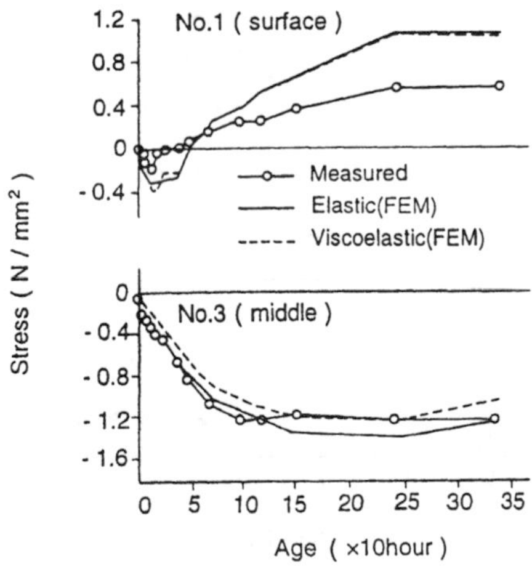

Fig. 8.19 History of thermal stress.

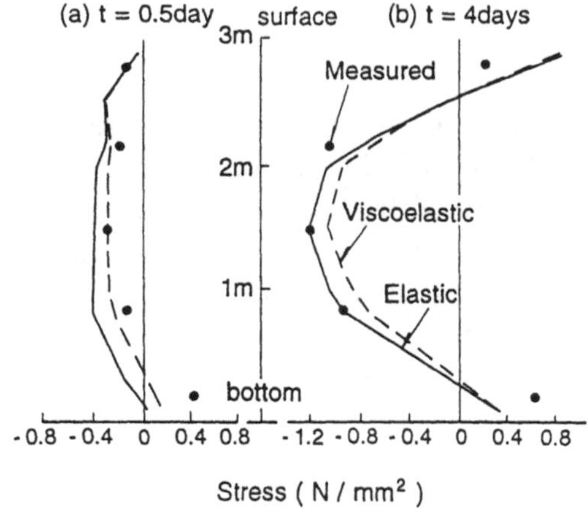

Fig. 8.20 Distribution of thermal stress.

Utilising the results from a concrete stress gauge and a strain gauge embedded at the same point in the footings, the actual Young's modulus, E_a,in the structure can be measured as a function of time. This may differ from the Young's modulus, E_c, obtained from standard compressive tests using cylindrical specimens. Fig. 8.21 shows the variation of E_a, E_c and E_a/E_c with time. It can be seen that, E_a, the average value of several measured locations, which have different stress and temperature histories, is smaller than the Young's modulus, E_c, obtained from standard tests. E_a/E_c decreases for the first week and then remains about 0.55.

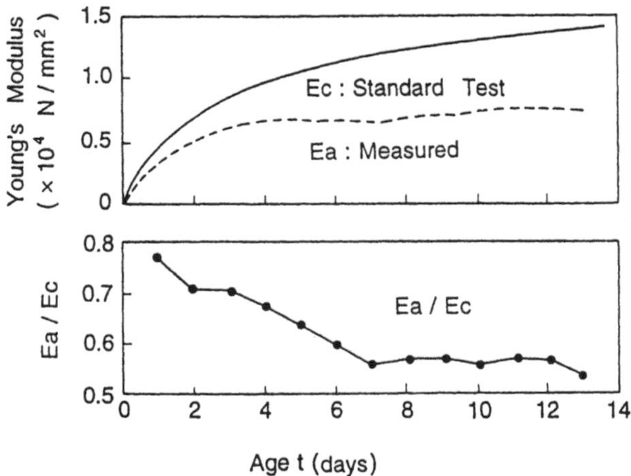

Fig. 8.21 Young's moduli by standard test and measured in the structure.

Example-2 Footing-2 [1]

A block foundation shown in Fig. 8.22 was constructed placing concrete layer by layer. In the second layer, stress gauges of Japan type and the strain gauges were buried at the same locations and thermal stresses and strains were measured and the comparison between the thermal stresses evaluated by the stress gauge and by the strain gauges were shown in Fig. 8.23 and in Fig. 8.24.

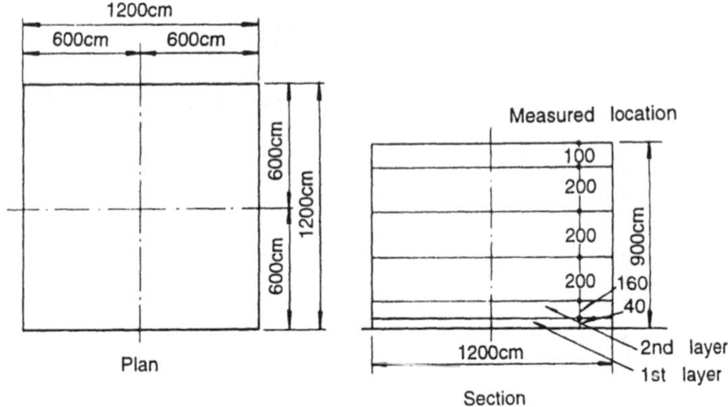

Fig. 8.22 Experimental block foundation.

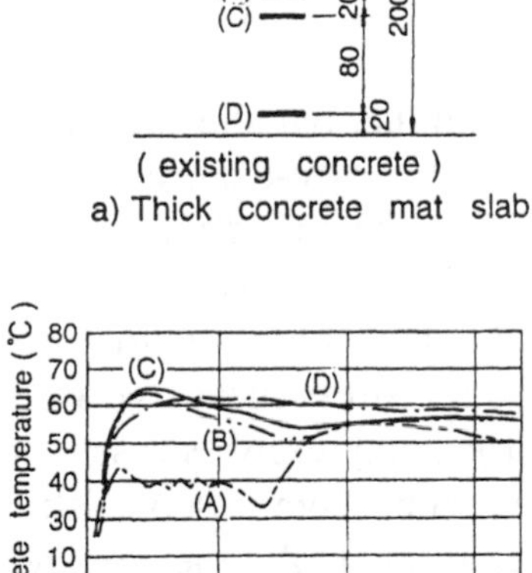

Fig. 8.23 Temperature rise.

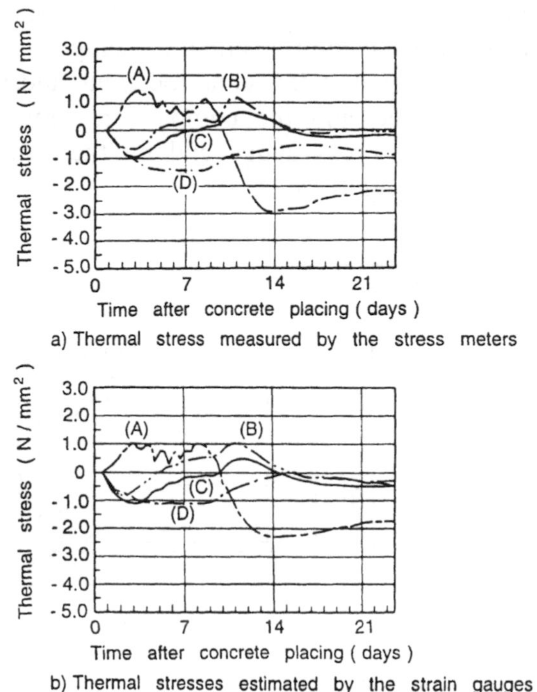

Fig. 8.24 Comparison of stress values measured by stress gauges and by strain gauges.

Example-3 Wall [6]

Temperature history and thermal stresses developed in a thick wall of 1.5 m thickness were measured by c-c thermocouple and the stress gauges of Japan type. The wall was the basement wall in a large testing laboratory shown in Figs. 8.25 (a), (b), (c). The temperature and the stress were measured at several points in one representative section.

The temperatures and the stress measured in the section are shown in Figs. 8.26 (a) and 8.26 (b). Compressive stress increased with temperature went up, and the maximum compressive stress of 1.6 N/mm² was observed two days after casting. As the concrete cooled, compressive stress decreased and tensile stress increased. On the sixth day, there was a sudden decrease in the measured value of tensile stress possibly because of the initiation of cracking at that moment in the area being monitored.

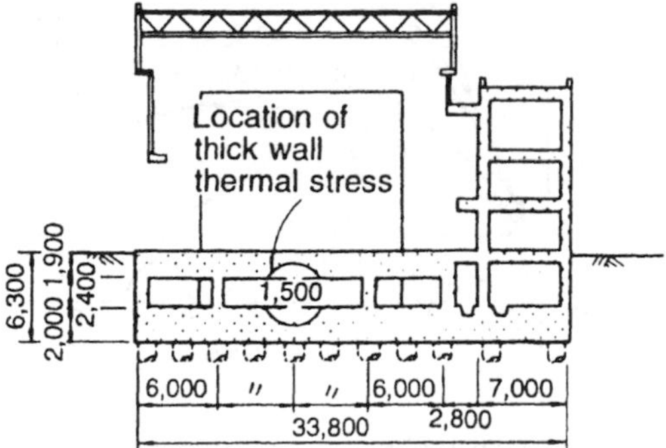

Fig. 8.25(a) Testing laboratory building section.

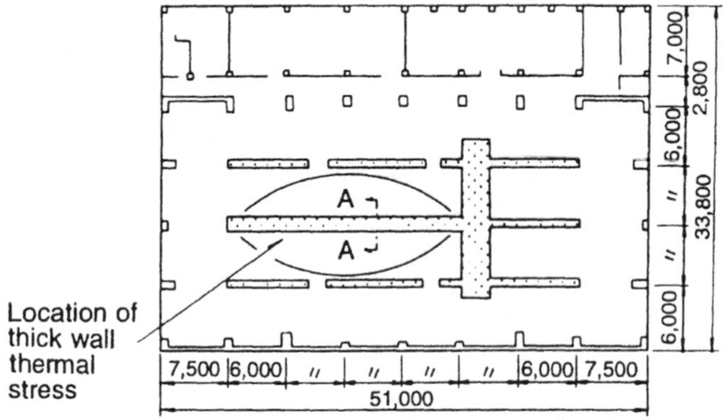

Fig. 8.25(b) Basement plan view of testing laboratory [mm].

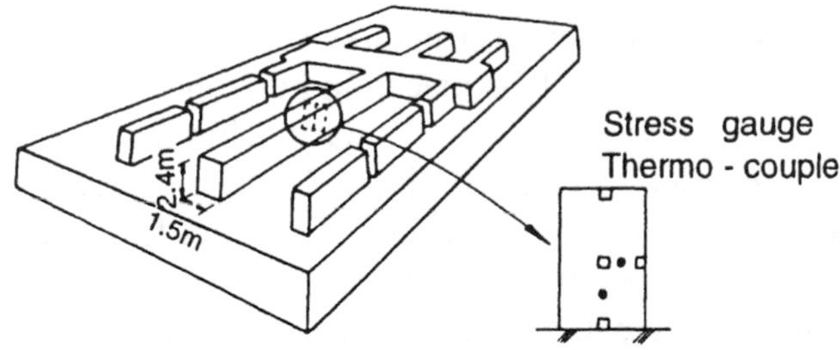

Fig. 8.25(c) A representative section where temperature changes and concrete stresses were monitored.

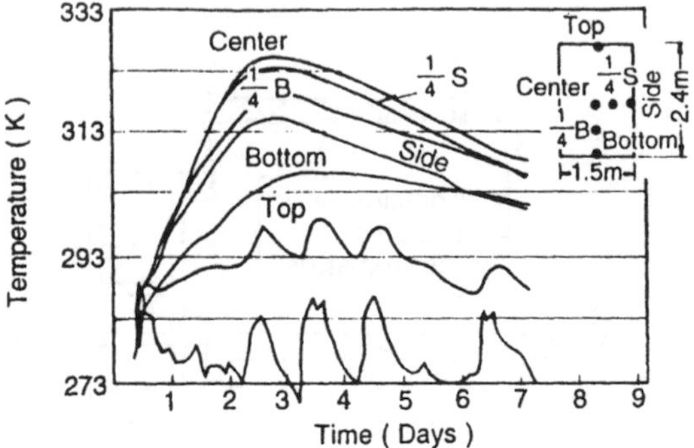

Fig. 8.26(a) Temperatures measured within the concrete section.

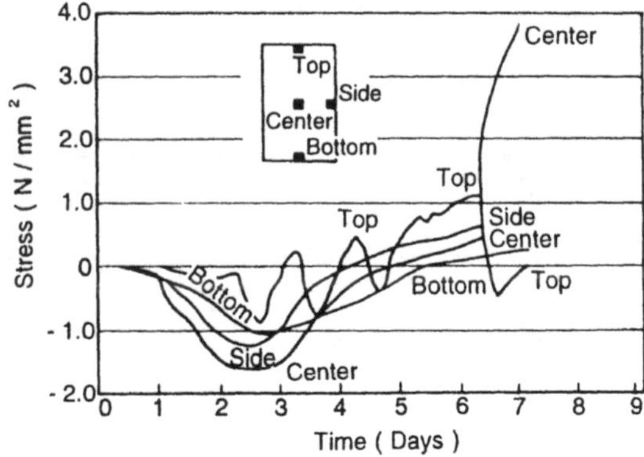

Fig. 8.26(b) Actual measured stresses within the section.

Example-4 Anchorage [7]

When constructing concrete structures such as foundations or anchorages of large bridges, concrete is placed in horizontal lifts (layers) of 0.7 - 2.5m thickness at intervals of 5 to 20 days. This method is also widely employed in constructing concrete dams. In such cases, it is understood that the temperature and stresses generated in any lift in each lifts are also influenced by the temperature and stress histories of neighbouring lifts. Fig. 8.27 shows a schematic representation of a large anchorage of a suspension bridge. The construction was carried out using lifts of 0.8 - 1.5m placed at an interval of 6 - 7 days. Blast furnace slag cement 280 kg/m³ was used in the construction. The stress resulting from temperature change of each lift were measured using thermocouples and the stress gauges of Japan type.

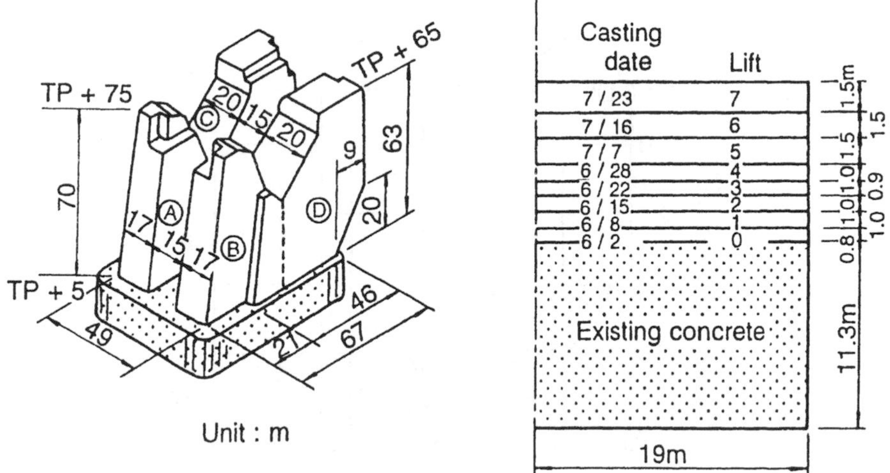

Fig. 8.27 Schematic representation and analytical model of the structure.

Though similar trends were obtained for all the lifts, a typical set of results (for lift No. 6) are plotted in Figs. 8.28 and 8.29. It can be seen that the measured stress for a particular lift was influenced by not only the temperature change in that lift but also the temperature histories of successive lifts.

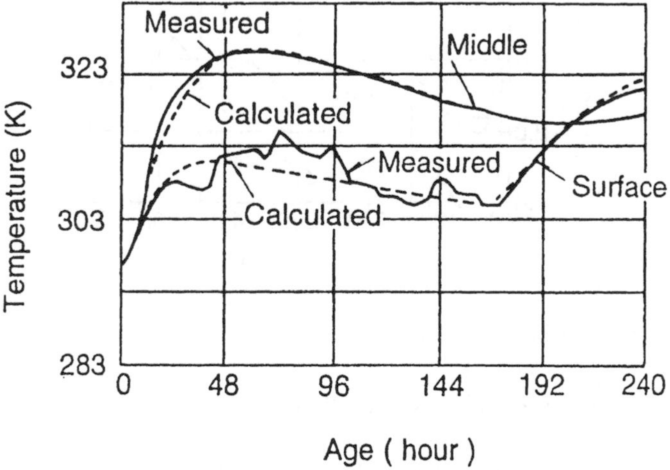

Fig. 8.28 Temperature history.

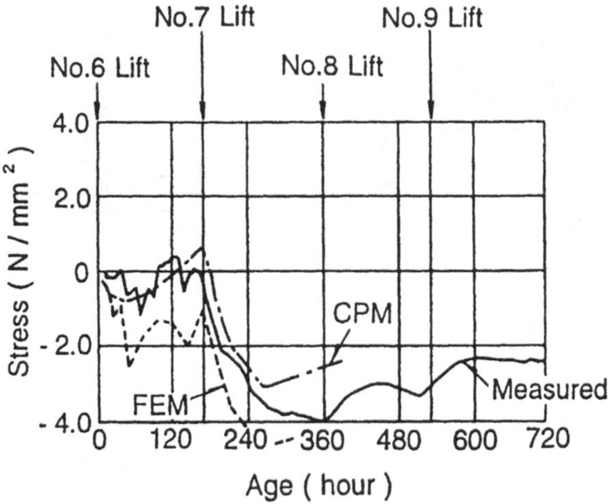

Fig. 8.29(a) Stress history at surface.

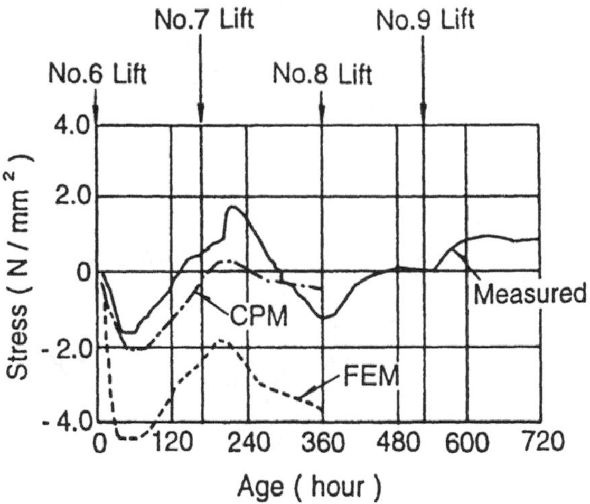

Fig. 8.29(b) Stress history at middle.

Example-5 Containment Vessel Wall [8]

The other example is shown in Fig. 8.30. These were measured in a cylindrical thick concrete wall. These figures show the thermal stress evaluated by a strain gauge has very good coincidence with that evaluated by a stress gauge.

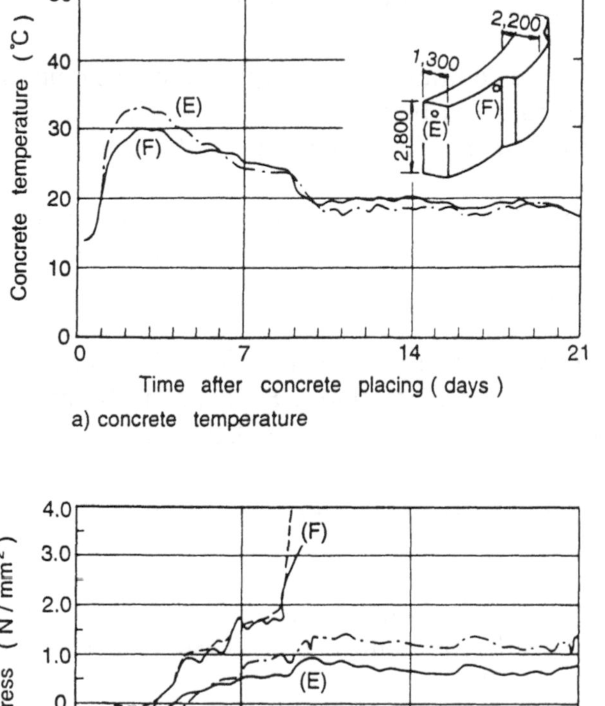

Fig. 8.30 Thermal stresses in thick concrete wall estimated by the strain gauge compared by the stress meter.

REFERENCES

[l] T. Kawaguchi and S. Nakase: Basic Investigation on the Determination of Thermal Stress in Massive Concrete Structure, *Proc. of the JCI*, **14**, No. 1, 1992, pp. 1087 1092.

[2] Y. Ganbara, T. Natsume, M. Sasaki and Y. Masamura: New Type Stress Meter for Concrete Structures, Annual Report of Kajima Institute of Construction Technology, **26** (1977), June 1978.

[3] J. Bonvalet: Le contraintemètre : Appareil pour la mesure directe des contraintes dans les bétons et matériaux sujets à fluage et retrait, *Revue Française de Mécanique*, No. 72-1979, pp. 45-51

[4] M. Ishikawa, T. Maeda and T. Nishloka: Improvement of Anchorage of Buried in type Strain Gauge, The 43rd Annual Conference of JSCE, 5th Section, 1988.

[5] H. Yoshikawa, M. Kusabuka and T. Tanabe: Visco-elastic Behaviour of Massive Concrete Structures, Proc. of Int. Symp. on Long Term Observation of Concrete Structures, Budapest, 1984, pp. 317 - 324.

[6] M. Yamazaki: Thermal cracking of a Thick Wall, *Concrete International*, Aug. 1986.

[7] JSCE Committee on Concrete: Documentation for Revision of JSCE Standard Specification on Concrete-1986, Concrete Library No. 61, Oct. 1986, pp. 80-82.

[8] T. Kawaguchi: Ohbayashi Corporation Technical Report, 1991.

9

Practical Measures to Avoiding Early Age Thermal Cracking in Concrete Structures

S. Bernander

9.1 INTRODUCTION

9.1.1 Historical comments

For reasons of function, durability and appearance early age cracking is a major concern in concrete construction. The increasing use of more cement for higher concrete qualities has rendered this problem more acute in recent years. For this reason more severe crack prevention requirements have been specified by future proprietors and normative bodies in the last decades - and that for an extended variety of structures.

The fact that, in particular, massive concrete structures tend to crack at early ages as a result of temperature induced stresses, is a problem as old as the use of concrete as construction material. The issue has thus been subject to the attention of civil engineers for almost a century.

In the past, crack control has in practice almost exclusively been implemented on the basis of temperature measurements. In the nineteen-fifties and earlier, cracking in maturing concrete was controlled mainly by simply limiting the temperature rise due to hydration. The degree of restraint, to which the young concrete might be subjected, was considered in a qualitative manner by suitable arrangements of construction joints and construction sequences. In the last decades it has become common practice to limit cracking by fulfilling certain more specific requirements with respect to *temperature differences* within the cast section or in rela-

Prevention of Thermal Cracking in Concrete at Early Ages. Edited by R. Springenschmid. RILEM Report 15. Published in 1998 by E & FN Spon, 11 New Fetter Lane, London EC4P 4EE, UK. ISBN 0 419 22310 X

tion to older adjoining concrete members. In some instances restrictions on maximum concrete temperatures have been set.

For temperature-based crack criteria to be technically justified a reasonable degree of proportionality between temperature differences and stresses has been presumed. However, recent research has shown that this assumption often does not hold true - particularly in respect of very early age stresses occurring in the heating or expansion phase of the hydration temperature cycle, cf fig. 9.5. (Rawhouser [1], Bernander, [2], [3], [4], van Breugel [5], [6], [7], Emborg [8], Eberhardt *et al.* [9]).

9.1.2 Recent developments

In the last decades researchers in the field of concrete engineering have become more acutely aware of the fact that temperature alone is not the sole crucial factor involved in the phenomena of temperature-induced cracking in young concrete. Hence, also the importance of the *degree of restraint,* to which the deformations in the early age concrete may be subjected as well as that of the *transient mechanical properties* of the hardening concrete, has been more generally recognized and actually considered in various pilot applications. Yet it is only of late that concrete engineers have begun to acquire a reasonable understanding of the complex phenomena involved in the early age cracking of concrete. By means of advanced computer simulations, based on pilot research on the transient material properties of hardening concrete, estimation of temperature stresses and prediction of crack risk in young concrete is today often actually implemented in design and construction of advanced concrete structures. (Bernander, [2], [4], [10], van Breugel, [7], Emborg, [9]). Further discussion about the practical application of strain and stress - related approaches to early age cracking is dealt with in sections 9.2 and 9.8 below. For more detailed information of analytic modelling and computation of early age stresses and risk of cracking see Emborg, Chapter 7.

Other important developments are the so called *cracking frames* and the *Temperature -Stress testing Machines* (TSM - devices). By means of cracking frames different concrete compositions can be qualitatively, but reliably, tested for their relative aptness to contribute to forestalling early age cracking. (Springenschmid, [12], [13]). See also fig. 9.1 and Springenschmid & Breitenbücher, Chapter 3.

The TSM-devices, which presently exist in many concrete laboratories, not only provide information about the relative ductility of concretes, but also furnish experimental data for calibration by back-analysis of basic mechanical properties and parameters of young concrete. (Emborg, [8], Breitenbücher, [14], Mangold, [15], Westman, [16]). Further progress can be expected from comparison between calculated stresses and in situ measurements using stress meters. See Tanabe, Chapter 8.

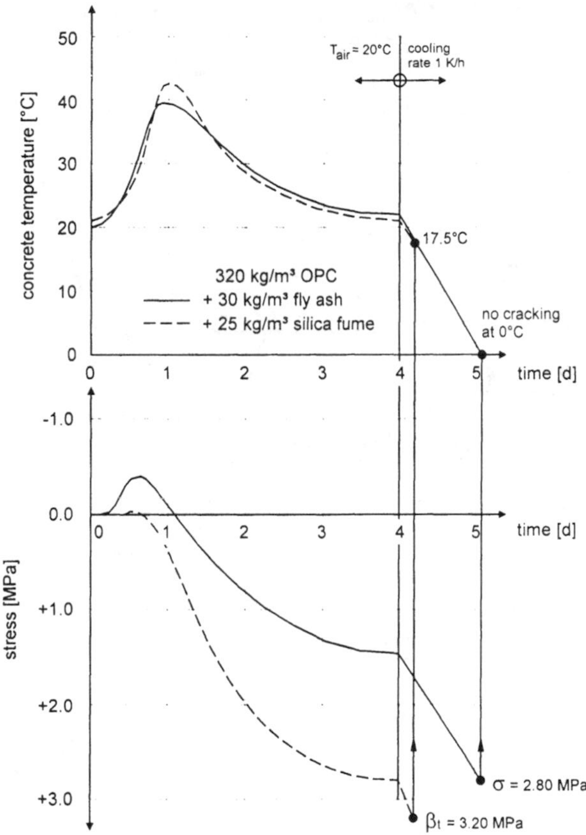

Fig. 9.1 Cracking frame tests on two different concretes - one containing silica fume and the other fly-ash. Limestone aggregate. Springenschmid *et al.* [11].

9.1.3 Implications of new development for construction engineering.

Temperature, restraint and properties of young concrete are, however greatly affected by preconditions, basic concrete data, curing conditions, environment, design and construction procedures.The widened perspective on early age cracking in concrete thus has numerous practical implications - not only for design - but also for all sorts of measures to be considered in connection with construction. The block diagram shown in figure 9.2 illustrates the complex interaction of important factors to be observed when assessing the risk of early age temperature cracking.

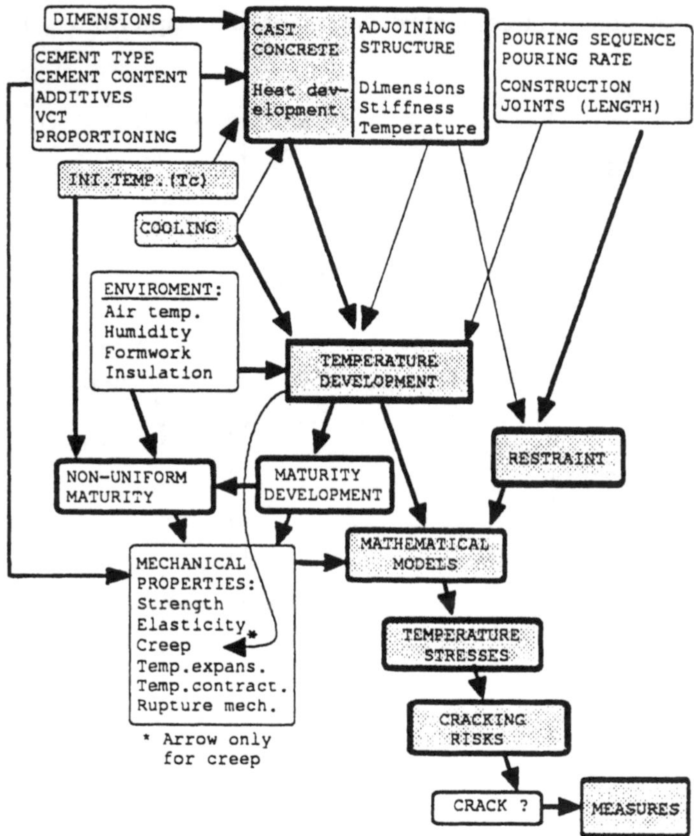

Fig. 9.2 Diagram showing interacting factors governing temperature-induced stresses and cracking in early age concrete, (Bernander [17]). See also Chapter 7.

9.1.3.1 Applicability of conventional temperature based crack risk criteria.

It would be an understatement to say that the recent laboratory testing and the new stress analysis procedures have shed some doubt as to the adequacy of estimating the risk of early age cracking solely on the basis of temperature differences as is still mostly being done. This applies in particular to cracking in the heating phase, where little or no affinity is found between temperature and stress development. A mere glance at figure 9.2 (or figure 9.5) should suffice to convince any firm believer in the conventional temperature based approach, that there is *no generally valid proportionality between temperature difference and stress*.

Additional information about this problem is given in paragraph 9.2.3 below. Another insufficiency of the purely 'temperature related' crack risk assessment is that it does not include - or readily make possible the incorporation of - the effects of likely simultaneous drying and / or chemical shrinkage. (See also Koenders *et al.* [18], [19]).

9.1.3.2 Non-uniform maturity

A novel feature, made possible by modern type stress calculations, is the ability to properly consider *non-uniform* maturity development due to varying temperature evolution in different parts of a cast section. The resulting non-uniform evolution of mechanical properties within a cast section has proven to have a most devastating effect on the reliability of conventional temperature-related crack criteria, which presuppose a *linear* correlation, or at least some degree of affinity, between the temperature field and the resulting stress field. (See also subdivision 9.2.3.2.)

Nevertheless, even today, the assessment of crack risk in hardening concrete is in practice predominantly based solely on temperature criteria - often even in connection with advanced civil engineering structures. However, it should be observed that this practice is *strictly* justified *only* if all influencing factors, i.e. not only temperature, are inherent in the reference experience on which the temperature-related estimations of crack risk are based.

9.1.4 Extended definition of the concept of 'mass concrete'

The problems in mass concrete consist importantly in the need to avoid cracking due to thermal strain by considerable reduction of cement content and to still satisfy other vital performance criteria. Now, the ongoing development towards higher requirements on strength and durability in concrete structures (e.g. off-shore platforms, nuclear reactor enclosures, tunnels, bridges etc.) has entailed - not only higher cement contents - but also growing concern with regard to early age cracking in a wide range of structures - i.e. not only in massive dams and the like.

However, the performance criteria are usually not identical for very large sized bodies (e.g. dam cores) and medium sized concrete elements such as tunnel walls, bridge girders and the like. For this reason a differentiation is made in this report between ordinary '*mass concrete*' and what is here termed '*medium mass concrete*' as explained in more detail in paragraphs 9.4.1, 9.4.2 and 9.5.1.

9.2 EARLY AGE CRACKING IN CONCRETE DUE TO HYDRATION VOLUME - GENERAL

9.2.1 The issue - definitions and notations

Cracking in plain *unreinforced* concrete is likely to have serious static and functional implications such as e.g. loss of strength and monolithic action. However, as far as *reinforced* concrete is concerned, cracking mainly affects functional and durability requirements (tightness, weathering, chemical resistance, frost resistance, appearance, etc).

In order to promote the understanding of this report, as well as the denotations to which are referred in the following, a brief review of the technical issue is presented. Figure 9.3 represents a condensed illustration of the phenomenon of early age *through* cracking due to hydration temperature rise. In the *heating* phase most of the total

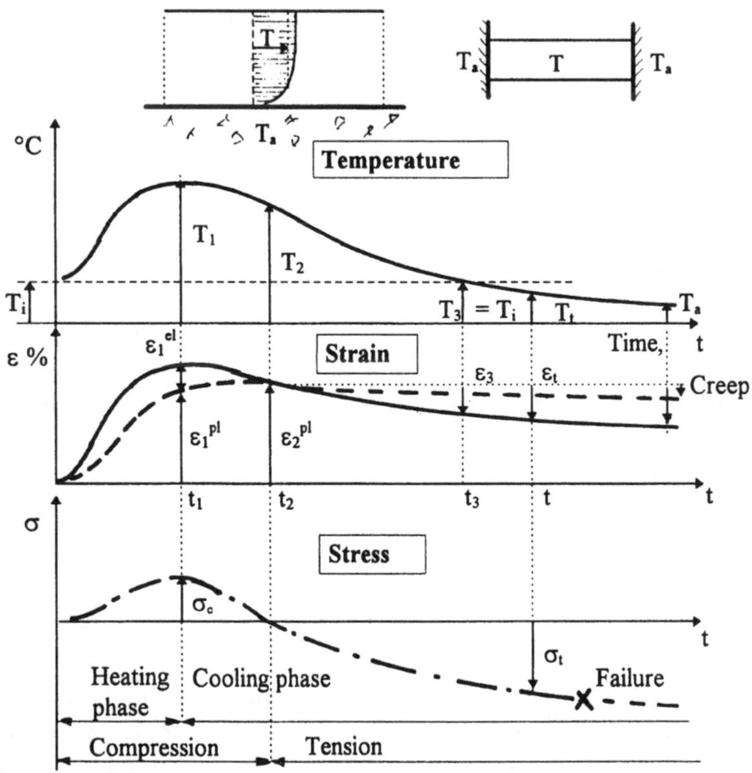

Fig. 9.3 Formation of through cracks due to temperature in early age concrete. Löfquist, [20], (modified by Bernander, [10]).

Notations:

- T_i initial fresh concrete temperature (placing temperature)
- T_1 maximum temperature
- T_2 'zero stress' temperature
- T_s surface temperature of concrete
- T_a mean temperature of *adjoining* restraining structure at time of pouring
- α_e coefficient of thermal *expansion* (heating phase)
- α_c coefficient of thermal *contraction* (cooling phase)
- ε_1 maximum thermal expansion
- ε_1^{el} elastic strain at maximum temperature (time t_1)
- ε_1^{pl} plastic strain at maximum temperature (time t_1)
- ε_2 maximum plastic strain at zero-stress (time t_2)
- $\varepsilon_{t.t}$ $= \varepsilon_t^{el} + \varepsilon_t^{pl}$ total deformation in tension (at time t)
- $\varepsilon_{t.u}$ ultimate tensile strain. ε_3 as defined in figure 9.3

volume expansion in the very young concrete is absorbed as plastic strain (ε_1^{pl}) and only a minor portion constitutes elastic compressive strain (ε_1^{el}) even at maximum temperature rise. In the initial stage of the cooling phase the beginning contraction of the now stiffer concrete soon results in compression stresses being reduced to zero at the time t_2. From then on, continued cooling induces increasing tensile strain and stress in the material. Using the denotations in figure 9.3, the crack criterion may be defined as:

$$\varepsilon_{t,t} < \varepsilon_{t,u} \tag{9.1}$$

$$\alpha_c(T_2 - T_t) < \varepsilon_{t,u} \tag{9.2}$$

If ε_1^{pl} is defined as $k \times \varepsilon_1$ where $\varepsilon_1 = \alpha_e \times [T_1 - T_i]$) it can be shown (Bernander, [10]) that the 'zero-stress' temperature T_2 can be be expressed with reasonable accuracy by the following simple relationship:

$$T_2 = T_i + (T_1 - T_i)[1 - (\alpha_e/\alpha_c)(1-k)] \tag{9.3}$$

or with equ. 9.3 substituted into equ. 9.2 the tensile strain at time (t) can be written as

$$\varepsilon_{t.t} = (T_1 - T_i)\,[\alpha_c - \alpha_e(1-k)] + \alpha_c(T_i - T_t) < \varepsilon_{t.u} \tag{9.4}$$

If $\alpha_e = \alpha_c = \alpha$, $t = \infty$ and $T_t = T_a$, then the eventual strain is:

$$\varepsilon_{t,t=\infty} = \alpha(T_l - T_i)k + \alpha(T_i - T_a) = \mathbf{k} \cdot \varepsilon_l + \alpha\,(\mathbf{T_i} - \mathbf{T_a}) < \varepsilon_{t,u} \tag{9.5}$$

The crack risk may of course readily be expressed in terms of stresses in which case the mobilized stress level ($\sigma_{t,t} / f_{ct}$) is an appropriate criterion.

Equation 9.5 pin-points the **'heart of the matter'** namely, that early age temperature stresses originate in part from the *plastic strain component* ($k.\varepsilon_l$) in the heating phase and partly from the *adaption* of the initial concrete temperature (T_i) to the temperature (T_a) of restraining adjoining structural members. Equation 9.4 can conveniently be used for quick estimations of the risk of through cracking, e.g in the context of actual construction on site, (Bernander, [10]). An approximate expression for 'k' and creep characteristics for some concretes are given in Bernander & Emborg [21] and Emborg & Bernander, [22].

9.2.2 Classification of cracks - cracks in *expansion* and *contraction* phases

In this report cracks are classified as proposed by the author at the RILEM symposium in Munich, 1994, (General Report regarding Practical Measures). This classification is believed to correspond better than before to the *genesis*, the *temporal behaviour* and the *functional implications* of early age cracks in concrete structures. Figure 9.4 illustrates how cracks in a *symmetrical* case may originate from either the heating or the cooling phases of the hydration temperature cycle. However, in this context it should be observed that the following classification embraces also all kinds of flexural cracks arising from un-symmetrical temperature fields and of bending restraint.

a) ***Early age cracks in the heating (or expansion) phase***.
These cracks appear shortly, (one to a few days) after pouring and tend to close at the end of the cooling cycle. (Cf. fig. 9.4 and 9.5). Very early age cracks of this type are usually - but not always - *surface cracks*. Temperature differentials within a foundation slab or over a closed box girder section with markedly different wall and slab thicknesses may e.g. generate *through cracks* also in the heating phase.Through cracks may also develop in the previously cast adjoining concrete, (Cf fig. 9.4 and 9.25).

As the cracks forming in the heating phase normally close in the cooling phase, the effects of these cracks on static capacity, function and durability are subject to some discussion and may therefore have to be deemed from one case to another. However, surface cracks may be expected to have deleterious effects on durability especially if simultaneous shrinkage, ambient temperature variations and aggressive environments are at stake. Surface cracks can also initiate through cracks, that may otherwise not have developed.

b) ***Cracks occurring in the cooling (or contraction) phase*** as described in article 9.2.1. These cracks are related to mean negative volume change and generally form '*through*' cracks as a result axial tension or flexure. Depending on dimensions and other prevailing conditions the cracks appear weeks, months and in extreme cases even years after a section has been poured. Cracks forming in the cooling phase tend to remain open permanently.

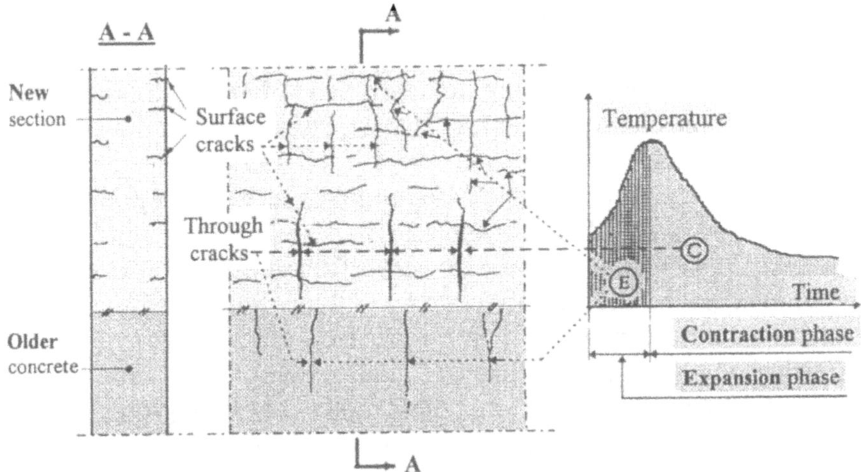

Fig. 9.4 Exemplification of early age expansion cracks and contraction through cracks in a symmetrical wall cast on older concrete. (Bernander, [17]).

Through cracks, therefore, often impair function and appearance but may also affect static capacity, monolithic behaviour and durability.

9.2.2.1 Alternative classification of early age cracks

Traditionally cracks have been classified as either

a) through cracks, arising from externally restrained mean volume change or

b) surface cracks originating from internally restrained non-uniform volume change.

In both cases the temperature field causing the volume changes as well as the restraint may be un-symmetrical thus giving rise to bending stresses and flexural cracks.

This classification has been adopted by Springenschmid, Breitenbücher and Mangold in various papers, where sharp distinction is made between 'eigenstresses' (due to what is defined as internal restraint) and longitudinal stresses. This approach has also been applied by Rostásy Tanabe and Laube, Chapter 6.

Differentiation between so called eigenstresses and other normal stresses are reasonable from the point of view that the eigenstresses are always 2- or 3- three dimensional and normally generating surface cracks whereas the longitudinal stresses, in at least elongated concrete pours, are often 1-dimensional causing through cracks. The differentiation between internal and external restraints may be justified in simple practical cases.

For a number of reasons the author of this report has not adhered to the above classification. Some of these reasons are:

The definition of 'eigenstresses' is confusing from a structural mechanics point of view, because *all* early age normal stresses due to restrained volume change are in fact eigenstresses i.e. even those generating through cracks and flexural cracks. Moreover, in general the 'eigenstresses' interact with the 'longitudinal' stresses and cannot - if determined separately - be superimposed on account of the non-linearities of the early age mechanical properties. Surface stresses/cracks often *initiate* through cracks and flexural cracks and thus often constitute a crucial feature of the failure mechanism in the process of through cracking.

Further, the concepts of 'internal' and 'external' restraints are arbitrarily linked with a simplistic analytical model addressing only the early age stresses in *one* section, i.e. the youngest pour in a construction sequence. However, in actual construction there is often a need to study concurrent stress developments in two or more adjacent sections that have been poured consecutively at short intervals. In such cases more general computational models are required, incorporating all the newly cast sections including the abutting the older concrete, into the analysis,. (See e.g. the example shown in fig. 9.25 and 9.24). Clearly in such generalized analyses all early age stresses are *internal stresses* and the term 'external restraint' does not in the authors opinion appear as a useful basic condition for classifying cracks.

9.2.3 Recent findings - '*beneficial*' side effects of hydration temperature development

9.2.3.1 Compression of concrete surface layers

Modern type analyses of early age temperature stresses have shown that the effects of hydration are not *always* deleterious.

Natural cooling of a massive concrete pour involves transmission of heat of hydration generating temperature differences between the interior and the outside surfaces. As regards the stresses induced by these temperature differences it is important to observe that - *provided cracking does not take place or is provided against* - tensile stresses occurring at the concrete surfaces in the heating phase generally evolve into compressive stresses or contribute to reduced tension at the same surfaces in the cooling stage. (fig. 9.5).

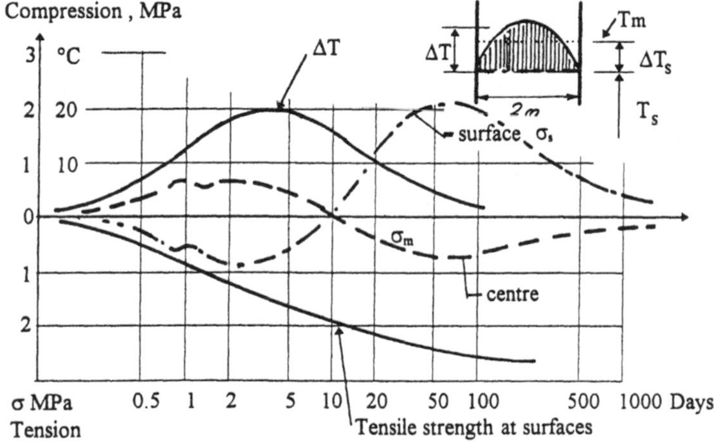

Fig. 9.5 Stresses due to temperature difference over a 2 m thick concrete wall section. Mean temperature rise ΔT_{ave}= 25°C. Maximum temperature difference ΔT= 20°C. Concrete: Std Portland cement 315 kg/m^3, w/c-ratio = 0.5. (Bernander, [4])

At locations, where there is a simultaneous tendency to through cracking, the tensions in the surface layers may be duly reduced, (Figure 9.6).

Practical implications:

The phenomenon illustrated in figures 9.5 and 9.6 has some important practical implications:

a) there is, as shown, no affinity whatsoever between the temperature differential $\Delta T = (T_c - T_s)$ and the corresponding development of stresses.

When the peak value of ΔT occurs, the risk of cracking in terms of $(\varepsilon_{tt}/\varepsilon_{tu})$ has already well passed the critical stage. Then for instance, when ΔT on cooling is still about 50 % of the maximum value, considerable compression (instead of tension) has developed at the surfaces.

b) the critical stage with regard to surface cracking normally takes place already 1 to 2 days after pouring, implying that form stripping at this time is definitely not recommendable.

c) another practical aspect of the said beneficial surface compression is that it sheds some doubt as to the usefulness of current practices of reducing temperature differentials by means of external insulation.

This is because redundant insulation reduces *unnecessarily* the eventual surface compression. More importantly still, too efficient insulation also prevents or lessens the vital *early age adaptation of concrete surface temperatures to that of the environment,* thus dramatically increasing tensile stresses (impact of temperature shock) at the inevitable final exposure to ambient conditions. The use of extra insulation i.e. other than normal formwork and protective canvases or mats should therefore always be preceded by qualified analysis. Both theory and practice indicate that haphazard application of high performance insulation usually does more harm than good.

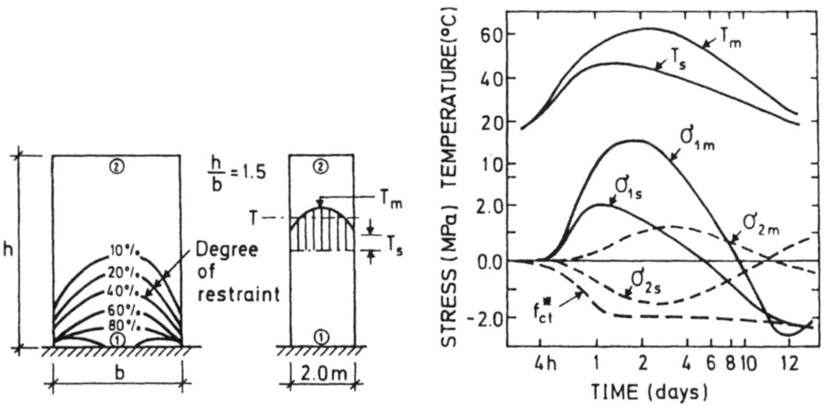

Fig. 9.6 Wall cast on a rigid foundation. Typical stress developments close to a construction joint and at some distance from the joint. Full adhesion at the joint interface is presumed. T_s and T_m denote temperatures at surfaces and mid-section respectively. σ_{1s},σ_{1m},σ_{2s} and σ_{2m} are corresponding stresses at locations (1) and (2). $T_i = T_{air} = 15°C$. Cement std Portland 350 kg/m^3. Form 12 mm plywood, (Emborg & Bernander, [23]). ***Note***: Full adhesion in construction joints cannot always be presumed. (Cf section 9.7.)

The universal ***proneness*** of concrete bodies - subjected to early age volume change due to heat of hydration and subsequent natural cooling - of developing a tendency to being compressed at the surfaces surely constitutes a factor of great - but, until recently in literature not very well recognized - importance to the statistical performance of concrete as a construction material. It is probably *difficult to overestimate the significance of this phenomenon with regard to the durability of concrete surfaces and to the corrosion resistance of superficially embedded reinforcement.*

Compression in the surface layers at the end of the hydration temperature cycle has also been reported by Mangold, [15], Springenschmid *et al.* [11], Mangold & Springenschmid [24], Eberhardt *et al.* [9], Solovyanchik *et al.* [25], and others.

9.2.3.2 Importance of considering non-uniform maturity development

Another important outcome of modern stress analysis entailing crucial practical implications is the - hitherto not much studied - effect of 'non-uniform maturity' development in a hardening concrete cast. The temperature differences due to heat transmission inexorably result in non-uniform development of the mechanical properties in different parts of a hardening concrete section. In addition, notable phase differences in the time domain arise in the evolution of temperatures in the different parts - often causing untimely, restrained relative deformations between these parts.

Since the *placing temperature* (T_i) of the concrete and *the ambient air temperature* (T_{air}) *strongly* influence the temperature distribution in the concrete, the relationship between these two parameters may be extremely decisive for - in particular - very early age cracking in the expansion phase. (Bernander, [26], Bernander & Emborg, [21], [17], Mangold & Springenschmid, [24]).

Consideration of non-uniform maturity has *important bearing* on practical measures in connection with decisions on placing temperature when concreting. Cf. art. 9.5.3.

Thus, suitable choice of placing temperature in relation to ambient temperature presents possibilities of optimizing possible prestress in the surface layers (dealt with in the previous article) as well as of avoiding adverse conditions in respect of surface cracking.

9.3 GENERAL PRINCIPLES AND PRACTICAL MEASURES OF CONTROLLING EARLY AGE CRACKING

9.3.1 General aspects

Although temperature, as has been stressed earlier in this report, is not the only important parameter in the phenomenon of early age cracking in concrete, it still remains the primary cause of the problems at stake. Experience from research and practical engineering indicates that temperature predictions in hardening concrete can be performed with relatively high accuracy, provided they are carried out with adequate knowledge and insight. It must, however, be emphasized in this context that prediction of hydration temperature rise is subject to a complex of interrelated and interacting parameters, that may vary widely for different concrete compositions.

It is therefore of vital importance to ascertain, that the chosen basic parameter values really apply to the specific concrete used in an analysed structural situation. For information about heat generation and prediction of temperature development in hardening concrete see the report by van Breugel, Chapter 4.

Methods to determine the heat of hydration are found in the report by Morabito, Chapter 1.

The problems inherent in mass concrete consist, as mentioned previously, importantly in the need to eliminate cracking due to thermal strain in the early ages while still satisfying the performance criteria with regard to compressive (tensile) strength, impermeability to fluids, chemical resistance, frost resistance and other aspects of durability.The difficulties often rest therein, that measures designed to avoid early age cracking to a major extent are directed towards reducing the cement content, i.e. measures, which may jeopardize the mentioned material requirements with regard to strength and durability. As the demands on strength and durability may vary with type of structure and with environment, a wide and diverse set of measures to controlling cracking is called for. For instance, the strength requirements for dams are normally quite different from those applying to bridge girders.

Careful constructional planning involving optimization of - sometimes even contradictory or incompatible - relevant measures has to be made. The outcome of this optimization may for large projects be of great economic consequence.

9.3.2 Crack criteria and crack control on working site

This topic is specifically and extensively dealt with in section 9.8.

9.3.3 Practical measures in construction related to the design phase

There are a number of courses of action related to the activities on the construction site, which are decided upon already in the design stage of a project such as dimensions, specification of the concrete to be used, location of construction and expansion joints, casting sequences, limitations regarding maximum allowable temperature rise and temperature differentials etc. However, as all basic data and site conditions are not known or not yet decided upon in the design and bidding stages of a project many of the measures dealt with in here also apply to paragraph 9.3.4 below.

a) *Structural dimensions*
Unnecessarily large thicknesses and volumes should be avoided. The option of replacing massive structures by hollow constructions is to be considered. See e.g. figure 9.7.

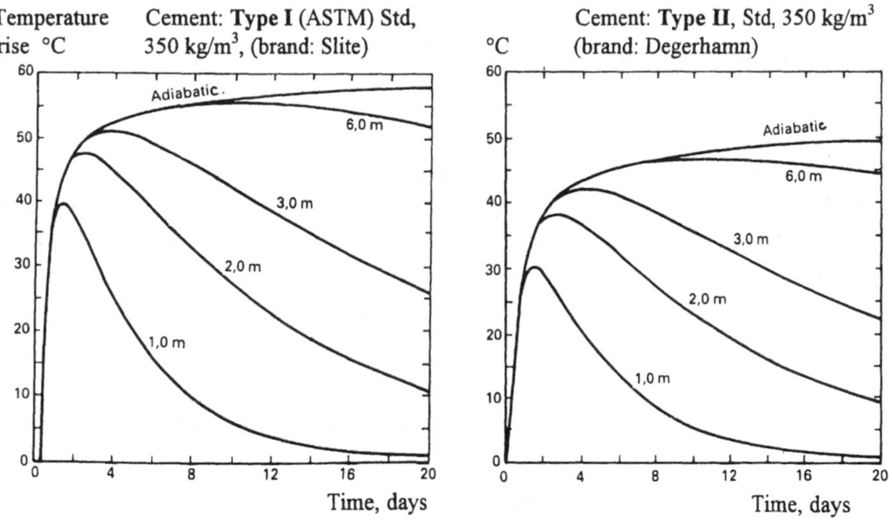

Fig. 9.7 Computed maximum temperature rise in concrete walls of different thickness. Cement: types I and II, ASTM, 350 kg/m^3, form: un-insulated plywood, $T_i = T_{air}$.

b) *Design of concrete mix*
As most requirements on the concrete are usually specified at an early stage of a project many of the practical measures related to the design of the concrete mix are often conceived well before construction is begun.

An effective measure to control cracking is the use of cement with a low cracking sensitivity, or if this property cannot readily be defined, the use of *low heat* type cements, usually type II or IV according to ASTM. For further information see item d) below and Springenschmid & Breitenbücher, Chapter 3.

However, since maximum temperature rise is roughly proportional to the amount of cement, some of the most effective measures to control cracking are associated with reduction of the cement by utilizing measures as listed below:

The *cement content* (the cement factor) can be reduced by:

- using *chemical admixtures* (water reducing agents, superplastizers, air entraining agents, etc) .The over all purpose of using these admixtures is to minimize the required amount of cement paste while still retaining desired workability and specified target water/binder-ratios.
- the use of *cement replacement* materials such as fly-ash, pozzolana, ground furnaceslag, silica fume, calcareous filler, pulverized rock etc. In some projects it may be favourable to use different concretes for construction in summer and winter. In 'summer concrete' up to 25 % of the cement may be replaced by fly-ash.
- the use of *large* maximum aggregate size
- using as high *water/binder*- ratios as possible while observing specified performance criteria. This applies in particular to mass concretes in dam cores and the like where w/b-ratios often range from 0.55 to 0.70, (Cf table 9.2).
- *roller compaction*, i.e. applying efficient compacting machinery enabling placing of 'dry' concretes with minimum cement paste content.

c) Measures directed towards reduction of restraint from neighbouring structures:

Restraint can be optimized by:

- Suitable choice in design and location of *construction joints*.
- Careful planning of construction and *pouring sequences*.

d) Choice of cement type and concrete composition with regard to its cracking sensitivity. Concrete compositions can be designed to produce concretes which are less sensitive to early age cracking. This is done by investigating applicable concrete compositions with regard to their intrinsic relative lack of ductility or so called 'cracking sensitivity' (Springenschmid *et al.*) using cracking frame and/or TSM tests. Depending on a variety of parameters such as cement type, aggregate mineral composition, water/binder- ratios, admixtures and cement replacements (by fly-ash, pozzolana and/or silica) different concrete compositions exhibit varying degrees of proneness to cracking. By means of the mentioned tests, available and/or economically suitable concrete mixes can be *reliably* tested with regard to their *cracking sensitivity* i.e. their innate potential of promoting a concrete structure free of cracks. For more information, see Chapter 3.

The results of cracking frame tests and TSM-tests can also be used for calibrating basic mechanical properties of the young concrete (tensional strength, visco-elastic behaviour, thermal coefficients for volume change etc) by back-analysis using simulation procedures. Often, direct tests of specific mechanical properties are called for. The object of securing these basic parameters is of course to provide necessary input data for stress simulations. (See article 3.4 g).

9.3.4 Measures related to the construction stage

This section will deal with steps taken in the construction stage in order to comply with specifications or requirements laid down in the design or to accommodate necessary modifications in relation to the design.

a) Control of the placing (initial) temperature (T_i) of the fresh concrete. See fig. 9.8.

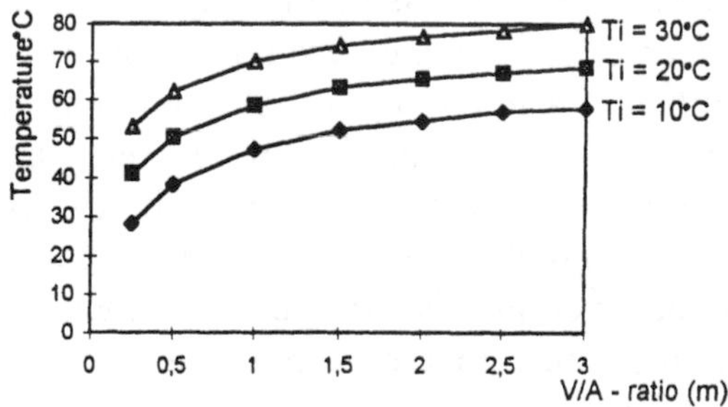

Fig. 9.8 Maximum temperature in concrete as a function of placing temperature (T_i) and volume/surface ratio, V/A (m). Cement type I (ASTM), Slite std,350 kg/m^3 , T_{air} is assumed to be = T_i, (Computed).

Lowering the placing temperature reduces the maximum temperature rise (T_1-T_i) as well as having a direct effect on the temperature difference between new and older concrete (T_i-T_a). T_i can be reduced by precooling of *cement*, *aggregates,* cooling of *mixing water* and replacement of water by *crushed ice*. Another current practice is using *liquid nitrogen,* which is either blown into the concrete at the batching plant through holes in the mixing shovels or at the site via lances directly into the concrete in a rotating transport container.

b) Artificial control of temperature in *hardening* concrete.
The overall objectives of the following methods of artificial temperature control are to reduce maximum temperature (T_1) and temperature differentials (T_i-T_a).

Temperatures in the *new* hardening concrete can be artificially regulated by means of:

- internal cooling using water circulating through embedded pipes,
- internal air cooling by flushing air through ducts of embedded steel sheeting,
- external cooling by sprinkling of water,
- insulation. (Note comment under '*attention*' below).

Special considerations - ***Attention!***
It is important to realize that some of these measures to controlling temperature are not always compatible with regard to the objective of forestalling cracking. For instance, insulation of free surfaces of hydrating concrete will on one hand effectively reduce temperature differentials, but will also inevitably bring about higher maximum temperatures (T_1) within a cast section - a condition, which is potentially deleterious from the point of view of *through* cracking due to external restraint.

Another important consideration is that insulation of the very young concrete counteracts early age adaptation of concrete surface temperatures to those of the environment. Excessive or superfluous insulation may well reduce temperature differentials within a cast section but will in most cases *aggravate* the over all risk of surface cracking (see comments regarding insulation of 'new concrete' in article 9.2.3 and about external cooling in article 9.5.4.1c).

A further discrepancy of this kind is that a low placing temperature (T_i), which is often effectively beneficial in respect of avoidance of through cracks, may under circumstances - usually hot summer conditions

- *dangerously* increase the tendency to surface cracks being generated as previously explained in the comments on non-uniform maturity development in art. 9.2.3 above.

c) Artificial control of temperatures in 'old concrete'.
Temperatures in older adjoining concrete can be regulated or modified by

- *internal heating* using e.g. pipes or ducts, previously used for cooling - or electric heating cables.
- *external heating*
- *insulation*

Insulation of exposed surfaces of the previously poured concrete is motivated by the benefit of retaining the heat of hydration in neighbouring casts until pouring of the section of interest is made. As soon as maximum temperature in the newly poured concrete is reached the insulation of the older concrete should be removed, while observing the effect of possible temperature shock. (Fig. 9.14 and 9.15). It should be noted that heat exchange between new and older concrete may have considerable impact on the temperature field, a condition that may readily be taken into consideration in the stress simulation procedures dealt with in section 9.8.

d) Curing measures
Cracking in early age concrete may be seriously enhanced or aggravated by simultaneous exposure to low ambient temperatures, high surface temperatures (solar radiation) and/or drying shrinkage. As shown in fig. 9.5 the risk of cracking due to interior expansion is very acute in the heating phase. Avoiding exposure to lower temperatures and drying out *at this stage* is a necessary requirement if surface cracks are to be warded off. It is therefore of utmost importance always to observe current practices of curing hardening concrete such as:

- time to form stripping
- protective insulation of stripped surfaces when exposed to low ambient temperatures
- water sprinkling or application of sealing agents to mitigate shrinkage effects
- protection from direct sunshine (summer time).

e) Optimization of restraint.
Considerations with regard to location of construction joints, casting sequences, detailing of reinforcement etc are as mentioned normally determined in the design stage. However, in practice, for a wide range of reasons, design considerations and intentions often have to be revised or modified in the course of construction. Thus, measures pertaining to restraint as mentioned above may have to be reconsidered during construction. For further information on this topic see chapter 9.7.

f) Trial castings - or 'Mock-up:s'.
Experiments on the working site involving trial castings or so called '*mock-up:s*' are often implemented in order to verify workability and other aspects of concrete construction. Such 'mock-up:s' are very useful, and sometimes indispensable, also for the determination or verification of the hydration and thermal properties of the concretes intended for use as well as for checking the efficiency of planned temperature control arrangements (e.g. with regard to cooling systems).

g) Numerical stress analysis.
Modern types of stress analysis or stress risk simulations - referred to in paragraph 9.1.2 and section 9.8 - have initiated an evolution towards a better understanding of early age cracking phenomena. Such analysis may conveniently be used for investigating the relative efficiency of different measures taken on site with regard to crack control or when design provisions in this respect have been changed or modified, (See section 9.8). In connection with larger projects the concretes intended for use should therefore be tested for their basic early age mechanical properties. See Chapter 7.
Crack risk assessments can also be made on the basis of data derived from cracking frames and/or TSM- devices. Cf. Springenschmid *et al.* [11], Fleischer & Springenschmid, [27]. See Chapter 3.

9.4 SPECIFIC MEASURES RELEVANT TO MASS CONCRETE

9.4.1 Introduction

The tendency in recent years to setting increasingly higher performance requirements on concrete has markedly aggravated the early age temperature cracking problem in medium sized concrete elements (0.5- 3 m thick) such as are common in e.g. infra-structural building. This fact has brought about a development towards more interest in early age crack phenomena in these types of structures and, gradually, a better understanding of this topic on the part of construction engineers.

In the chapter on 'Mass concrete' of the Manual of Concrete Practice, ACI Committee 207 report, [28], concrete in a structure or a portion thereof is considered as "*mass concrete, if measures are necessary to control thermal behaviour to alleviate cracking*".

By this standard, according to the author's practical experience, it could be argued that most 'high quality concrete' for advanced structures should be denoted as *mass concrete.*

However, the objectives and the requirements are not identical for, on one hand,

- large size - *low stress* - voluminous structures without or with little reinforcement (such as dam cores and the like) and on the other
- medium sized *highly stressed*, highly performing concrete elements such as e.g. heavy bridge girders.

It has therefore been found appropriate in this report to differentiate between three categories of concrete applications as follows:

a) Massive concrete structures (Section 9.4)
b) Medium sized massive concrete structures or structural components (Section 9.5)
c) Slender concrete applications (Section 9.6.)

9.4.2 Massive concrete structures - *definition of mass concrete*

Mass concrete is thus in this report defined as concrete for very large sized portions of a structure, where the level of *concrete stress* and/or *environmental exposure* - on account of the very ample presence of concrete - is not a *primary issue.* Such constructions are for instance dams, massive bridge piers and associated foundation slabs, foundation blocks for pylons and anchors of suspension or cable stay bridges etc. The problems related to mass concrete for these types of structures are high maximum temperatures (T_1) and large differences in temperature ($\Delta T_s = T_1 - T_s$) between the interior - where almost adiabatic conditions prevail - and the surfaces more or less exposed to the environment. The cooling phase of the hydration cycle is by nature extended in time and may, if unattended, range from months, years to decades and even centuries (massive straight gravity dams). Specific measures for mass concrete are thus directed towards *reducing maximum temperatures* and *temperature differences* and to *speeding up* the cooling process.

However, a typical feature in mass concrete production is that many of the measures taken to accomplish these objectives require that the projects are large enough for them to be economically viable. Hence this section (9.4) is essentially focused only on procedures, which are appropriate for large size projects such as dams and the like.

9.4.3 Reduction of temperature rise by appropriate concrete mix design

As strength is not a first priority, there is in mass concrete a marked potential for minimizing the temperature rise by reducing the cement content merely by raising the water/binder ratio up to acceptable limits, while still observing all specified properties in the concrete. Practically all of the measures mentioned under article 9.3.1 are being used separately or in combination in order to lower the maximum temperature rise (T_l-T_i) while maintaining appropriate workability. The current application of superplastifiers and/or air entraining agents in conjunction with cement replacement materials, large maximum aggregate stone size and new placing techniques have virtually revolutionized the cement contents required to meet the functional requirements for dams and similar structures. Table 9.1 illustrates the effect of maximum aggregate size on cement content. Table 9.2 reflects the overall progress made in this respect in some spectacular dam projects.

Table 9.1 Effect of maximum aggregate size. (Example, Wischers, [29])

Max. aggregate size [mm]	*Required cement content [kg /m³]*
30	270
50	240
70	220
150	175

In a paper to the RILEM symposium 1994, Iwata *et al.* [30] report having used a superplasticizer admixture (a poly-hydroxy carbolic acid ester) in combination with selected anion type surfactants for a sub-ground railway tunnel project. This so called 'high range water reducing type heat control agent' engendered - apart from its water reducing effect - not only an adiabatic temperature reduction but also a markedly lower rate of adiabatic temperature rise. This lead to a significant reduction of maximum temperature in the actual structure.

9.4.3.1 Roller compacted concrete

A specific method of reducing the cement content is the - in the U.S. initiated - and later in Japan and other countries adopted placing procedure called 'roller compacted concrete'. By this method the cement content is effectively cut down by minimizing the amount of cement paste in making virtually a 'no slump concrete' from carefully graded aggregates. The 'dryish' concrete is compacted by means of efficient vibratory equipment. The working surface being horizontal, the concrete is placed by dumpers and levelled using bulldozers. The lift is then compacted by means of vibratory rollers. The ACI Committee 207, [28] recommend placing of the concrete to be compacted in layers of 250 - 300 mm. However, when constructing the Shimajigawa dam in Japan Suzuki, Iisaka *et al.* [31] report having placed roller compacted concrete in 700 mm lifts in cycle intervals of 4 days (summer) and 3 days (winter). Concrete composition was chosen as per table 9.2.

Table 9.2 Mass concrete data for some major dam projects, (ACI Committee 207, [28] and other sources)

Name of project, country	*Year*	*Dam type*	*Cement factor [kg/m^3]*	*Cement replac. (p) [kg/m^3]*	*Max. aggr. size [mm]*	*w/c+ p = wbr*
Norris dam, USA	1936	Straight	200 (II)*	-	152	0.67
Grand Coulee,USA	1942	gravity	224 (II+IV)	-	152	0.60
Warragamba,Australia	1960	"	196 (II)	-	152	0.53
Krasnoiarsk,Russia	1970	"	230 (II)	-	100	0.55
Dworshak, USA	1972	"	125 (II)	fly-ash, 42	152	0.59
Lower Granite, USA	1973	"	86 (IV)	-	152	0.71
Rich.B Russel, USA	1982	"	134 (II)	fly-ash, 35	152	0.57
		"	103 (II)	", 43	152	0.67
Hoover dam, USA	1936	arch	225 (IV)	-	130	0.58
Pieve di Cadori, Italy	1949	gravity	150 (IV)	nat.poz. 50	120	0.63
Glen Canyon, USA	1963	"	111 (II)	pumicite 53	152	0.54
Itaipu,vBrazil-Paragu.	1982	grav butress	108 (II)	fly-ash, 13	152	0.70
Rossens, Switz.land	1948	arch	250 (I)	-	79	0.53
Salamonde, Portugal	1949	thin ar.	250 (II)	-	200	0.54
Shimajigawa, Japan	1978		90	fly-ash,	(40)	
Xiaolangdi dam project	1995	concr. D5	170	fly-ash, 70	150	0.58
People Rep.of China		concr. A2	465	micro silica, 30	80	0.28

* Cement types according to ASTM classification.

9.4.4 Lowering of fresh concrete temperature as placed

The initial fresh concrete temperature (T_i) has a direct impact on the maximum temperature (T_l) as well as on the temperature differences (T_l-T_a) relative to the environment. Lowering of the placing temperature is therefore a sought objective in mass concrete construction. (cf fig. 9.8). Current measures to that end are precooling of cement and/or aggregates, cooling of mixing water or replacement of the water by crushed or flaky ice. Table 9.3 shows the effect on the temperature of fresh concrete by precooling of each - respectively all - of the components by a certain measure, (10°C).

The use of liquid nitrogen as a means of lowering placing temperature is a rather costly method and is therefore normally restricted to treating smaller volumes of mass concrete.

Table 9.3 Temperature reduction in fresh concrete when decreasing the temperature of the constituents by 10 °C for two concretes. *Concrete A*: w/c = 0.55, cement: 320 kg/m^3. *Concrete B*: w/c = 0.6, cement: 160 kg/m^3.

Temperature reduction by 10 °C of:	*Specific heat [kJ/kg. °C]*	*Effect on fresh concrete [°C]*	
		Concrete A	*Concrete B*
Cement	0.88	1 °C	0.6 °C
Water	4.18	3 °C	1.7 °C
Aggregate	0.85	6 °C	7.7 °C
Total	1.00	10 °C	10.0 °C

If, in this example, all added water in concrete A were to be replaced by crushed ice the resulting additional reduction of the placing temperature would be 25 °C. However, normally it is not practical to administer all water in the form of crushed or flaky ice.

9.4.5 Artificial temperature control

Modern methods of design and construction procedures for mass concrete result today, as indicated in table 9.2, in cement factors that are sufficiently low to eventuate in 'interior dam concretes' being free of cracks. For structures, or parts of structures, where functional requirements call for higher cement contents (e.g. exterior dam concretes) internal cooling with embedded pipes presents a technically very appropriate expedient. Internal cooling affects maximum and mean temperature rise as well as the temperature differentials, the latter effect directly reducing the expansion of the concrete in the interior of the cast section and its due implications on surface cracking. In addition, pipe cooling provides a means to speed up the cooling of the integral concrete mass, thus shortening the time required to reach joint closure temperature for block joint grouting. This is often a necessary consideration in mass concrete construction, e.g. in dams and massive barriers in under-ground storages for hydro-carbons. For more particulars about *pipe cooling*, see corresponding item in the following section 9.5.

9.5 SPECIFIC MEASURES RELEVANT TO 'MEDIUM MASS CONCRETE' - APPLICABLE TO MEDIUM MASSIVE STRUCTURES

9.5.1 General

Many medium sized structures or structural components are also prone to early age cracking on account of the considerable cement/binder contents associated with high demands on strength, tightness and resistance to deleterious environmental exposure. This category of '*medium massive structures*' represents a wide range of applications of concrete construction such as bridge girders (frames), pylons and piers, tunnels, quays, jetties, off-shore concrete platforms, water storage tanks, purification plants, foundation slabs etc. In many of these applications normal performance concretes (NPC s) up to C60 are currently used. Further, there is at present a marked tendency to increased utilization of high performance concretes (HPC s).

The use, in medium massive structures, of water/cement (binder) - ratios even less than 0.45 and the simultaneous requirement on good workability (high slump or flowing concrete) normally entails high cement contents, which in turn tend to generate problems with elevated hydration temperatures and early age cracking. In addition, low water /binder-ratio concretes tend to be less compliant (stiffer) in the contraction phase and hence, also on this account, more prone to cracking.

These circumstances justify, in the author's opinion, classification of '*medium mass concrete*' as a special, *separate category* of mass concrete The high demands on medium mass concrete thus *preclude* many of the current practices related to 'conventional' mass concrete , described in section 9.4. E.g. methods of cement reduction such as the use of high water/binder-ratios, utilization of maximum aggregate sizes of 80 to 200 mm or roller compaction are no longer practicable. Further, as medium mass concrete applications may range in size from minor projects (even parts of a structure) to very large projects (such as km-long sub-ground or trench tunnels), many other measures applying to mass concrete construction may or may not be economically feasible in specific projects. Thus, although most of the measures listed in sections 9.3 and 9.4 are applicable in principle only such measures, that are particularly appropriate for 'medium mass concrete' will be dealt with in this section. For *crack criteria* see section 9.8.

9.5.2 Reduction of temperature rise by concrete mix design

The high cement content, normally required in medium mass concrete, presents a potential for cement reduction using chemically reactive fines and/or chemical admixtures (water reducing and air entraining agents, superplasticizers etc). Cement replacement by silica fume, fly-ash, blast-furnace slag, separately or in combination, is today normal practice in construction of off-shore platforms, large bridges and tunnels. (See table 9.4). In fact the very circumstance, that HPC strengths of C105 are these days specified in several national Codes of Practice and that strengths of C130 have been utilized in practical applications, point at the potential of making medium mass concretes (NPC s) in the range of C35 to C60 using *moderate cement contents* while still retaining excellent *workability*. (Cf. table 9.4)

Table 9.4 Examples of binder compositions in some major projects

Project	*Strength*	*Cement [kg/m³]*	*Fly-ash [kg/m³]*	*Silica fume [kg/m³]*	*Blast furnace slag [kg/m³]*	*wcr wbr*
Hibernia, (C-75),Canada	C75	450	-	38	-	0.34
Troll Floater,(C-75), Norway	C75	405	-	20	-	0.37
Heidrun (LC-60, γ=1.95), Nor.	LC60	420	-	20	-	0.37
Store Baelt, West bridge, DK	C65	335	40	20	-	0.35
Veda br., main span: 1210 m S	K40	360	-	-	-	0.45
Öresund Bridge, DK-S	C45	360	-	13	-	0.37
Arche de la Défense, France	K60	425	-	30	-	0.40
Storm Surge Barrier, Holland			-	-	350	0.45
Rudolphstein highw. bridge, D	C45	280	80	-	-	0.51
Watertight foundation, D	C35	280	60	-	-	0.49

9.5.2.1 Concrete mix design with regard to reduction of cracking sensitivity

As discussed in article 9.3.3 cement type and brand may be selected and concrete composition can be designed with regard to innate ductility and propensity to mitigating early age cracking problems. By means of cracking frame and/or TSM tests different concretes can be reliably tested for their intrinsic behaviour under defined conditions. This approach is particularly prolific for recurring projects and structures involving a high degree of repetition, where conditions are not too variable and known in advance. In a paper to the RILEM Symposium, 1994, Springenschmid, Breitenbücher and Mangold [11] have reported how cracking frame tests were used to optimize concrete compositions suitable for construction of recurring railway tunnel projects. The following measures were found to promote the requirement of *low cracking tendency* of the concrete:

- Low fresh concrete temperature (8 to 12°C)
- Low early strength development
- Low cracking temperature ($T_c < 10$ °C) of cement (std test, Chapter 3)
- Partial cement replacement by fly-ash
- Use of aggregates with low coefficient of thermal expansion (contraction)
- Air entrainment, about 4% by volume
- Addition of silica fume should be *avoided,* (See fig. 9.1)

9.5.3 Lowering of fresh concrete placing temperature

Lowering of the placing temperature (T_i) is an efficient and commonly used expedient for the prevention of *through* cracks also in medium mass concrete. A decrease of T_i by e.g. 10 °C (acc. to table 9.3) corresponds to a reduction of the mobilized tensile strain by some 0.07 % which constitutes a considerable proportion of the total failure strain of 0.10 to 0.15 %. In principle all the measures listed in articles 9.3.4 and 9.4.4 are applicable also to medium mass concrete. However, in minor applications many of the steps taken in regular mass concrete are not economically viable. On the other hand, cooling by 'ventilating' the fresh concrete batch with liquid nitrogen is normally only feasible when placing smaller volumes of concrete.

As mentioned in section 9.2, a too strong reduction of the placing temperature is not always recommendable. Consider for instance concrete being cast at a placing temperature of 10°C in a hot environment of say 25°C. The concrete in the surface layers will then harden quickly while the maturity development in the interior concrete will be delayed. When, eventually, the greater part of interior expansion does take place the surface concrete - having already acquired the mechanical properties of a stiffer material - will be extremely prone to cracking.

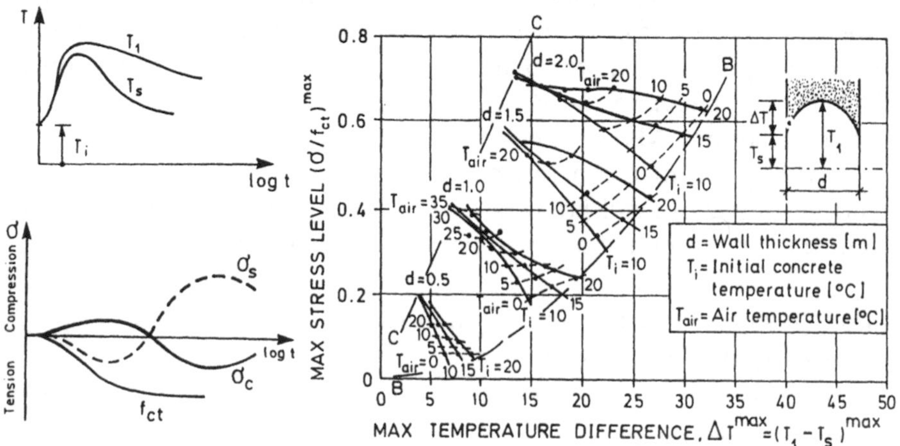

Fig. 9.9 a) Temperature differences over a wall section causing early age tensile stresses at the surfaces. b) Maximum tensile stress levels as a function of maximum temperature difference between core and surface for different placing and air temperatures. Computations by means of TEMPSTRE-N. Concrete: Std Portland cement, type I, 331 kg/m^3, w/c = 0.55, (Bernander & Emborg, [21], and Emborg & Bernander, [32]).

If, in contrast, the placing temperature is say 20°C while the ambient air temperature is 5°C (such as conditions may be in winter in temperate climates), then the situation will be reversed. When the major proportion of the interior expansion takes place the concrete in the surface layers is still in its early age extremely plastic state and will comply readily to forced deformation. Then, as the surface temperatures in due course tend to rise faster than those of the interior, tensions will be liable to decrease already *before* the occurrence of maximum temperature. The advantage of using warm concrete in a cold environment as compared to the opposite scenario with low placing temperature in hot climate is drastic as may be interpreted from figure 9.9.

Hence, with respect to the risk of surface cracking, warm concrete with moderate insulation (i.e. enough to prevent freezing) is recommendable in winter construction while a warning relevant to *surface cracking* should be issued against considerable reduction of placing temperatures (T_i) in hot summertime and hot climates.

9.5.4 Artificial temperature control in the hardening concrete

9.5.4.1 Cooling of concrete in the heating phase

For *medium mass concrete* cement reduction as described above is often not sufficient to comply with specifications, irrespective of whether the crack criteria are purely temperature related or based on stress analysis. In such cases methods involving artificial temperature control may be indispensable.

Of all the measures listed in article 9.3.4 b, internal cooling by pipe-born water is technically the most effective. Internal cooling by air ventilation through ducts of embedded steel sheeting has recently also been applied. Internal cooling alleviates early age cracking in three ways, namely by

- reducing the *mean* maximum temperature (T_l) thus mitigating strength loss due to high curing temperatures,
- reducing temperature differentials in relation to abutting older concrete (T_l-T_a), hence generally alleviating the risk of *through* cracking,
- reducing temperature differences between interior and surface concrete and duly also the tendency to formation of surface cracks as well as through cracks along edges in thick slabs.

Considering the propensity of temperature differences (due to natural cooling) for generating compressive (or markedly reduced tensile) stresses in the surface layers - as explained in article 9.2.3 - forestalling surface cracking by internal cooling thus provides a means to create beneficial states of stress in the concrete cover to reinforcement. *In fact, internal cooling, in conjunction with modern early age stress analysis, presents a systematic method of procedure by which concrete structures with exceptionally high performance characteristics in respect of durability can be produced.*

a) Internal air cooling.

Air cooling using larger ducts (100 - 500 mm) of embedded steel sheeting may be looked upon as an effective way of reducing the volume to surface ratio (V/A - ratio) of a thick concrete section. Other parameters being the same, air cooling is obviously not as effective as water/pipe cooling, but provided there is available space, this can be compensated using wider ducts. Air cooling is *easy* to implement and is especially advantageous in connection with slip-forming or when using climbing shuttering - in the contexts of which pipe cooling using water is impractical.

Figures 9.10 shows a section of the pylons for the Tjörn cable stay bridge, where air cooling was implemented on the author's initiative in order to mitigate the effects of large temperature differentials over the asymmetric section. The mean maximum temperature rise in the thick part of the section was reduced by some 12°C. Fig 9.11 shows recordings from another project, in which air cooling has been successfully applied to forestall, otherwise unavoidable, extensive surface cracking in the 3 m thick bridge piers of the Igelsta bridge in the vicinity of Stockholm.

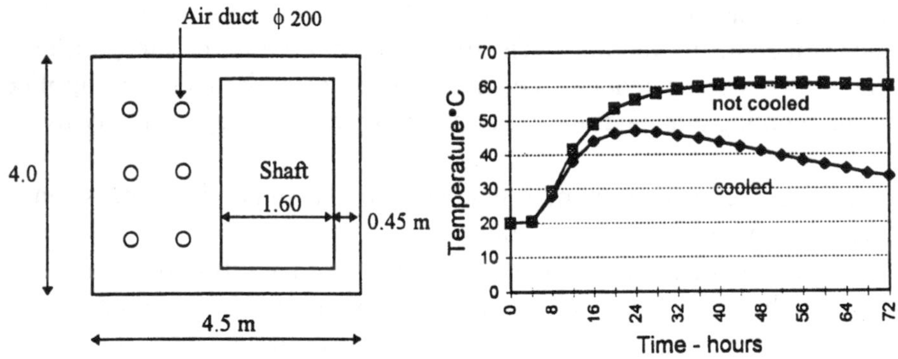

Fig. 9.10 Example of internal air cooling in pylons for the Tjörn cable stay bridge, Sweden, (1980). *Mean maximum* temperature rise is shown. Cement: type II: 400 kg/m^3, micro-silica: 40 kg/m^3, w/b = 0,47. (Bernander, not. publ.).

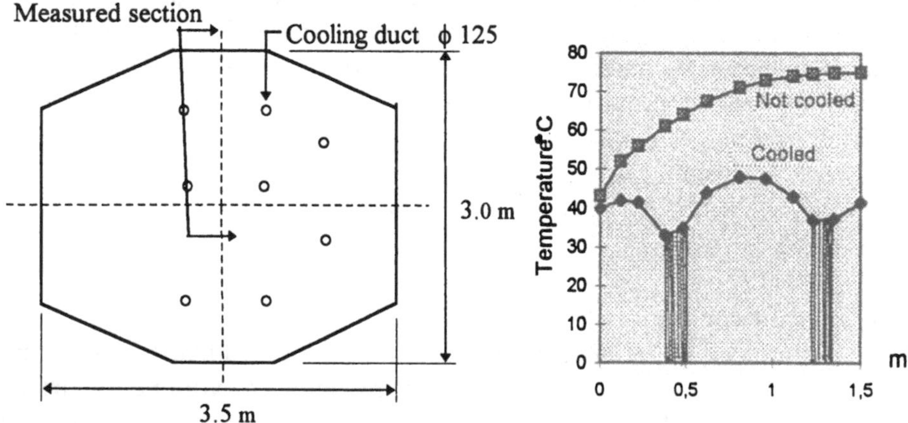

Fig. 9.11 Results from air cooling of bridge piers for the Igelsta bridge near Stockholm, 1992, Compiled from Groth & Hedlund, [33].

b) Internal water cooling
Cooling with the intent of reducing the hydration temperature rise and the temperature differentials should be designed to be as effective as possible throughout the heating phase. Normally, this intense cooling is to be discontinued once the concrete temperature has surpassed its maximum value by some measure. Further cooling may be detrimental and must not be undertaken unless the consequences of prolonged cooling have been carefully studied by means of appropriate stress analysis. However, recent *two-dimensional* stress analyses in connection with ongoing projects (e.g. the Öresund bridge) seem to indicate that cooling beyond the maximum temperature can in many instances have a beneficial effect on the overall risk of cracking.

In structures, where the objective of the cooling measures also is to shorten the time required to reach the temperatures specified for closure of block joints by grouting, cooling may be continued (or resumed later) at a lower rate using e.g. cooling water of suitable temperature i.e. water not as cold as is recommended for the initial intense cooling phase.

Termination of cooling
Abrupt interruption of pipe cooling - like removal of efficient external insulation - radically changes the states (the distribution) of temperature and stress in the concrete, sometimes generating harmful tensional stresses. (See e.g. fig. 9.12 and 9.24). It is therefore generally recommended, that the cooling efficiency be reduced in steps or decreased gradually.

Heat transmission capacity
The capacity of a pipe cooling system is essentially dependent on the difference between concrete temperatures and the temperature of the cooling water as well as on the distance on centers of the cooling pipes. If e.g. the pipe distances are doubled the cooling efficiency is in principle reduced by a factor of ¼ , provided the pipes are not too close. The distances to be used depend of course on the aspired cooling effect but vary typically between 0.5 and 1.5 m.

Effect of pipe size
Smaller variations of the pipe diameter has little impact on the rate at which heat is carried away. For instance, in order to double the cooling rate the pipe diameter would have to be increased by a factor of about 10, provided the ratio of pipe diameter to pipe distance is small. Normally used pipe diameters range from 25 to 50 mm.

Conductivity of pipe material

The conductivity of the pipe material is, however, very important. A study by the author revealed, for example, that steel pipes at 0.6 m on centers gave roughly the same mean cooling effect as plastic pipes of same size at 0.5 m on centers - i.e. the number of plastic pipes to replace steel pipes should be increased by about 45 to 50 %.

Cooling by means of embedded piping is traditionally associated with typical mass concrete structures (dams and the like). Experience over the last decades shows, however, that pipe cooling constitutes a potent aid to avoid cracking also in medium mass concrete. Pipe cooling was e.g. systematically implemented in the construction of the Tingstad tunnel (1964-66) resulting in virtually total absence of early age cracks. (See section 9.7, figure 9.18). Figure 9.13 presents another eloquent example of the striking benefit of selective pipe cooling of a 0.5 m thick wall of a subground railway tunnel (500 m long), which also proved to be devoid of cracks and leakages.

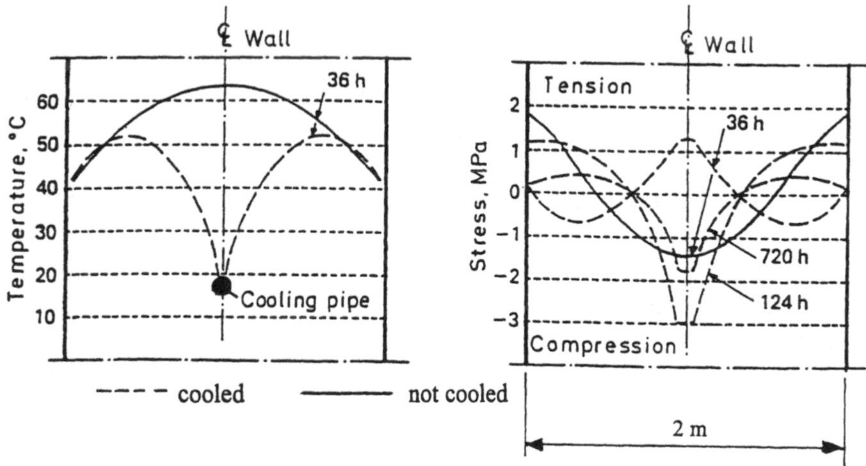

Fig. 9.12 Temperature and stress distribution at a time of 36 hours after pouring in a cooled and a non-cooled section. Furthermore, the stress distributions in the cooled wall are shown at 124 and at 720 hours. Cooling was stopped at 120 hours. Note how tensions around the pipes at 36 hours abruptly change into compression at 124 hours i.e. immediately after cooling is terminated. Concrete: Cement type I, Std, 350kg/m^3, w/c = 0.55, form:12 mm plywood.

Useful nomograms (alignment charts) for defining the cooling time in concrete bodies with and without cooling pipes were published by Rawhouser [1]. The charts were later also relayed by Wischers [29] and the ACI Committee 207, [28].

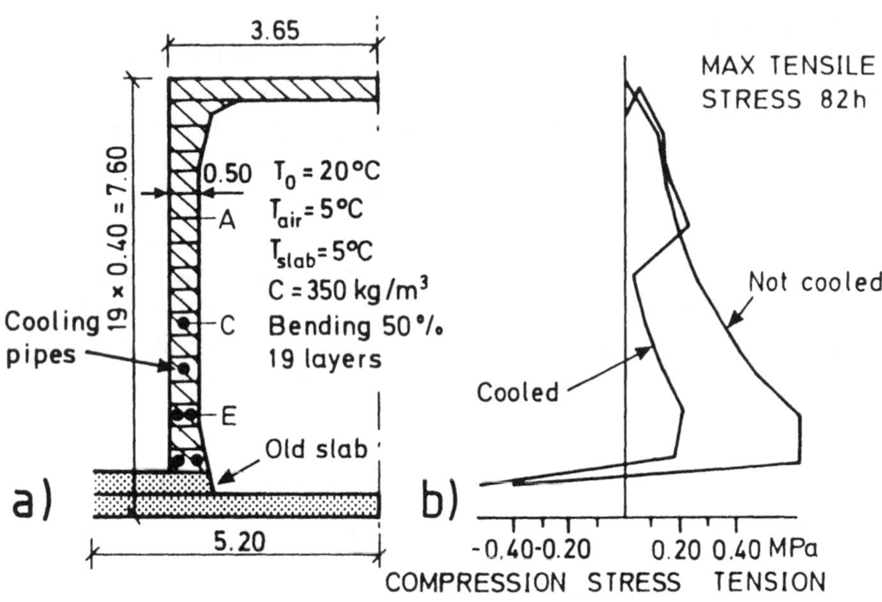

Fig. 9.13 a) Box tunnel element cast on a cold slab. b) Thermal stress analysis depicting the strong alleviation of crack risk in a *cooled* section as compared to a *non-cooled* section. Concrete: cement type II (ASTM), 350 kg/m^3,w/c=0.4, form: 25 mm wood. (Emborg & Bernander, [23]).

c) External cooling.
From a conventional point of view, external cooling by sprinkling water on metal shuttering would appear to be an inappropriate measure as it is normally bound to increase the temperature differential in a concrete body. However, having in mind what has earlier been said about the 'beneficial' compression of the surface layers in articles 9.2.3 and 9.3.5 it is evident that external cooling may well contribute to avoidance of surface cracking. External cooling naturally reduces the mean temperature of a section and slows down the maturity development at the surfaces, but the most important impact of surface cooling is the early age adaptation of surface concrete temperature to that- or even below that - of the environment. In consequence the subsequent decrease of the mean concrete temperature tends to generate surface compression. In a paper especially dedicated to external cooling, Mangold, [15], has for instance demonstrated how the temperature induced prestress of the concrete surfaces can

be controlled by the *duration* and *timing* of the sprinkling procedure. See also paper by Yamazaki, [34].

9.5.4.2 Insulation of hardening concrete

Insulation of concrete in the heating phase is an effective way of cutting down temperature differentials across a section. However, as has been stressed previously insulation also entails the draw-backs of not only raising the mean temperatures in the cast section (adversely affecting through cracking) but also of increasing the impact of temperature shock at the later unavoidable exposure of the concrete surfaces to a colder environment. It is therefore recommended, that high performance insulation - if to be used at all - be applied in several layers so that these layers may be 'peeled off' at time intervals, thus mitigating the intensity of temperature shock. The cautions issued in article 9.3.4 b should be considered in this context.

9.5.5 Artificial temperature control in previous pours (old concrete)

An obvious way to reduce the temperature difference between two adjacent concrete pours is by maintaining or raising the temperature of the previous older cast. Figure 9.14 illustrates how the heat in pour no 1 is retained by insulation so that the temperature in the cooling phase of pour no 2 is more or less synchronized with the post peak temperature development in the bottom slab in order to minimize the differentials.
Further, the concrete in pour no 3 is heated (e.g. by electric cables) as shown, thus substantially reducing temperature difference between pours 3 and 4 (ΔT_{34}) to a minimum. In cases, where heating of the older pour is applied it is important to observe that the expansion of the structural member in question must not be restrained to the extent that the measure undertaken is rendered ineffective.

In figure 9.15 it is exemplified how insulation can be used to reduce the risk of *through* cracking by minimizing the temperature differences (ΔT_{12}) and (ΔT_{23}) between the pours 1, 2 and 3.

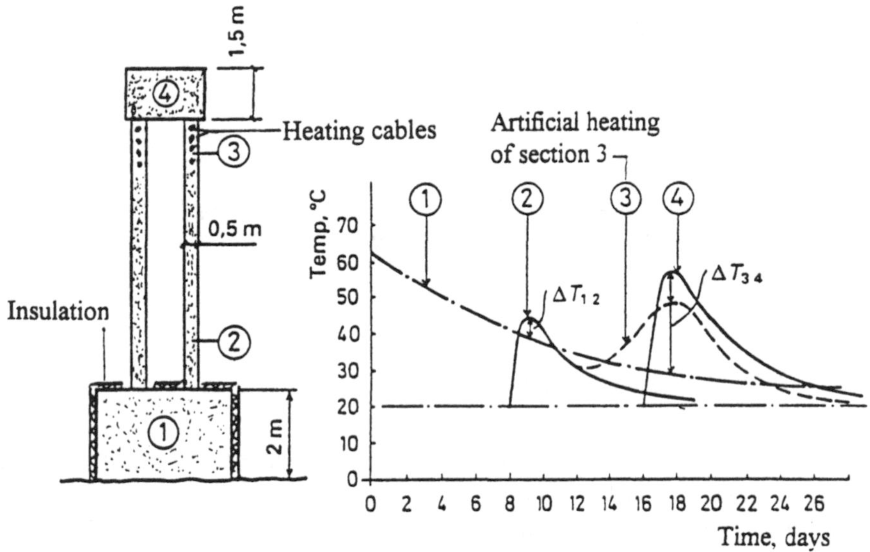

Fig. 9.14 Bridge pier shaft - examples of temperature control by a) retaining heat in previous cast (pour n° 1) by insulation and b) by heating older concrete, i.e. the upper portion of wall 3.

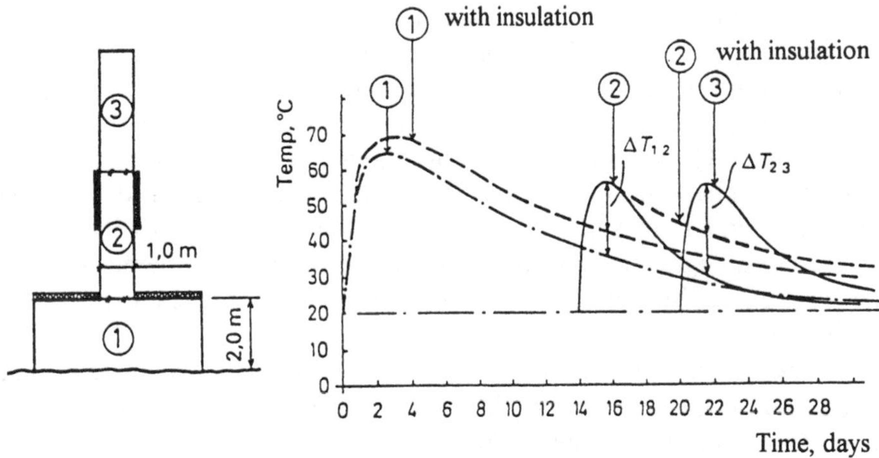

Fig. 9.15 Example of working procedure by which insulation is applied on the concrete surfaces in connection with form stripping thus engendering slower cooling of the pours 1 and 2.The temperature differences ΔT_{12} and ΔT_{23}, liable to generate through cracks, are substantially reduced. Concrete: Cement Std type I, 330 kg/m^3 .

The examples given above underline the necessity of studying each situation on its own merits as conditions will vary widely from one case to another. An important factor in this context is of course the time interval between two consecutive casts

9.5.6 Restraint

The fact, that the degree of restraint is a factor in early age cracking as important as temperature itself, has justified dedicating a separate section (n° 9.7) to this topic.

9.6 SLENDER STRUCTURES

Also for this category of concrete structures or elements thermal early age *through* cracking may be of concern, in particular when water tightness is a crucial requirement. Early age *surface* cracking due to hydration temperatures does *not* usually occur in these cases unless there are concurrent influences of shrinkage or of temperature shock due to premature form stripping.

The primary crack generating factor in slender structures is not so much the hydration temperature rise - which is normally insignificant - as the *initial difference* (T_i- T_a) between the concrete placing temperature and the temperature of older bordering concrete. (Cf. paragraph 9.2.1). Further, in thinner structures, the restraining effect of construction joints is often very high - primarily because of higher reinforcement ratios across joints. (Cf. section 9.7). In consequence, through cracking is not an uncommon phenomenon in thin container walls, bridge decks, road pavements and floor slabs. The early age temperature cracking is often initiated and aggravated by simultaneous drying and/or chemical shrinkage as well as by temperature gradients imposed by external weather conditions.

Crack prevention measures are therefore in slender structures normally focused on the temperature difference (T_l-T_a). Lowering of the initial concrete temperature by the methods described in the previous sections is thus often an effective expedient, the crucial temperature differential being directly affected. The use of concrete with known low cracking sensitivity may be considered. Optimizing the cement content is another step that may be taken, but which may not prove to be so effective as the hydration temperature rise is not a main concern.

It is, however, important to observe current methods of controlling drying shrinkage (water sprinkling, sealing membranes, plastic foils etc)

also in this context. Early exposure to low temperatures by form stripping should, as always, be avoided. At low w/c- ratios the effect of chemical shrinkage should be considered. Steps to minimizing the restraint should be contemplated when deciding on lengths of construction joints and construction sequences. See also Koenders [18, 19].

Concrete road- and airstrip pavements belong to this category of structures. Because of the slenderness of the slabs, the concrete is greatly influenced by ambient temperature and humidity conditions. The temperature gradients induced by e.g. solar radiation and temperature changes may give rise to deleterious flexural stresses. The impact of solar radiation and drying shrinkage can be effectively balanced or mitigated by water sprinkling. Covering of the concrete with plastic foil is another measure to retain humidity, but which should be avoided in summertime because of possible high temperatures resulting from 'greenhouse' effects.

9.7 RESTRAINT

9.7.1 General

It has been emphasized in the beginning of this report that restraint is an element in early age cracking as important as temperature volume change itself. For restraint in general see Rostásy, Tanabe and Laube, Chapter 6. In the following some practical aspects are given as well as a few important considerations focused on construction joints.

The degree of restraint may be defined as a percentage of 'full restraint', i.e.= 100 %, signifying no deformation in the restraining structure. Complete freedom of movement is thus in this context attributed a value of 0 %.

9.7.2 Construction measures relevant to restraint

Restraint depends primarily on the configuration of the restraining structure and the freedom of movement that it permits. Figure 9.17 indicates qualitatively various types of restraint that may result from alternative pouring sequences.

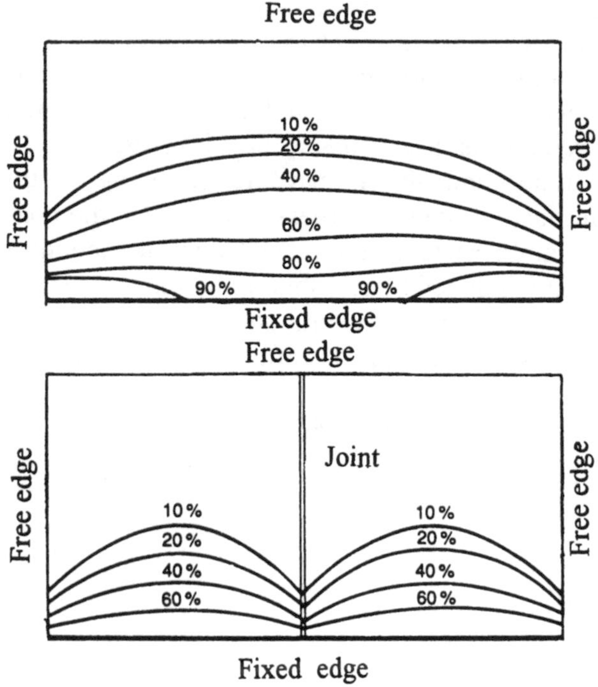

Fig. 9.16 Walls subjected to temperature reduction. Numbers signify percentage reduction of volume change because of inhibited movement thus constituting a measure of the restraint in different parts of the walls. (Copeland, [35]).

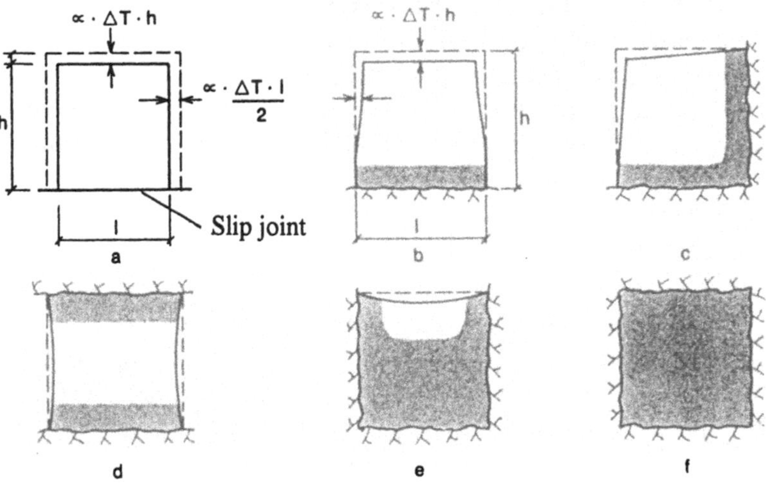

Fig. 9.17 Examples of different types of restraint. Shaded parts indicate areas subjected to considerable restraint, implying potential proneness to through cracking. (Bernander, [2]).

The degree of restraint varies spatially within a cast section and depends on many factors such as the dimensions and geometry of the section, pouring sequences in construction, stiffness of the restraining structure, length and nature of construction joints, freedom of movement of the integral structure e.g. in bending etc. In order to model and quantify restraint it is necessary to resort to normal procedures in structural mechanics analysis, evaluating current restraint from one case to another. The case record shown in figure 9.18 may serve as an illustrative example of how pouring sequences can affect the degree of restraint and the measures required to cope with the varying restraint.

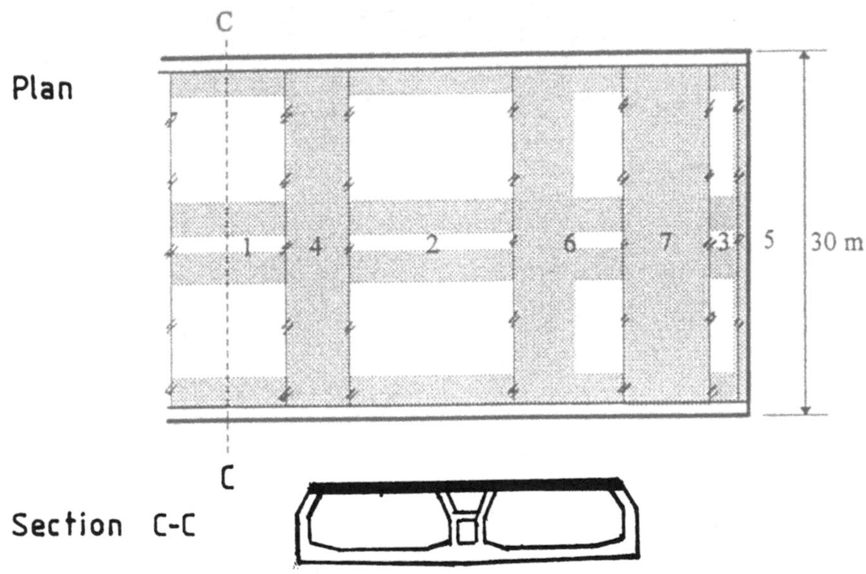

Fig. 9.18 Pouring sequence for the roof slab of the Tingstad tunnel (1964-66). Shaded parts, being areas of high restraint, were water- cooled by means of embedded piping. Estimation of crack risk was systematically based on strain criteria according to equation 1b of article 2.1 above, thus duly considering early age plastic behaviour. The cooling effect was adapted to a target crack risk, to the degree of restraint and to seasonal variations of the temperature of fresh concrete and cooling water. (Bernander, [10, 2]).

9.7.3 Construction joints

More often than not restraining action on a poured section is transmitted by shear and normal stresses across a construction joint - a condition that, as we shall see, deserves special consideration. Figure 9.19 a shows a case similar to that in fig. 9.16 where the magnitude and distribution of principal stresses in a wall - also subjected to evenly distributed reduction of temperature - are shown. Full adhesion in the joint is assumed.

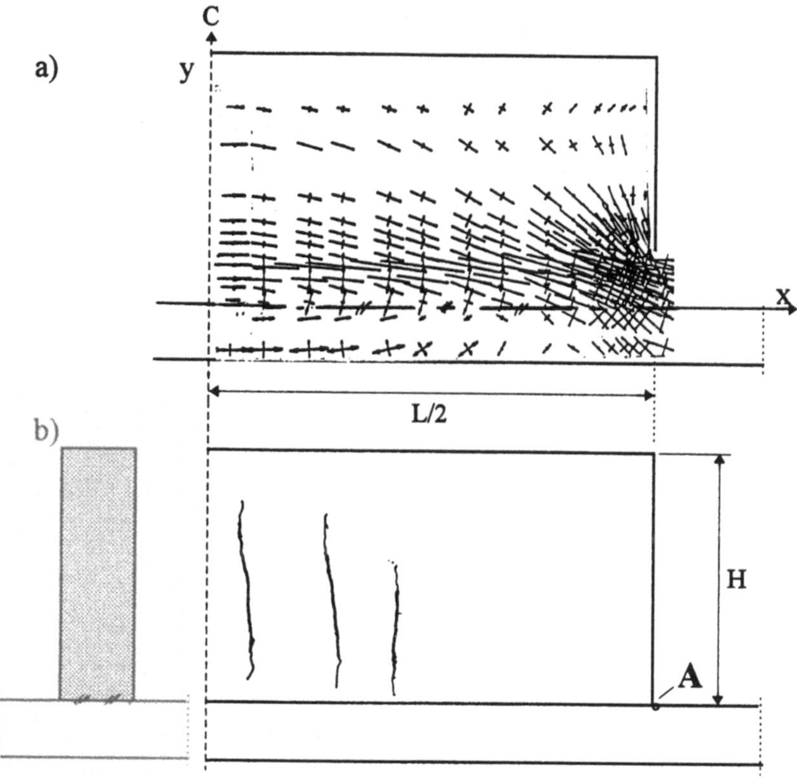

Fig. 9.19 The figures show a) the distribution and relative magnitude of principal stresses in a wall subjected to a temperature reduction assuming full adhesion between the joint interfaces. b) Through cracks as they generally appear in reality in a corresponding scenario.

As can be seen in the diagram the principal stresses are, not unexpectedly, highest in the corner portion at the extreme end of the construction joint (point A). However, in practice cracks generally tend to occur in the central parts of such walls as indicated in figure 9.19 b. The inevitable conclusion from this discrepancy between theory and practice is that full adhesion in the joint *cannot* be present, evidently due to a 'slip' failure taking place - starting from the far end (A) of the joint.

Figure 9.20 shows a result from an investigation illustrating the high shear stresses and the typical *peak tensile stresses* at the extreme end of a construction joint - in this case as a result of a temperature reduction of

10°C. The combined effect of high shear stresses and the concentrated tensions perpendicular to the joint constitute a very difficult state of stress - inexorably producing at least a local failure. In fact, a failure analysis at this location demonstrated, that even if the cohesion strength of the concrete were assumed not to be affected by the innate weakness of construction joints, a local failure would occur, resulting in a slip between new and older concrete. Further, even in a joint with through reinforcement, this failure is likely to propagate along the joint as a progressive 'shear band' failure until shear forces in the joint balance the total tension force (N_o) at the mid-section as indicated in figure 9.29. The associated slip entails a reduction of the restraint exercised by the joint. In fact, the only way in the example to avert this progressive failure would have been to eliminate the splitting action of the *'end peak tension'* by applying sufficient concentrated transverse reinforcement at the corners(A).

Hence, the restraining action of a construction joint will depend on its *length,* amount of *through reinforcement* and the *friction characteristics* of the joint interfaces.

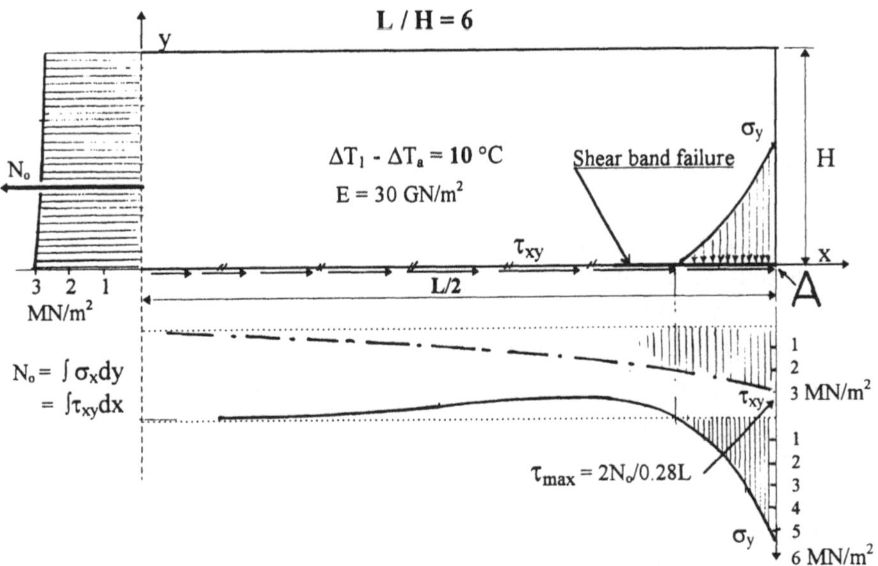

Fig. 9.20 Result from a study of the behaviour of construction joints in respect of their restraining action. Stresses have been calculated assuming *full adhesion* at the joint. The important shear stresses and the high peak tensile normal stresses at point A will, however, initiate a progressive shear band failure as indicated in the figure. (Bernander, april 1994, not yet publ.).

The above mentioned study thus indicates that through cracks occurring in the central part of a wall (as shown in fig. 9.19 b) *must be preceded by*, at least, *a partial failure* with an associated slip along the joint - *because*, if this were not so, the cracks would first occur in the corner portion near the end of the joint at (A).

Practical implications

The above considerations have far-reaching practical implications on design, construction and economy in respect of early age through cracking.

Firstly, it is evident that the slip in the joint between new and older concrete is bound to have a decisive effect on the restraining action to which the young concrete is subjected. If, for instance, the length L in the example shown in figure 9.20 is 10 meters, the total contraction at each end of the wall, corresponding to the 10° C, will only be roughly 0.5 mm, implying e.g. that a minute slip in the joint at one end of only 0.25 mm would reduce the restraining effect by about 50 %.

Secondly, a *necessary prerequisite* of the formation of the centrally located cracks shown in fig.9.19 b is that the construction joint must have *sufficient length* and/or *sufficient reinforcement* to mobilize enough shear to build up the tension force (N_o) required to provoke tensile failure in the wall section. The *mobilizable shear forces* in the joint will thus depend on 'post-failure' friction, dowel action of reinforcement and resulting compression perpendicular to the joint. Such compression will ensue from the balancing of the lifting forces at the joint-end by reinforcement, from dowel action of re-bars (minor effect) and as a result of deviations of the joint alignment from a straight line.

The described phenomenon of the accumulative build-up of the wall tension (σ_x) along the joint involves the relationships between the slip failure, through reinforcement, friction and joint length. It offers a reasonable explanation to the factual observation in practical engineering, that through cracks tend to occur when the lengths of construction joints exceed 8 to 10 meters. This is, namely, a circumstance that challenges theoretical prediction (when assuming full adhesion) according to which tensile strain and crack risk ought to be independent of size and scale effects.

A third rather odd implication is, that a poorly cleaned construction joint will be less prone to produce cracks in the newly poured section than a carefully treated one. This may e.g. be a circumstance worth contemplating in connection with water-tight structures, where construction joints in any case are fitted with sealing profiles for tightness.

The structural situations shown in figures 9.16, 9.19 and 9.29 presume a stiff, *inflexible sub-base*. It should be noted that the stresses in the joint will be somewhat different if the wall is only partially restrained in bending due to flexibility of th esub-base or because of week connection of the same.

The ***main conclusion*** to be drawn from the reasoning above is, however, *that extensive future research in this field must be focused on the modelling and quantification of the true restraining effect of construction joints*. The current common practice of assuming full adhesion in joints is, in many cases, evidently much too conservative - in particular for joints with moderate lengths (L < 6 m) and with insignificant reinforcement.

9.8 CRACK CRITERIA - MONITORING, CONTROL AND FOLLOW UP OF SPECIFIED REQUIREMENTS DURING CONSTRUCTION

In major projects the owner or the designer will normally specify the requirements with regard to early age cracking. However, the owner may sometimes just demand a crack-free structure, thus leaving it to the contractor to decide on means and methods to accomplish this target. In any case the precise circumstances that will prevail on the future construction site or the detailing of production procedures are in general not yet known or defined in the design stage.

Therefore, and also for contractual reasons, it is normally the responsibility of the contractor to finally decide on what measures are to be taken in order to fulfil the specified requirements. Hence, also in the construction stage, assessment of crack risk based on temperature measurements and /or follow up stress analysis have to be performed.

9.8.1 Crack risk assessment based solely on temperatures

Although recent research in early age concrete engineering has clearly demonstrated the inaccuracy and, in many applications the irrelevance, of purely temperature-related crack predictions, specifications are even today (1998) predominantly concerned with only temperature requirements when aspiring to control the risk of cracking in young concrete. Normally the specified limitations apply to maximum temperatures, temperature differentials within the cast section and mean temperatures differences between the new and the older restraining concrete. Table 9.5

lists temperature criteria that have been applied in some major projects in Europe.

Table 9.5 Examples of specified crack criteria applied in some major projects in Northern Europe.

Project		T_l [°C]	ΔT_s [°C]	ΔT_a [°C]	*Requirement:* Crack-free ?	Stress analysis ?
Tingstad tunnel, Sweden	1964	-	-	-	Yes	Yes 1)
Faroe bridges, Denmark	1982		20 ?	?	-	No
Alssund bridge, Denmark	1978	-	20	-	-	No 1)
Helsingborg sub-ground tunnel,* S.	1989	T_i+5	-	-	Yes	Yes 2)
Store Baelt West bridge, Denmark	1990	-	15^{Δ}	12	-	Yes 1
Igelsta railway bridge piers, Sweden	1992				Yes	Yes 3)
Veda susp.bridge,1210 m span, Swe.	1994	70	20	20	Yes	Yes 4)
Öresund bridge, (length 7.9 km) S.	1996	50	15^{Δ}	15(20)#	Yes	Yes 5)
Öresund tunnel, (length 3.7 km), DK	1996	50	"	"	Yes	Yes 5)

1) Stress analysis was applied 'unsolicitedly' by the contractor for his own planning.
2) When beginning construction, temperature criteria were replaced by stress analysis on contractors initiative.
3) Temperature criteria gave rise to serious surface cracking. These cracks were later avoided when applying stress analyses, which were then used to decide on the necessary measures.
4) Originally temperature criteria according to Swedish bridge norms (BN 88,44.411) *or* stress criteria were specified but later the stress approach was adopted.
5) *Both* temperature- and stress related criteria were specified by the proprietor.

* Contract n° I, (Spåret)

\# On the initiative of the contractor it was later convened that ΔT_a could be adjusted with respect to the current degree of restraint - e.g. at 80 % restraint the accepted value of ΔT_a would be 20 /0.80 = 25°C.

Δ The internal temperature difference is defined as the difference between mean temperature and surface temperature.

It should, however, be born in mind - as already stated in article 9.1.3 - that *temperature differences alone are no measure of stress level or crack risk* unless all other factors affecting early age cracking are constant and included in the reference experience on which the assessments are made. For instance, a certain temperature differential obviously does not induce the same tensile stresses if the degrees of restraint in two, otherwise identical, cases are not the same (Cf. fig. 9.6 & 9.23 c). It is therefore surprising that most specifications - even to this day - do not relate the temperature field to the spatial configuration of the degree of restraint - for instance in order to allow for the fact, that even high temperature differentials at locations with little restraint will obviously not contribute much to the risk of through cracking.

Temperature related crack criteria may well seem convenient and practicable to apply in construction, but it should be recognized that, *using such criteria, the reliability of prediction of the risk of early age cracking will be poor.*

9.8.2 Crack risk assessment based on stress analysis

Considering the above mentioned shortcomings of the temperature based assessment of early age crack risk it may seem astonishing that a problem, which has attracted the attention of concrete engineers for so long, has been dedicated so little qualified structural analysis. Any experienced engineer in structural mechanics should not have to contemplate this problem for very long in order to realize that volume change due to temperature is not the sole factor - not even the only crucial one - generating early age temperature cracks. Other important parameters to be considered are:

- the degree of restraint on the cast section imposed by the stiffness and geometry of adjoining structures, length and nature of construction joints etc,
- the transient mechanical properties of hardening concrete and the dependence of these properties on cement type and brand, cement content, water/binder-ratio, additives, maturity development and many other conditions.
- non-uniform temperature and the consequent non-uniform maturity development, which is significantly affected by the relation between (T_i) and (T_{air}).
  conditions related to fracture mechanics.
- artificial temperature control - cooling, heating, insulation etc.

Note: Non-uniform maturity development is here referred to as a separate factor because of its decisive impact on the risk of surface cracking. (See item 9.2.3.2).

All of these factors are in turn affected by other basic conditions of which some important ones are shown in figure 9.2 (See section 9.1).

Research, especially in the last decade, has resulted in various types of computer software for the simulation of stress development based on structural and material modelling of thermal stress build up in early age concrete. In order to qualify in this context the analysis should - apart from the thermal and chemical volume change - at least also consider the timing of the spatial mechanical properties of the concrete in the cast section (i.e. non-uniform maturity development), and the restraining effect of resistance to deformations in adjoining existing structural members.

As may be deduced from table 9.5 such stress analysis has been applied partly or in full in quite a number of projects. The Swedish National Road Administration has for instance issued provisory specifications for avoidance of early age cracks in high quality construction using stress related crack control. As shown in table 9.5 stress related crack control is mandatory for the entire 20 km long connection across Öresund between Denmark and Sweden. Here, however, to be really on the safe side, temperature criteria are also applied, which in the opinion of the author is a costly and an *utterly superfluous* precaution.

9.8.2.1 Practical importance of stress related crack risk evaluation

In conclusion it may be asserted, that the computer-aided theoretical approach to estimating crack risk presents a useful working tool with great potential when evaluating the aptness and cost efficiency of different practical measures in construction. The computer simulations have the capacity to deal with the transient materials properties of young concrete, composite structural elements, complex static and environmental relationships. Computer-aided simulations are also speedy and flexible to implement and enable investigation of optional construction scenarios as well as of parameter studies on the construction site. For instance, which may be the most favourable option: reducing fresh concrete temperature, internal cooling, cement reduction by water reducer or by larger aggregate size, use of low heat cement, modifying the lengths of construction joints or any arbitrary combination of these measures? Experience shows that the need to investigate many alternative construction procedures of this kind is usually considerable and, in the authors experience, this need tends to grow fast with our capacity to furnish relevant answers.

Figure 9.21 through 9.24 may serve to demonstrate the potential of applying stress simulations in construction engineering:Thus, figure 9.21 shows the results from an investigation of the stress development in a 6 m thick foundation slab for the BFG building in Frankfurt a. M. Such a stress diagram may e.g. be used in practice for a rational design of a cooling system, lest the risk of cracking should turn out to be too high.

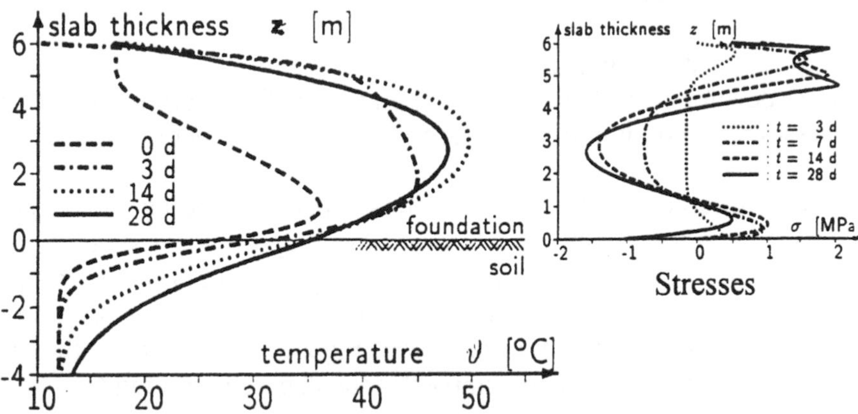

Fig. 9.21 Temperature - stress simulation for the 6 m thick foundation slab of the BFG building in Frankfurt a. M. Selected and compiled diagrams from Huckfeldt, Duddeck & Ahrens, [36].

In figure 9.22 optimum cooling systems in terms of cost are shown for different wall thickness and varying lengths of a construction - a type of investigation that may be very useful when deciding on suitable measures to accommodate specified crack criteria.

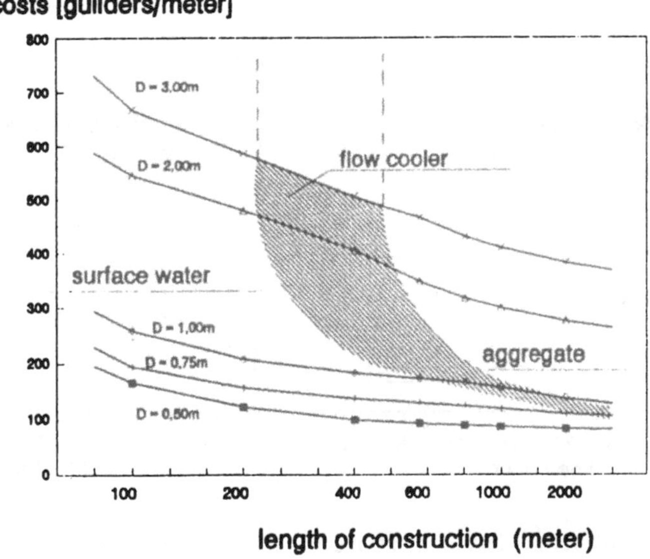

Fig. 9.22 Optimum cooling systems for different wall thickness and construction lengths, Wagenaars & van Breugel, [37].

In figure 9.23 the results of another stress simulation related to a concrete wall are given illustrating:

a) temperature development
b) stress development due to *only* temperature differentials across the wall
c) stresses on account of external restraint on the cast section

The discontinuous behaviour of the stress curves at the age of 48 hours reflects the radical influence of form stripping on the risk of surface cracking. The previously highlighted tendency (art. 9.2.3) of temperature differences from natural cooling to generate beneficial compression in the surface layers is evident from fig. 9.23 b. Figure 9.23 c features the obvious, but not always considered, impact of the degree of restraint. Cf fig. 9.2).

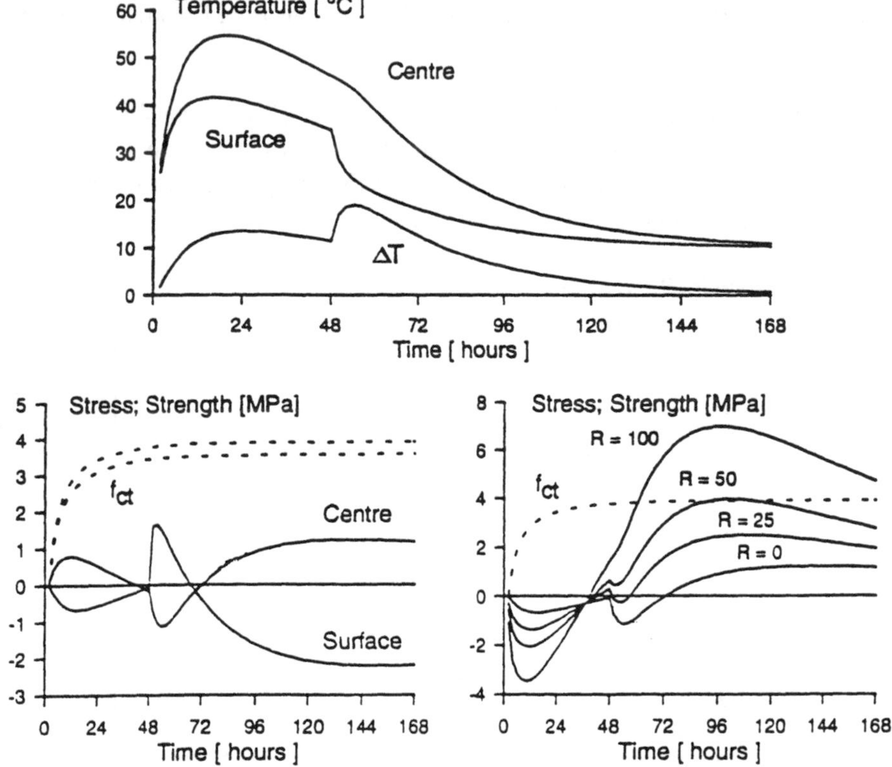

Fig. 9.23a Temperature development in a 1.0 m thick concrete wall. Strength and stress in a concrete wall. **b**) only eigenstresses. **c**) effect of the degree of restraint (R) . Eberhardt, Lokhorst & van Breugel, [9].

In paragraph 9.2.2 it was pointed out that cracking may result from the *integral* effect of the hydration processes in *two or more consecutive pours* (or lifts) in a construction sequence. The last two figures in this context highlight the importance of *extending* the analyses of stresses and crack risk to *previous casts* in the sequence or to the *restraining older concrete.*

The case record shown in figure 9.24 represents a result from an investigation in connection with the erection of the Store Baelt West bridge. The stress simulations were actually used by the contractor when planning optional locations of construction joints and when designing alternative cooling arrangements. As may be seen in the diagram, the existing stresses in pour n° 2 - due to the restraining action of the caisson wall (pour n° 1) in the previous construction stage - are superimposed by the effects of the hydration process in pour n° 3. However, the stresses in this last pour are in turn affected by the added effects of the simultaneous *restrained volume change* in both of the two preceding casts.

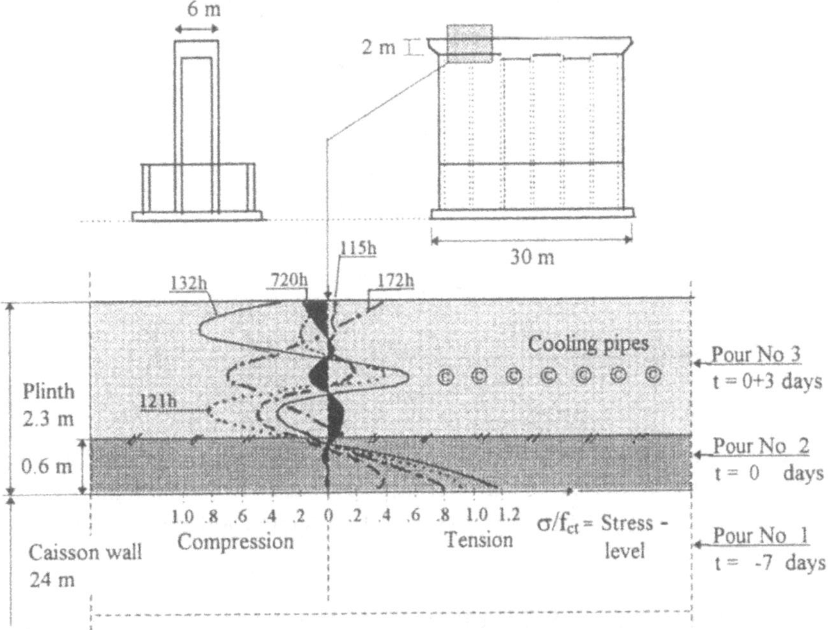

Fig. 9.24 Analysis in the construction phase of a pouring sequence with regard to early age thermal cracking in the roof slab of a pier caisson for the Store Baelt West bridge. The curves show the stress fields at four different ages. The 2 m thick slab is poured in two steps with an interval of three days. Note the radical change of the stress field between 120 h and 172 h due to the termination of cooling, (Bernander, [26]).

Figure 9.25 depicts another case record , where the expansion due to hydration in a bridge pier shaft provoked serious cracking in an underlying tremie-concrete foundation slab. In this case the cracks in the *older restraining concrete* resulted from the joint action of *internal stresses,* originating from the hydration process in the tremie-concrete, and of the *superimposed stresses* due to the expansion in the shaft. The example highlights the importance of not limiting the early age crack analyses only to the latest step in the construction process.

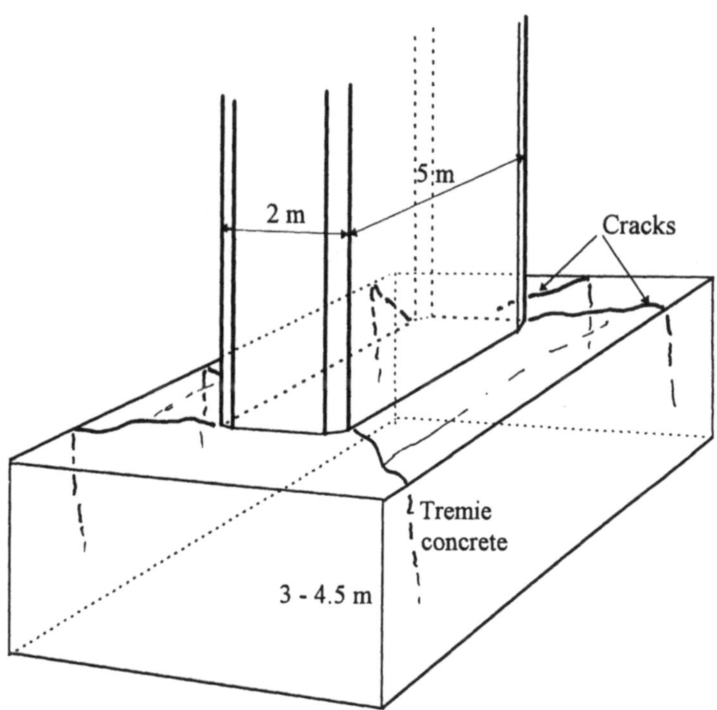

Figure 9.25 Pier shaft cast on a foundation slab. Serious cracking occurred in the sub-base of older tremie concrete (the Öland bridge). The case record highlights the necessity to consider the stress conditions also in restraining structural members of 'older concrete'.

The figures 9.21 through 9.25 reflect the present state of the art in respect of stress analysis in the field of early age thermal cracking - at least in its more sophisticated manifestations. The examples call attention to how stress simulations can be utilized in engineering also on the working site.

9.8.2.2 Benefits of application of the more advanced stress analysis in the assessment of early age crack risk

As has been stressed earlier in this report, crack risk prediction based only on temperature criteria is *inadequate in principle* and, in many contexts, *not sufficiently accurate* from an engineering point of view. This implies, that the considerable resources actually spent on crack control on the sole basis of temperature monitoring

are largely *squandered* because of the poor prospects of attaining the target crack risk or - if so specified - complete avoidance of cracks in a project. Sometimes costly measures undertaken on the working site prove to be insufficient, sometimes they are, in fact, superfluous.

In contrast the potential benefits related to the more sophisticated stress analysis approach, as described above, may be summerized as follows:

a) Improved accuracy in the prediction of crack risk - not to mention all the associated favours to function, durability and general credibility of concrete engineering.
b) Improved economy as a result of better over all precision of crack prevention measures. Accuracy in the spatial localization of crack risk in structures enabling selective and localized - and hence more economical - implementation of crack prevention measures such as cooling, heating, insulation etc.
c) The very fact, that more of the crucial factors affecting crack risk are considered, offers more options, greater flexibility and economic potential in the choice of crack prevention measures in design and construction.
d) The effects of simultaneous drying shrinkage can, at least approximately, be incorporated into the analysis if the shrinkage development in the time domain is known.
e) Instant repairs, injection and grouting work, embellishment, owner/contractor contentions etc may be largely avoided as a direct consequence of improved accuracy with regard to crack risk prediction.

It should, however, be emphasized in this context that, at present (1998), there is *no general acceptance* of the new stress related methods of evaluating of early age thermal cracking among users, bodies for the development of buildings or infrastructure and normative authorities. This is largely due to the following factors:

- The novelty of the approach to practising engineers.

- The recognized complexity of early age cracking phenomena and of how to model crack prediction by e.g. stress simulations.
- The lack of established, generally accepted know-how with regard to the modelling of the basic materials properties of the young concrete.
- The scarcity of experimental data to back up the modelling of these parameters.
- The notorious lack of understanding about the nature and importance of restraint as well as of established methods of quantifying it.

9.8.3 Crack risk assessment based on experimental data from cracking frame or TSM- tests

Cracking frame and/or TSM test data can be directly or indirectly applied to estimate the 'zero-stress' temperature (T_2) and the subsequent evolution of stresses. The advantage of this more experimental approach over the computerized intricate stress analysis procedures is, of course, that - provided the thermal development and the boundary conditions (e.g. restraint) in each point of interest are correctly simulated in the tests - the results should reflect the actual behaviour of the young concrete.

These techniques are demonstrated in two papers to the RILEM Symposium, (1994). In one paper by Mangold & Springenschmid it is shown how the so called eigenstresses due to differential temperatures over a wall section can be evaluated for different scenarios regarding the 'zero-stress' temperature, which, in a round-about way, reflect the influences of the relationship between the initial placing temperature and the ambient surface temperature - whether this be artificially controlled or not. (Cf article 9.2.3.2). The paper also confirms the aforesaid tendency of natural cooling to generate compression in the surface layers of a concrete body.

In another paper to the symposium Fleischer & Springenschmid demonstrate a direct application of results from cracking frame and TSM tests to the actual construction of a bridge deck.

However, for a number of reasons the potential of using results from cracking frame and TSM tests in complex structural contexts lies in the application to structures with a high degree of recurrence and repetitive procedures and where the nature of the early age cracking problem is fairly constant. (Cf article 9.5.2.1). In, perhaps, most practical engineering applications the cracking problem is extremely multifaceted. Conditions on the working site are complex, changing and variable exacting alternative measures and construction procedures - sometimes even provisional arrangements - to be investigated. This is likely to render stress assessments by the cracking frame method too slow and not sufficiently versatile in the process of construction.

9.9 CRACK PREVENTION BY MEANS OF REINFORCEMENT

9.9.1 Surface reinforcement ('shrinkage reinforcement')

Surface cracks due early age temperature development and shrinkage can be limited, and in borderline cases, prevented by providing exposed surfaces with superficial reinforcement. The amount of surface reinforcement - having no static function - is normally chosen not less than 10 mm bars 300 mm on centers (= 3.33 cm^2 /m) and rarely exceeds a steel content of 6 to10 cm^2 /m.

As surface reinforcement, in general, only mitigates cracking by limiting crack width the use of surface reinforcement may not be recommendable in chemically very aggressive environments. Cracks, although small, provide access of deleterious substances through the cover to superficial reinforcement, as a result of which serious corrosion may ensue. The writer has witnessed many a case, where - in hindsight - it could be concluded that the durability of the structure would actually have been substantially better *without* the surface reinforcement.

9.9.2 Reinforcement with regard to 'through' cracking

If investigations according to sections 9.4 through 9.8 indicate that through cracking cannot be avoided by the measures taken, then the tendency of the structure to crack may - as in the case of surface cracking - be limited by extra reinforcement. The primary objective is thus not to prevent cracking but to reduce the widths and the lengths of the cracks at the unavoidable expense of increasing the number of cracks. In some cases it has been observed that additional reinforcement has effectively mitigated or even eliminated cracking, which may be the result of several contributing factors. The steel bars can for instance transmit heat away from critical areas on account of their high conductivity, hence performing as 'cooling pipes'. Further, the high relative stiffness of steel bars tends to promote 'strain-controlled' ductile failures in concrete sections subjected to tension - a phenomenon that is readily understood if the principles of 'failure mechanics' are considered.

The amount of reinforcement required to reduce crack width may be estimated from theoretical considerations, whereby a common criterion is that the *elastic* tensile capacity of the steel bars should suffice to provoke tensile failure in the *relevant part* of the concrete section and that at the time when crack formation is critical.

When crack width is limited to a specified value, also the distribution of the reinforcement as well as diameters and distances between bars has to be taken into consideration. Figure 9.26 shows e.g. a diagram according to which the necessary reinforcement can be estimated on the basis of acceptable or specified crack width and the re-bar diameter.

Additional reinforcement, solely designed to mitigate cracking, may often prove to be very costly and should therefore always be contemplated along with the other optional measures to control early age cracking that have been dealt with in sections 9.3, 9.4 and 9.5.

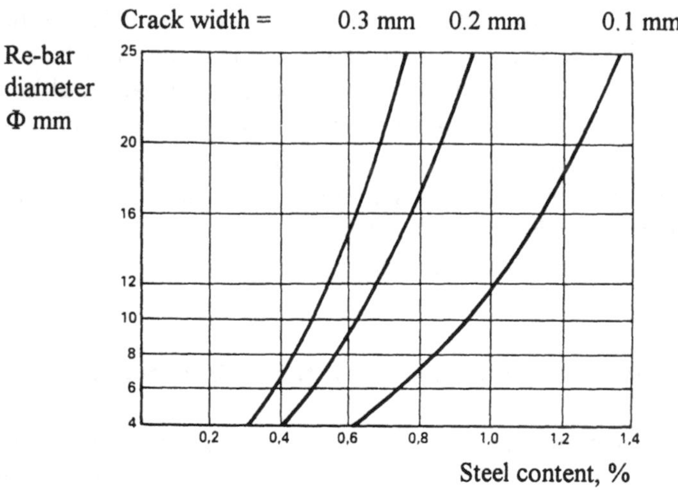

Fig. 9.26 Relationship between steel content, re-bar diameter and crack width for through cracks,(centric restraint). Concrete: K20 -. K40, Reinforcement: Dented bars K400 - K550. Falkner [38].

Accurate design of crack-related reinforcement necessitates detailed knowledge about the relationships between many factors such as crack width, amount and distribution of reinforcement, re-bar diameters, the transient properties of young concrete, restraint, volume change as well as magnitude and distribution of eigenstresses. Because of this complexity precise prediction of crack control by extra reinforcement cannot be made at the present state-of-the art. Detailed recommendations in this field will necessitate further research.

9.10 FUTURE DEVELOPMENT

The foreseeable development in the art of predicting the risk of early age thermal cracking is primarily an extension of our present knowledge of relevant basic concrete materials properties and of our abilities to model these parameters in numerical simulations. Improved modelling of the degree of restraint, to which the hardening concrete is subjected, will have to be an important target for future research.

A further improvement of crack risk prediction is to incorporate into the analysis also the effects of drying and chemical shrinkage.
The need of comprehensive experimental research on early age properties of different types of concrete is conspicuous. The results of this future research will have to be accumulated in suitable data-bases. The on-going development in the direction of numerical temperature-stress analysis of crack risk is expected to continue at a rapid pace, encompassing in due time also the related problems in high strength concretes (HPC:s).

9.11 NOTATIONS

T_i initial fresh concrete temperature (placing temperature)
T_1 maximum temperature
T_2 'zero stress' temperature
T_s surface temperature of concrete
T_a mean temperature of adjoining restraining structure at time of pouring
α_e coefficient of thermal expansion (heating phase)
α_c coefficient of thermal contraction (cooling phase)
ε_1 maximum thermal expansion
ε_1^{el} elastic strain at maximum temperature (time t_1)
ε_1^{pl} plastic strain at maximum temperature (time t_1)
ε_2 maximum plastic strain at zero-stress (time t_2)
$\varepsilon_{t,t}$ $= \varepsilon_{t,el} + \varepsilon_{t,pl}$ total deformation in tension (at time t)
$\varepsilon_{t,u}$ ultimate tensional strain.
ε_3 as defined in figure 1
c cement content
p cement replacement by puzzolanas
w water content
w/c = wcr water/cement ratio
w/b = w/c+p water/binder ratio (wbr)

f_{cc} cylinder compressive strength of concrete
f_{ct} tensile strength of concrete
σ_c concrete compressive stress
σ_t concrete tensile stress
V/A volume to surface ratio (m)
L ength of construction joint
k ratio between plastic deformation and total temperature-induced deformation at maximum temperature (T_1)

REFERENCES

[1] Rawhouser C. (1945): Cracking and Temperature Control in Mass Concrete, *ACI - Journal* 16: 4.

[2] Bernander S. (1980) Massive Concrete, Chapter 16, Manual for Concrete Technology, Work execution, (in Swedish).

[3] Bernander S. & Gustafsson S. (1981) Temperature induced stresses in concrete due to hydration. *Nordisk Betong*, no 2, 1981.(In Swedish with long summary and figure texts in English).

[4] Bernander S. (1982) Temperature Stresses in early Age Concrete. RILEM International Symposium on Early Age Concrete, Paris 1982.

[5] van Breugel K. (1980): Relaxation of young Concrete. Department of Structural Concrete, Faculty of Civil Engineering, Delft University of Technology, Research Report No 5-80-D8, Delft 1980, pp 140.

[6] van Breugel K. (1980): Artificial Cooling of hardening Concrete. Delft University of Technology, Research Report No 5-80-9.

[7] van Breugel K. (1982): Development of Temperatures and Properties of Concrete as a Function of the Degree of Hydration. Proceedings from RILEM International Conference on Concrete at Early Ages, Paris 1982, vol.1, pp 179 - 185.

[8] Emborg M. (1989): Thermal Stresses in concrete Structures at early Ages, Div of Structural Engineering. Luleå Univ. of Technology, Doctoral Thesis 1989:73 D, , revised edition, Luleå 1990, 286 pp

[9] Eberhardt M, Lokhorst S. J. and van Breugel K (1994): On the Reliability of Temperature Differentials as a Criterion for the Risk of early Age thermal cracking. Rilem Int. Symposium on Avoidance of thermal cracking in Concrete, Munich 1994.

[10] Bernander S. (1973) Cooling of hardening Concrete by Means of embedded cooling Pipes. Applications in connection with the construction of the Tingstad tunnel, Gothenburg. *Nordisk Betong*, no 2 - 73, pp 21 - 31. (In Swedish, but with long summary and figure texts in English).

[11] Springenschmid R., Breitenbücher R. and Mangold M.:Practical Experience with concrete technological Measures to avoid Cracking. Rilem Int. Symposium on Avoidance of thermal Cracking in Concrete, Munich 1994.

[12] Springenschmid R. & Nischer P. (1973): Untersuchung über die Ursache von Querrissen in jungem Beton, *Beton- und Stahlbetonbau*, Heft 9, Berlin 1973, p 221.

[13] Springenschmid R. (1984): Die Ermittlung der Spannungen infolge Schwinden und Hydratationswärme in Beton, *Beton- und Stahlbetonbau*, **79**, Heft 10, p 263.

[14] Breitenbücher R. (1989): Zwangsspannungen und Rißbildungen infolge Hydratationswärme, Dissertation, Baustoffinstitut, Technischen Universität, München1989, 206 p.

[15] Mangold M. (1994). Thermal Prestress of Concrete by Surface Cooling. Rilem Int. Symposium on Avoidance of thermal Cracking in Concrete, Munich 1994.

[16] Westman G: Thermal cracking in high performance concrete. Viscoelastic models and laboratory tests. Division of Structural Engineering, Luleå University of Technology, Licentiate Thesis, 1995:27L, Luleå 1995, 123 pp

[17] Bernander S. & Emborg M. (1994) Risk of cracking in massive concrete structures - New developments and experiences. Proceedings from Rilem International Symposium on "Avoidance of Thermal Cracking in Concrete", Munich Oct. 1994, 8 pp.

[18] Koenders, E.A.B, al. (1996): Design and Construction of an office Building in High Strength Concrete, 15 th IABSE Congress, Copenhagen, pp 583 - 588.

[19] Koenders, E.A.B, al. (1996), Cracking Risk and post-cracking Behaviour in High Performance Concrete, 15 th IABSE Congress, Copenhagen, pp 283 - 288.

[20] Löfquist B. (1946) Temperatureffekter i hårdnande betong (Temperature Effects in hardening Concrete) Technical Bulletins, Serie B, No 22, Royal Hydro Power Administration, Stockholm (in Swedish).

[21] Bernander S. & Emborg M. (1992) Temperature conditions and crack limitation in massive concrete structures. Chapter 27, *Handbook for Concrete Technology - Construction*, Stockholm, 1992 (in Swedish) 50 pp.

[22] Emborg M., Bernander, S. (1994): Thermal Stresses computed with a Method for manual Calculations. Proc. from Rilem International Symposium on "Avoidance of Thermal Cracking in Concrete", Munich Oct. 1994, 8 pp.

[23] Emborg M, Bernander S (1994): Avoidance of early Age thermal Cracking in Concrete Structures - Predesign, Measures, Follow-up. Proc. from Rilem Int. Symposium on "Avoidance of Thermal Cracking in Concrete", Munich Oct. 1994, 8 pp.

[24] Mangold M. and Springenschmid R. (1994). Why are temperature-related Criteria so undependable to predict thermal Cracking at early ages.Proc. Rilem Int. Symposium on Avoidance of early Age Cracking in Concrete, Munich 1994.

[25] Solovyanchik A. R et al (1994): Inherent thermal Stress Distribution in Concrete Structures and Method for their Control. Rilem Int. Symposium on Avoidance of thermal Cracking in Concrete, Munich 1994.

[26] Bernander S. (1992) Sprickrisk i grova betongkonstruktioner - Rön och erfarenheter. (Risk of Cracking in massive Concrete Structures - new experiences), *Nordisk Betong*, nr 3, 1992.

[27] Fleischer W and Springenschmid R (1994). Measures to avoid Temperature Cracks in Concrete for a Bridge deck. Rilem Int. Symposium on Avoidance of thermal Cracking in Concrete, Munich 1994.

[28] ACI Committee 207: Mass Concrete. ACI:1R-87. From: *ACI Manual of Concrete Practice 1990*, Part 1,Detroit, Michigan.

[29] Wischers G (1964): Betontechnische und konstruktive Massnahmen gegen Temperaturrissen in massigen Bauteilen, *Beton*, 1964: 1-2.

[30] Iwata M, Saito K, Ikuta K and Kawauchi T (1994): Countermeasure for thermal cracking of box culvert, Proc. from Rilem Int. Symposium on "Avoidance of Thermal Cracking in Concrete", Munich Oct. 1994, 8 pp.

[31] Suzuki N. and Iisaka T. (1994): Establishment of a new Crack Prevention Method for Dams by the RCD Method. Rilem Int. Symposium on Avoidance of thermal Cracking in Concrete, Munich 1994.

[32] Emborg M. & Bernander S. (1994): Assessment of the Risk of thermal cracking in hardening Concrete, *Journ. of Struc. Eng (ASCE)*, New York, **120**, No 10, October 1994, pp 2893 - 2912.

[33] Groth P. and Hedlund H. (1996): Luftkylning av betong med ingjutna kylrör (Air cooling of Concrete with embedded cooling pipes), Technical Report 1996:07 T, Technical University of Luleå.

[34] Yamazaki M (1994): A large Beam cooled with Water Shower to prevent Cracking. Proceedings from Rilem Intenational Symposium on Avoidance of thermal Cracking in Concrete, Munich 1994.

[35] Copeland R.E (1957): Shrinkage and Temperature Stresses in Masonry, *ACI - Journal* No 28 : 8.

[36] Huckfeldt, Duddeck and Ahrens (1994): Numerical simulation of crack-avoiding measures, Proc. from Rilem Int. Symposium on

"Avoidance of Thermal Cracking in Concrete", Munich Oct. 1994, 8 pp.

[37] Wagenaars W.G.L. and van Breugel K. (1994): Watertight Design, Artificial Cooling or extra Reinforcement. Proc.Rilem Int. Symposium on Avoidance of thermal Cracking in Concrete, Munich 1994.

[38] Falkner H, (1969): Zur Frage der Rißbildung durch Eigen - und Zwangspannungen infolge Temperatur in Stahlbetonbauteilen. Deutscher Ausschuß für Stahlbeton, Heft 208, Berlin 1969.

10

RILEM Technical Recommendations

RILEM Technical Committee 119 - TCE

Membership: Chairman: R. Springenschmid, Germany; **Secretary:** M. Plannerer, Germany; **Editorial Secretary:** J.-L. Bostvironnois, Germany; **Members:** P. Acker, France; S. Bernander, Sweden; R. Breitenbücher, Germany; K. van Breugel, Netherland; M.J. Coole, Great Britain; M. Emborg, Sweden; H. Grube, Germany; H. Hamfler, Germany; M.J. Hammons, USA; H. Huber, Austria; C. Jaegermann, Israel; F. Jung, Austria; M. Mangold, Germany; P. Morabito, Italy; P.K. Mukherjee, Canada; F.S. Rostásy, Germany; A.R. Solovjantchik, Russia; T. Tanabe, Japan; P.J. Wainwright, Great Britain.

10.1 ADIABATIC AND SEMI-ADIABATIC CALORIMETRY TO DETERMINE THE TEMPERATURE INCREASE IN CONCRETE DUE TO HYDRATION HEAT OF THE CEMENT

Foreword

This Technical Recommendation was drafted by P. Morabito, Italy, and revised after discussion in the Technical Committee 119-TCE and its sub-committee on hydration heat.

10.1.1 Scope

One of the most important factors associated with thermal cracking in concrete is the evolution and distribution of the temperature increase throughout the section at any time after casting. The temperature increase is a direct result of the heat evolved from the hydration of the cement.

Prevention of Thermal Cracking in Concrete at Early Ages. Edited by R. Springenschmid. RILEM Report 15. Published in 1998 by E & FN Spon, 11 New Fetter Lane, London EC4P 4EE, UK. ISBN 0 419 22310 X

The majority of standard tests currently in use for measuring the heat of hydration of cement are carried out at a constant temperature. The prediction of temperature increase of concrete from these results can be difficult because these isothermal tests do not take account of the change in reactivity of the cement with changing temperature and do not therefore reflect the conditions in the real structure where the temperature changes continually.

The alternative approach is to use adiabatic or semi-adiabatic calorimeters in which concrete specimens of the same mix as will be used on site are tested.

Such methods aim at determining the evolution of the adiabatic temperature increase under conditions which are very similar to those at the centre of a large pour.

The present recommendation gives details of the apparatus and describes the procedure to determine the evolution of the adiabatic temperature in a specific concrete.

10.1.2 Definitions

- Adiabatic calorimeter: a calorimeter is considered to be adiabatic if the temperature loss of the sample is not greater than 0.02 K/h.
- Semi-adiabatic calorimeter: a calorimeter where the maximum heat losses are less than 100 J/(h·K).
- Coefficient of temperature loss a [K/h]: Decrease in temperature of the sample for unit time.
- Time constant τ [h]: Parameter of the exponential cooling curve.
- Coefficient of heat loss α [J/(h·K)]: Quantity of heat lost from the sample for unit of time and for a unit temperature difference between sample and environment.
- Temperature increase of the concrete sample θ_s [K]: Temperature increase of the sample measured during hydration in a calorimeter.
- Adiabatic temperature increase θ_{ad} [K]: True adiabatic temperature increase of concrete.
- Temperature increase θ_{HH} [K]: Intrinsic temperature increase calculated from tests in a semi-adiabatic calorimeter.
- Temperature T_s [°C]: Temperature of the sample.
- Boundary temperature T_a [°C]: Temperature of the sample surroundings.

- Apparent heat capacity of calorimeter C_{cal} [J/K]: Heat capacity exhibited from the calorimeter.
- Heat capacity of the sample C_s [J/K]: Specific heat of concrete multiplied by the mass of the sample.
- Total heat capacity C_T [J/h]: Amount of the total heat capacity exhibited from the calorimeter and the sample ($C_T = C_{cal} + C_s$).

10.1.3 Principle of measurement

The measurement method consists of introducing into the calorimeter a sample of fresh concrete just after the mixing and measuring the temperature of the specimen. Adiabatic calorimeters rely on the principle that at any time during the test the temperature of the sample surroundings must be equal to the temperature of the concrete. This condition requires that additional heat must be supplied from outside.

Semi-adiabatic calorimeters rely only on some form of insulation around the sample to slow down the rate of heat loss.

The hydration heat of the tested concrete will be divided into three parts:

- heating of the specimen
- heating of the calorimeter
- heat loss.

The heating of the calorimeter must be regarded as an apparent heat capacity of the apparatus including the sample container, and has to be determined for each type of calorimeter.

The heat losses play a different role for the two kinds of equipment. In adiabatic calorimeters they are usually small and almost constant throughout the duration of the test and give rise to a residual correction to determine the adiabatic temperature from the measured one. In semi-adiabatic calorimeters they are not constant throughout the run time of the test as they increase with the increasing temperature of the sample and give the major contribution to the calculation of the adiabatic temperature increase of the specimen.

From the point of view of the reliability in determining the adiabatic temperature increase, the main difference between the two measurement methods is that the temperature of the sample in an adiabatic calorimeter is at any time very close to the adiabatic temperature so the influence of the change in reactivity of the cement with the temperature is directly taken into account. In a semi-adiabatic calorimeter the heat losses give rise to a temperature of the sample lower than the adiabatic temperature and a maturity function must therefore be assumed in order to take account of the change in reactivity of the cement. This can give rise to some approximation in determining the adiabatic temperature increase of the tested concrete.

10.1.4 Apparatus

10.1.4.1 Adiabatic calorimeter

Fig. 10.1 shows a schematic view of an adiabatic calorimeter.

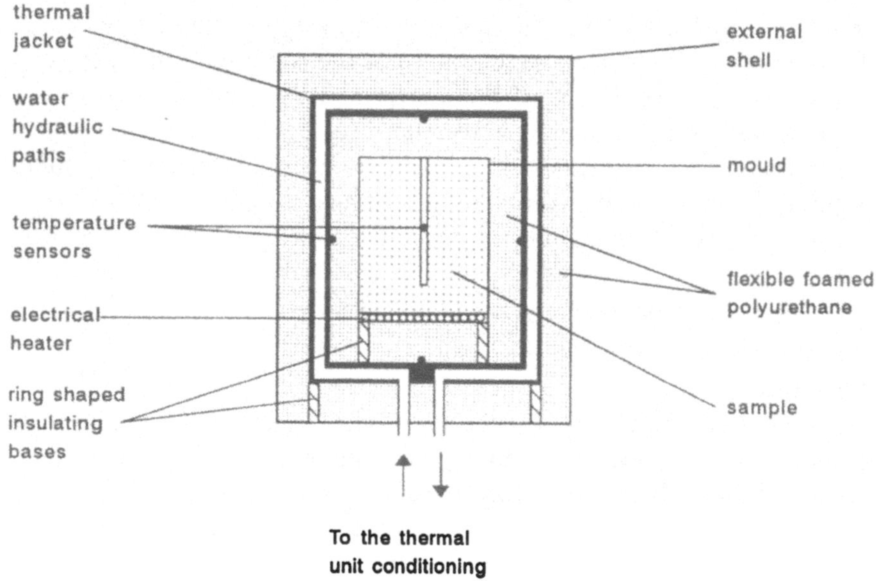

Fig. 10.1 Sketch of a typical adiabatic calorimeter.

The sample is cast in a cylindrical container fitted with a cover. The container should be tight against water vapour, which will be checked by weighing the sample before and after the test. The reduction of mass of the sample should be less than 0.1%.

The dimensions of the container should take into account the thermal capacity effects of a maximum aggregate size of around 32 mm. To incorporate this aggregate size the volume of the sample must be about 4,000 cm^3 and the smallest dimension of the sample must be at least three times the maximum size of the aggregate. The ratio between the apparent heat capacity of the calorimeter to the heat capacity of the sample must be limited to 0.1; such a condition is achieved by performing tests on large volume samples using thin containers. If the condition $C_{cal}/C_s \leq 0.1$ is not satisfied a thermal compensation should be adopted (see 10.1.10.1), otherwise the test method should be considered as semi-adiabatic.

The casing of the sample container should have a sufficiently wide base to ensure good stability against falling over. It is advisable to provide the base with an electric heater to supply external heat for purposes of calibration or thermal compensation of the calorimeter.

The temperature of the specimen is measured with a PT100 resistance thermometer or a thermistor inside a tube placed along the axis of the sample. To ensure good thermal contact the tube should be filled with oil.

The temperature of the calorimeter is controlled by a thermal jacket which should be capable of producing an isothermal surface around the sample whose temperature must follow the sample temperature. The thermal jacket will have the same cylindrical shape as the sample and be made of steel or aluminium (the latter is to be preferred because of its lower unit weight). Inside the jacket water is circulated the temperature of which is controlled by an auxiliary thermal unit conditioning whose working temperature should range from 5°C up to 80°C in order to be able to simulate the hydration of concrete with the same initial temperature as on site. Some form of insulation, such as flexible foamed polyurethane sheet 3 cm thick, should be provided between the sample and the thermal jacket; this improves the equalisation of the temperature around the sample, will minimise the accidental risk of supplying heat directly to the sample and will make the degree of sensitivity of the temperature controller less critical. A second layer of thermal insulation inserted between the jacket and the external shell will minimise any effects that the environmental temperature might have on the test. The temperature of the jacket shall be regulated by a controller whose sensitivity influences the amount by which the temperature of the jacket cycles around the set point and consequently will affect the amount of heat that is lost or gained. Since the temperature of the jacket must never be allowed to exceed the temperature of the sample under test, the temperature of the jacket is set slightly lower than the sample temperature and the difference between the two is carefully controlled. A small amount of heat loss is thus unavoidable.

The temperature difference must be such that temperature loss of the sample is less than 0.02 K/h and to achieve this temperature sensors having a sensitivity of ± 0.01 K should be used (PT100 or thermistors).

10.1.4.2 Semi-adiabatic calorimeter

A semi-adiabatic calorimeter is essentially made up of an insulating vessel filled with foam rubber and of an external shell to protect the thermos vessel from damage (fig. 10.2).

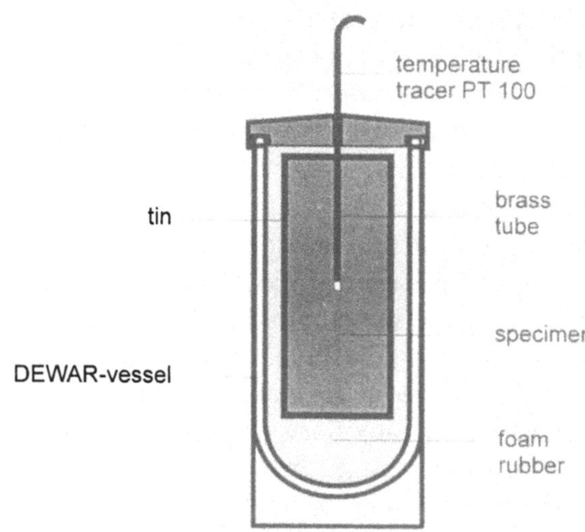

Fig. 10.2 Example of a semi-adiabatic calorimeter using a commercial thermos vessel.

The dimensions of the sample and the method of measuring the temperature are the same as for an adiabatic calorimeter. The ambient temperature where the vessel is placed is fixed at 20°C ± 1°C. The coefficient of total heat loss shall not exceed 100 J/h/K. The vessel should be placed away from any heat sources and strong air currents. The speed of the ventilation air around the calorimeter should be less than 0.5 m/s. When several tests are being carried out simultaneously the minimum distance between vessels should be over 10 cm.

10.1.5 Calibration

The calculation of the adiabatic temperature increase from the measured curves is given in 10.1.10.2 and requires the knowledge of C_s, α, τ and C_{cal}; the last three coefficients depend on the characteristics of the calorimeter therefore must be determined through a calibration.

The calibration sample shall be of distilled water (specific heat equal to 4,186 J/kg/K in the range from 0°C to 100°C) which is to be subjected to the following steps:

- heating of the sample
- spontaneous cooling of the sample.

10.1.5.1 Adiabatic calorimeter

First the total thermal capacity C_T of the calorimeter containing the calibration sample is determined then that of the calorimeter alone is deduced.

The sample container is filled with distilled water and left for a period of about 10 hours to stabilise the system. After this time the sample is heated at constant power P [W] under adiabatic conditions for a time $\bar{t}$ of about 20 h. The level of power should be such as to give a rise in temperature θ_s equal to 30-40°C, the rate of temperature increase should be maintained constant (α = const.). After this time the power is switched off and the sample is kept under adiabatic conditions for 10 h to stabilise the system.

The power supplied is expressed as:

$$P \cdot \bar{t} = C_T\left(\vartheta_s + a \cdot \bar{t}\right) = C_T\left[\vartheta_s + \frac{(T_s - T_a)}{\tau} \cdot \bar{t}\right] \tag{10.1}$$

In order to determine τ, the temperature of the thermal jacket is lowered to a temperature T_{jo} of about 10-20°C, the sample will then cool spontaneously according to the following expression:

$$-\frac{d\left[T_{jo} - T_s(t)\right]}{T_{jo} - T_s(t)} = \frac{dt}{\tau} \tag{10.2}$$

whence:

$$\tau = \frac{t}{\ln \dfrac{T_{jo} - T_{so}}{T_{jo} - T_{st}}} \tag{10.3}$$

where: t = elapsed time ln = natural logarithm
T_{so} = temperature of the sample at starting time of cooling
T_{st} = temperature of the sample at point in time t

If the temperature controller does not allow a steep descent of the temperature of the jacket to be achieved, an alternative way is to remove the sample from the calorimeter and keep it at about 50°C in an oven, reduce the temperature of the jacket to T_{jo} and when this temperature is reached replace the sample in the calorimeter. The cooling period should run for about 30 h.

The apparent heat capacity is given by:

$$C_{cal} = C_T - 4186 \cdot m_w \qquad [J/K] \tag{10.4}$$

m_w being the mass in kg of the water sample.

10.1.5.2 Semi-adiabatic calorimeter

The coefficient α is determined by measuring in steady state conditions the equilibrium temperature T_s for different levels of constant electrical power P provided to the calibration sample. When stable conditions are achieved the heat supplied is completely dissipated into the environment. The equation for the losses is expressed as:

$$P = \alpha \cdot \vartheta_s \tag{10.5}$$

α should be independent on the temperature; however, this should be checked at 5 points in the temperature range observed during the hydration tests (for example for temperature increase of 10, 17, 24, 31, 38 K) and a mean constant value will be assumed.

The measurement of the thermal capacity C_{cal} is carried out by the method of spontaneous cooling of the water sample. For this purpose the sample is disconnected from the electrical power supply after reaching the steady state condition for the last point of calibration to determine α. During cooling the heat exchange equation is expressed as:

$$-C_T \cdot d\vartheta = \alpha \cdot \vartheta \cdot dt \tag{10.6}$$

whence:

$$C_T = \frac{\alpha \cdot t}{\ln \frac{\vartheta_o}{\vartheta_t}} \tag{10.7}$$

with: t = elapsed time since disconnecting the power supply

θ_o = temperature increase at starting time
θ_t = temperature increase at point in time t

10.1.6 Evaluation of the heat capacity of concrete

10.1.6.1 Experimental evaluation

After the heat capacity of the calorimeter is known, the heat capacity of the concrete specimen can be determined according to the same procedures used to determine C_T with the difference that the calibration sample is replaced by a concrete sample which has been allowed to completely hydrate; so:

$$C_s = C_T - C_{cal} \tag{10.8}$$

10.1.6.2 Calculated evaluation

The heat capacity of concrete is temperature dependent and also depends on the hydration rate during the adiabatic or semi-adiabatic tests. Consequently, the experimental values are not better than the calculated values obtained from a knowledge of the composition of concrete and the specific heat of the components.

Calculation of the heat capacity of the sample is performed according to the following formula:

$$C_s = m_s \sum g_i \cdot c_i \tag{10.9}$$

with: m_s = mass of the concrete sample in kg
g_i = content by mass of mix components
c_i = specific heat of mix components:
aggregate: c_a = 0.7-0.9 kJ/kg/K
cement: c_{ce} = 0.84 kJ/kg/K
water: c_w = 4.186 kJ/kg/K

10.1.7 Testing procedure

The testing procedure is performed through the following two steps:

- hydration of the concrete sample;
- calculation of τ.

The volume of the batch should be more than the volume of the sample to be tested and all the constituents will be kept at the constant starting temperature for one day prior to mixing. The temperature of the fresh concrete should differ not more than ± 2 K from the temperature of the calorimeter. The concrete is placed inside the container, vibrated and weighed. The measurement of the temperature increase starts no later than 10-15 minutes from the start of the mixing. The measuring period is usually over seven days and calculation of τ is carried out for each concrete sample tested by spontaneous cooling at the end of the hydration period. Finally, data reduction is performed according to the procedure given in 10.1.10.2.

10.1.7.1 Adiabatic test

The difference $T_s - T_j$ is set to a value such that the coefficient of the temperature loss is less than 0.02 K/h. The temperature of the thermal jacket is regulated continuously or at a rate of no more than every 30 s. Over the whole period of hydration, data recording of T_s and T_j is performed, the experimental evaluation of τ is carried out as described in 10.1.5.1.

10.1.7.2 Semi-adiabatic test

The temperature of the thermos vessel is adjusted to the ambient temperature. The temperature of the fresh concrete should be equal to the ambient temperature within ±2 °C. After the measuring period the specimen is removed from the vessel and stored to let hydration cease. Then the specimen in the closed container is heated up in an oven to about 50°C and inserted in the thermos vessel. By means of the measured temperature loss the time constant τ is determined.

Because of the semi-adiabatic conditions, data reduction leads to a temperature increase θ_{HH} which is a lower estimate of the adiabatic temperature increase; an example of stepwise calculation is given in appendix 10.1.10.3.

The estimate can be improved by considering the influence of the temperature on the hydration by means of a maturity function. In general, the temperature factor k_T of Arrhenius gives a good approximation for the temperature dependence of the hydration of cement:

$$k_T = e^{\frac{E_A}{R}\left(\frac{1}{293} - \frac{1}{273+T}\right)} \qquad (10.10)$$

with: T temperature in °C
E_A activation energy in J/mol
R universal gas constant = 8.314 J/(mol·K)

$$\frac{E_A}{R}=\begin{cases}4000 & \text{for} \quad T\geq 20°C \\ 4000+175(20-T) & \text{for} \quad T<20°C\end{cases} \text{ for Portland cements} \qquad (10.11)$$

$$\frac{E_A}{R}=6000 \text{ for slag cements} \qquad (10.12)$$

The assumed maturity function is

$$M=\int_0^t k_T \, dt \qquad (10.13)$$

The correction is done stepwise according to the following equations:

$$\vartheta_{ad}(t)=\vartheta_{HH}(M) \qquad (10.14)$$

with time $t=\int_0^M \frac{1}{k_T}\,dM$

and $T_s=T_o+\vartheta_{HH}(M)=T_o+\vartheta_{ad}(t)$, to being the initial temperature of the specimen. An example for the calculation scheme is given in 10.1.10.4.

10.1.8 Test results

The test report will document the characteristic parameters of the apparatus and the relevant data of the test.

For the apparatus the following data will be given:

- working principle (adiabatic or semi-adiabatic)
- volume of the sample
- apparent heat capacity of the calorimeter
- coefficient of heat losses α
- date of the last calibration
- ambient temperature T_a for semi-adiabatic calorimeter
- the assigned value of T_s-T_j for adiabatic calorimeter

For each test the following data will be documented:

- concrete composition
- initial temperature of the mix
- initial weight of the sample
- duration of the measuring period
- time, temperature of the sample, ambient temperature (for semi-adiabatic test), jacket temperature (for adiabatic test) at steps not greater than 0.5 h: these data can be tabulated or plotted
- final weight of the sample
- heat capacity of the sample (specify if calculated or measured)
- time constant τ
- total heat quantity liberated by the unit weight of concrete

10.1.9 Some remarks on the choice of adiabatic or semi-adiabatic test

Although both types of testing method make it possible to determine the adiabatic temperature increase of a given concrete sample, semi-adiabatic tests are generally easier to conduct. On the other hand, hydration curves from adiabatic calorimeters are determined quite accurately.

The choice of the test type will depend on the degree of accuracy with which the laboratory test should simulate the hydration in the real structure. For thermal analysis of mass concrete structures where the temperature increase is expected to be almost adiabatic or when the true adiabatic temperature increase and heat of hydration are required, the use of an adiabatic calorimeter is recommended.

In structures of smaller size where heat losses are expected, greater precision in the prediction of the temperature increase can only be achieved from a knowledge or estimate of the expected temperature loss. This of course is an extremely difficult figure to calculate but assuming it was possible then either the calorimeter could be designed to reproduce this loss or adjustments could be made to the measured curve. In both cases a semi-adiabatic test may be more favourable: the semi-adiabatic hydration curve could be in relatively close agreement already with the hydration process as it actually develops in the structure; inaccuracy of any adjustment that allows for minor deviations of the actual temperature in the structure from the semi-adiabatic curve can be kept to a minimum.

10.1.10.1 Appendix A: Thermal compensation for the apparent heat capacity of the calorimeter

The thermal compensation technique consists of supplying by the electrical resistor an external heat source equal to that accumulated from the apparent heat capacity C_{cal} while the hydration test is in progress.
For this, the testing time is divided in intervals Δt_i equal to 0.5 h. Starting from the second interval from the beginning of the hydration, the heat to be supplied in each interval Δt_i is calculated according to the following expression:

$$E_i = \left[T_s(t_{i-1}) - T_s(t_{i-2})\right] \cdot C_{cal} \tag{10.15}$$

The power to be supplied is given by

$$P_i = \frac{E_i}{\Delta t_i} \tag{10.16}$$

The corresponding temperature increase at the end of the test is corrected for the apparent heat capacity of the calorimeter so data reduction will take into account the heat losses alone.

A check of the reliability of the thermal compensation technique should be made at the end of the test through the following verification:

$$\frac{C_{cal}}{C_T} = \frac{E_T}{\vartheta_s \cdot C_s + E_T} \tag{10.17}$$

being $E_T = \sum E_i$ the total heat supplied during the test.

10.1.10.2 Appendix B: Calculation of the adiabatic temperature

The adiabatic temperature at point in time t is expressed as:

$$\vartheta_{ad}(t) = \left(1 + \frac{C_{cal}}{C_s}\right) \cdot \left[\vartheta_s(t) + \int_0^t a(t) \cdot dt\right] \tag{10.18}$$

The following relationships exist between the coefficients of temperature loss a, heat loss α and the time constant τ of the calorimeter with the sample to be tested inside it:

$$a = \frac{\alpha\,(T_s - T_a)}{C_T} \tag{10.19}$$

$$\tau = \frac{C_T}{\alpha} \qquad (10.20)$$

$$a = \frac{T_s - T_a}{\tau} \qquad (10.21)$$

Equation (10.18) can be written in terms of heat of hydration Q given off by a unit weight of concrete:

$$Q(t) = \frac{C_T \cdot \vartheta_s(t)}{m_s} + \frac{C_T}{m_s} \cdot \int_0^t a(t)\,dt = \frac{C_T \cdot \vartheta_s(t)}{m_s} + \frac{1}{m_s}\int_0^t \alpha\left[T_s(t) - T_a(t)\right]dt \qquad (10.22)$$

where m_s is the mass in kg of the tested sample.
For adiabatic calorimeters T_a is equal to the temperature of the thermal jacket T_j and the difference $T_s(t) - T_a(t)$ should be constant at any point during the test.
In semi-adiabatic tests T_a is the temperature of the environment in which the apparatus is placed; T_a shall be kept constant (20°C) so the difference $T_s(t) - T_a(t)$ is equal to $\theta_s(t)$.

10.1.10.3 Appendix C: Example of a stepwise calculation to determine the intrinsic temperature increase from a semi-adiabatic test

The temperature θ_{HH} of the concrete specimen according to the defined semi-adiabatic conditions is given by:

$$\vartheta_{HH} = \left(1 + \frac{C_{cal}}{C_s}\right) \cdot \left[\vartheta_s + \sum a \cdot \Delta t\right] \qquad (10.23)$$

Table 10.1 shows an example for the calculation scheme.

Table 10.1 Example of a stepwise calculation to determine the intrinsic temperature increase from a semi-adiabatic test

$C_s = 4040$ J/K $\quad C_{cal} = 800$ J/K $\quad \tau = 24.8$h

t	Δt	T_S	θ_S	T_a	T_S-T_a	$a\Delta t$	$\Sigma a\Delta t$	$\theta_S+\Sigma a\Delta t$	θ_{HH}
h	h	°C	K	°C	K	K	K	K	K
0		19.9	0.0	19.9			0.00	0.00	0.0
	1.0				0.3	0.01			
1		20.5	0.6	20.0			0.01	0.61	0.7
	1.0				0.6	0.02			
2		20.7	0.8	20.0			0.03	0.83	1.0
	1.0				0.9	0.03			
3		21.0	1.1	20.0			0.06	1.16	1.4
	1.0				1.3	0.05			
4		21.5	1.6	20.0			0.11	1.71	2.1
	1.0				2.0	0.08			
5		22.4	2.5	19.9			0.19	2.69	3.2
...		...	...	...			...	...	...
15		40.1	20.2	20.1			5.14	25.34	30.4
	1.0				20.2	0.81			
16		40.4	20.5	20.1			5.95	26.45	31.7
	1.0				20.4	0.82			
17		40.5	20.6	20.1			6.77	27.37	32.8
	1.0				20.4	0.82			
18		40.5	20.6	20.1			7.59	28.19	33.8
	1.0				20.4	0.82			
19		40.4	20.5	20.1			8.41	28.91	34.7
	1.0				20.3	0.82			
20		40.2	20.3	20.0			9.23	29.53	35.4
...		...	...	...			...	...	...
80		22.9	3.0	20.0			33.08	36.08	43.3
	4.0				2.8	0.44			
84		22.6	2.7	20.0			33.52	36.22	43.5
	4.0				2.5	0.40			
88		22.3	2.4	19.9			33.92	36.32	43.6
	4.0				2.2	0.35			
92		22.0	2.1	20.0			34.27	36.37	43.7
	4.0				1.9	0.31			
96		21.9	2.0	20.1			34.58	36.58	43.9
	4.0				1.8	0.29			
100		21.7	1.8	19.9			34.87	36.67	44.0

10.1.10.4 Appendix D: Example of a stepwise calculation to estimate the adiabatic temperature by means of the maturity function

Table 10.2 Example of a stepwise calculation to estimate the adiabatic temperature by means of the maturity function

semi - adiabatic							$\theta_{HH} = \theta_{ad}$	adiabatic				
t	Δt	T_S		K_T	ΔM	M		$T_0 + \theta_{ad}$		K_T	Δt	t
h	h	°C		-	h	h	K	°C		-	h	h
0		19.9				0.0	0.0	20.0				0.0
	1.0		20.2	1.009	1.01				20.4	1.017	0.99	
1		20.5				1.0	0.7	20.7				1.0
	1.0		20.6	1.028	1.03				20.9	1.041	0.99	
2		20.7				2.0	1.0	21.0				2.0
	1.0		20.9	1.040	1.04				21.2	1.057	0.98	
3		21.0				3.1	1.4	21.4				3.0
	1.0		21.3	1.060	1.06				21.7	1.084	0.98	
4		21.5				4.1	2.1	22.1				3.9
	1.0		22.0	1.094	1.09				22.7	1.130	0.97	
5		22.4				5.2	3.2	23.2				4.9
...		...				...	...	...				...
15		40.1				23.0	30.4	50.4				13.1
	1.0		40.3	2.417	2.42				51.1	3.700	0.65	
16		40.4				25.4	31.7	51.7				13.8
	1.0		40.5	2.437	2.44				52.3	3.876	0.63	
17		40.5				27.9	32.8	52.8				14.4
	1.0		40.5	2.422	2.44				53.3	4.031	0.61	
18		40.5				30.3	33.8	53.8				15.0
	1.0		40.5	2.437	2.44				54.2	4.173	0.58	
19		40.4				32.7	34.7	54.7				15.6
	1.0		40.3	2.422	2.42				55.1	4.300	0.56	
20		40.2				35.2	35.4	55.4				16.2
...		...				...	...	...				...
80		22.9				130.2	43.3	63.3				34.1
	4.0		22.8	1.135	4.54				63.4	5.815	0.78	
84		22.6				134.8	43.5	63.5				34.9
	4.0		22.5	1.120	4.48				63.5	5.846	0.77	
88		22.3				139.3	43.6	63.6				35.6
	4.0		22.2	1.105	4.42				63.6	5.865	0.75	
92		22.0				143.7	43.7	63.7				36.4
	4.0		22.0	1.094	4.38				63.8	5.898	0.74	
96		21.9				148.0	43.9	63.9				37.1
	4.0		21.8	1.087	4.35				64.0	5.935	0.73	
100		21.7				152.4	44.0	64.0				37.9

10.2 METHOD FOR IN SITU MEASUREMENT OF THERMAL STRESS IN CONCRETE USING THE STRESS METER

Foreword

Thermal and other restraint stresses in young concrete can not be determined by the use of traditional methods based on a measurement of deformations. The stress meter enables a good approximation of thermal stresses in concrete. The design concept of the stress meter was developed by the Kajima Technical Research Institute. The Technical Recommendation was drafted by T. Tanabe, Japan.

10.2.1 Scope

This recommendation describes a method for in situ measurement of thermal stresses during the hardening of concrete using a stress meter which was especially developed for this purpose and which is embedded in concrete. The recommendation briefly describes the principle of the measurement, including the details of the stress meter used and its installation, etc.

10.1.2 Principle of measurement

The measurement of thermal stresses in concrete at early ages cannot easily be carried out using conventional means of only monitoring the strains, mainly because the other properties like the Young's modulus, etc. are not accurately known.

The principle of measurement described in this recommendation using the embedded stress meter is shown schematically in Fig. 10.3.

The method uses essentially a load cell in series with concrete cast in the stress meter (hereinafter referred to as a concrete "prism") at the time of casting the surrounding mass concrete. There is no mechanical bond between the prism and the mass concrete along the four walls.

The stress in the concrete prism is obtained by dividing the force (as measured by the load cell) by the cross-sectional area of the prism. Also, since the length of the load cell (l_s) is much smaller compared to the length of the concrete prism (l_c), it can be assumed that the latter is approximately equal to the gauge length ($L = l_s+l_c$) of the surrounding concrete. Furthermore, the method ensures that the properties of prism, such as the Young's modulus, creep, etc. and those of the surrounding concrete are similar, by allowing a free exchange of water across the walls.

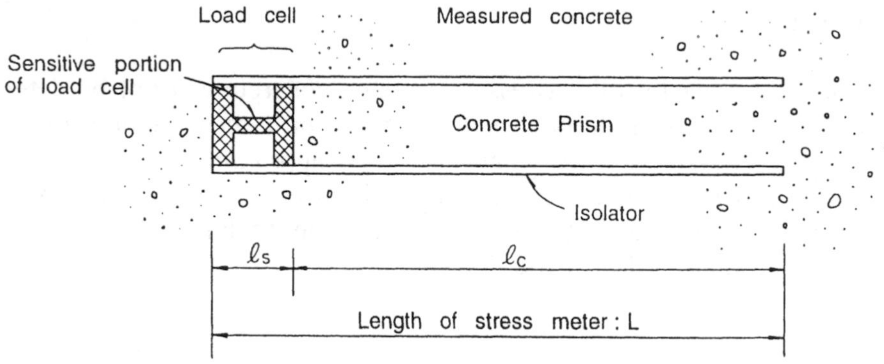

Fig. 10.3 Basic structure of the stress meter

Essentially, the thermal stress in the surrounding concrete is related to the stress in the prism concrete (concrete within the stress meter) as follows :

$$\sigma_c = \frac{1}{(K_g / K_c)} \cdot \sigma_g \tag{10.24}$$

where

σ_c stress in the surrounding concrete
σ_g stress in the concrete prism
K_g rigidity of the overall system comprising of the load cell and the prism
K_c rigidity of the surrounding concrete.

Thus, in order to ensure accurate measurement of σ_c, the rigidities of the stress meter system (load cell and prism) and the surrounding concrete must have as far as possible the same value.

A more detailed discussion on the effect of the length and stiffness of the load cell, etc. on the accuracy of the measured thermal stresses is given in 10.2.7.

10.2.3 Stress meter*

Fig. 10.4 shows schematically the stress meter used. The device consists of a load cell which is fixed at one end of an open box made of wire mesh. The prism and the outside concrete are "monolithic" through the anchor that is provided at one end. A similar anchor is provided at the load cell. The lid of the box is also made of wire mesh and is used to cover the concrete prism after casting. The thermal stresses generated in the concrete due to restraint, etc. are directly obtained using the data from the load cell.

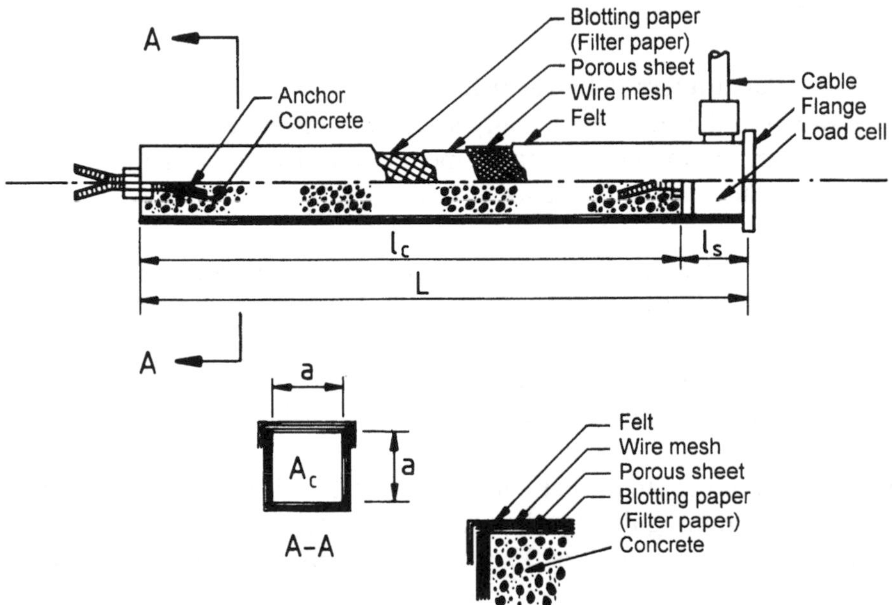

Fig. 10.4 Schematic drawing of the stress meter

Size of the typical stress meter
(Exemplary valid for one specific type of stress meter)

Length of the stress meter (L)	50 cm
Length of the load cell (l_s)	5 cm
Length of the concrete box (l_c)	45 cm
Mean value of sectional area of the load cell (A_s)	15.3 cm²
Sectional area of the concrete box (A_c)	25 cm²

* Addresses of suppliers of the stress meter may be obtained at the Institute for Construction Materials, Technische Universität München, Baumbachstr. 7, D-81245 Munich, Germany, Fax: +49 89 289-27064, E-Mail: Baustoffinstitut@baustoffe.bauwesen.tu-muenchen.de

The walls of the stress meter, including the lid, are lined with blotting paper, pour sheet and felt. This arrangement allows a free exchange of water between the concrete within and outside the device, ensures that there is no mechanical bond between the concrete inside and outside the meter along the four walls, and provides continuity between the concrete prism and the surrounding concrete through the anchors.

Thus, the properties of concrete within the device, e.g. the Young's modulus, creep coefficient, etc. are almost the same as those of the bulk concrete and any thermal stresses can be accurately measured.

10.2.4 In situ measurement

10.2.4.1 Parameters to be recorded

In addition to the stresses (actually strains) as measured by the embedded stress meter, the following measurements may also be carried out to facilitate better analysis of the data.

- Variation of temperature within the concrete: This measurement should be carried out at a location close to the one where the stress is being monitored.
- Variation of strain within the concrete: This measurement should also be recorded at a location close to the one where the stress is being monitored.
- Variation of the atmospheric temperature: This is an important parameter in determining the development of thermal stresses in concrete and should be recorded at 2 or 3 suitable locations in the neighbourhood of the structure.

10.2.4.2 Installation of stress meter

The stress meter can be installed within concrete in a horizontal, vertical or any other position, to measure the thermal stresses in that direction. The stress meter needs to be fixed rigidly in place for accurate measurement of the stress in the desired direction. The steps in the installation of the stress meter can be briefly outlined as given below.

(1) Make a suitable holder which can be used to support the stress meter. Reinforcing bars, or any other suitable material can be used for this purpose, provided the meter can be securely held in place during concreting and that the stand provides the minimum restraint to the movement of the concrete in the direction of measuring.

(2) Secure the holder within the formwork before the concrete is cast. During casting, when the level of concrete reaches the level at which it is desired to install the stress meter, remove a sample of the concrete and fill the stress meter with it. It must be established that the following conditions are satisfied:

- The width of the stress meter is more than twice the size of the maximum size of the aggregate,
- The blotting paper and the felt lining of the stress meter have been moistened thoroughly before pouring concrete into it.
- The concrete placed in the meter (prism) should be compacted using a tamping rod or by vibrating.

(3) Once the meter has been completely filled with concrete , cover it with the upper lid and securely tie it using binding wires.

(4) Wrap the assembly using cloth and moisten with water until the stress meter is completely enclosed by concrete. Secure it to the holder within the formwork.

(5) The vibration in the neighbourhood of the stress meter may be carried out using mechanical needle vibrators. Care should be taken not to disturb the stress meter from its position.

Note: The above method for the installation of the stress meter may be suitably modified depending upon the type of the structure in question or the location chosen for monitoring the thermal stresses.

10.2.4.3 Frequency of measurement

The interval and duration of measurement depend upon the aim of the measurement, the type of the structure, etc. The following may, however, be used as a general guideline in conducting the measurements:

Age after placing	Frequency of collecting data
Up to 1 day	Every hour
1-3 days	Every 6 hours
3-7 days	Every 24 hours
After 7 days	Every 48 hours

It is recommended that the measurements be continued till the temperature of the concrete reaches the ambient temperature.

It may however, be noted that the stresses can be accurately measured only so far as no cracks are formed in concrete in the neighbourhood of the stress meter.

10.2.5 Results

A typical example of the results obtained using the stress meter in terms of the stress history in a 1.5 m thick concrete is shown in Fig. 10.5. The variation of temperature recorded is also given in Fig. 10.6 for reference.

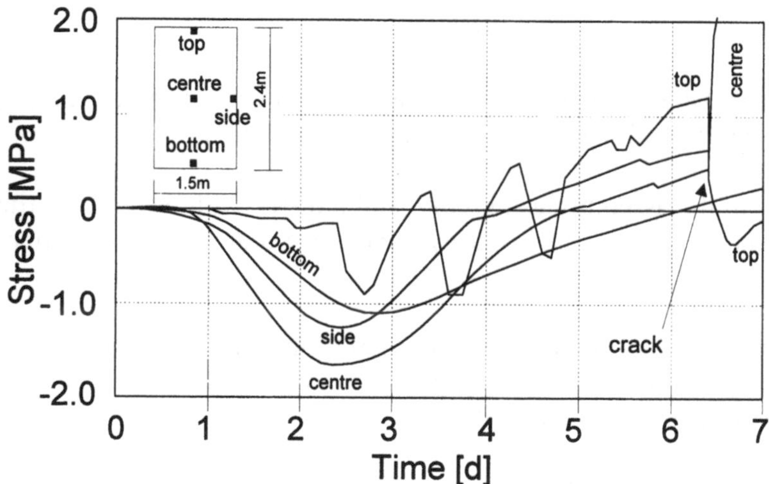

Fig. 10.5 Development of measured stress in concrete structure

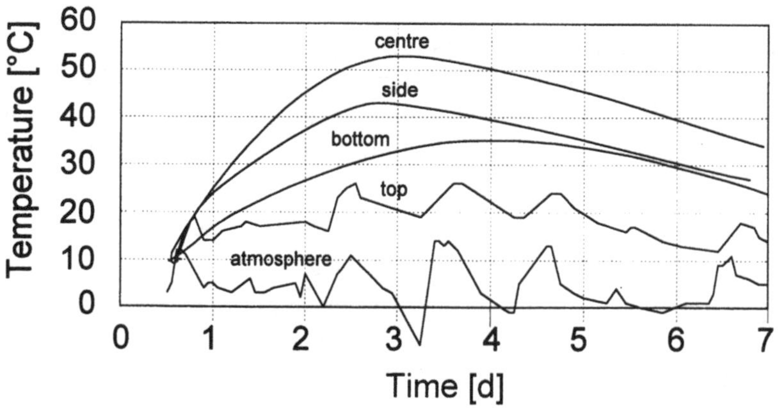

Fig. 10.6 Development of temperature

The figures show the initial compressive stresses generated during the period when the internal temperature rises, and the subsequent transition to tensile stresses on account of the "cooling" of the concrete. A sudden change in the measured stress values indicates possible formation of cracks in the neighbourhood of the meter and may be taken to mean that subsequent the stress values recorded are not accurate.

10.2.6 Report

A report giving the details of the measurement of thermal stress carried out as outlined in this recommendation must give the following details:

10.2.6.1 Basic information

- Details of the stress meter used including the size and type of model used, the range of the load cell and the calibration constant
- Type of structure, including the size and details of the member
- Location and direction of the embedded stress meter
- Method of fixing the stress meter within the concrete
- Exact mix design (including the type of cement, cement content, maximum size of aggregate, etc.)
- Temperature of fresh concrete when placed
- Variation of the ambient temperature

10.2.6.2 Additional information

- Compressive strength development using laboratory specimens, including the detail of measurement
- Tensile strength development using laboratory specimens, including the details of measurement
- Variation of temperature of concrete at the location of stress monitoring
- Variation of strain within concrete at the location stress monitoring

10.2.7 Appendix E

10.2.7.1 Rigidity and Error in Measurement

From the schematic representation of the measurement of thermal stresses as given in Fig. 10.3, the rigidities of the stress meter and surrounding concrete depend upon their length and the modulus of elasticity of the load cell, etc. Their relationship can be expressed as follows

$$\frac{K_g}{K_c} = \frac{l_s + l_c}{\frac{E_c}{E_s} \cdot \frac{A_c}{A_s} \cdot l_s + l_c} \tag{10.25}$$

with

K_g rigidity of the load cell and the prism
K_c rigidity of the surrounding concrete
E_c, A_c Young's modulus and cross-sectional area of the concrete prism
E_s, A_s Young's modulus and the cross-sectional area of the load cell
l_c and l_s lengths of the concrete prism and load cell (see Fig. 10.3).

Based on Eq. (10.24) and using the Young's modulus of steel (210 GPa) for E_s, the variation of K_g/K_c with the Young's modulus of concrete for different values of l_s, l_c, A_s and A_c is given in Fig. 10.7.

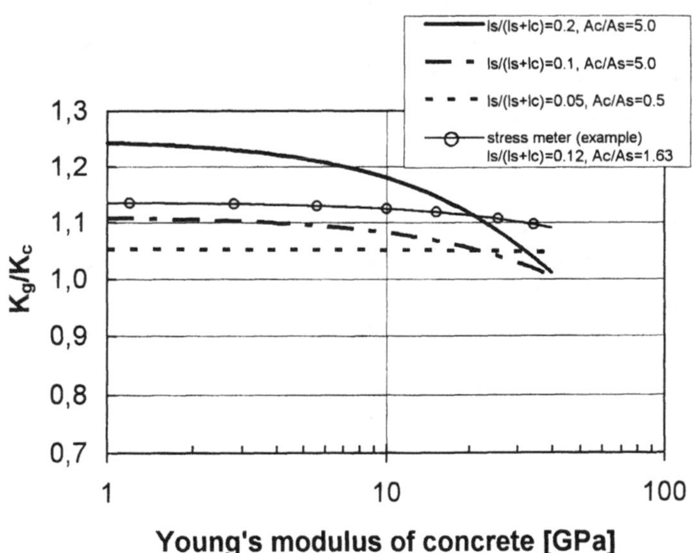

Fig. 10.7 Effect of the rigidity of the stress meter and of the Young's modulus of concrete on K_g/K_c

From Fig. 10.7, it can be seen that by having a large enough l_c compared to l_s, the ratio K_g/K_c approaches unity, a condition where the measurement ensures the best results.

10.2.7.2 Effect of creep and drying shrinkage

Because of the special feature of the stress meter, in which the prism is cast with the same concrete as the surrounding concrete, drying shrinkage and creep characteristics of the concrete in the stress meter are similar to the surrounding concerte and therefore the effect of these factors on the measured values of thermal stress is insignificant.

10.2.7.3 Effect of temperature

The coefficient of linear expansion of the stress meter (load-cell and the concrete prism) can be represented as follows :

$$\frac{\alpha_g}{\alpha_c} = \frac{\frac{\alpha_s}{\alpha_c} l_s + l_c}{l_s + l_c} \tag{10.26}$$

with

α_g coefficient of linear expansion of the stress meter
α_s coefficient of linear expansion of the load cell
α_c coefficient of linear expansion of the surrounding concrete

This equation also suggests that more accurate results can be obtained by increasing the length of the prism in relation to the length of the load cell. As an example, if α_s is taken to be 10.10^{-6}/K, l_s/l_c is assumed to be 0.1 and the range for α_c is taken as $9\text{-}13.10^{-6}$/K, it is found that α_g/α_c is in the range 1.01-0.98. Thus it can be seen that the changes in temperature do not much affect significantly the measured thermal stresses.

10.3 TESTING OF THE CRACKING TENDENCY OF CONCRETE AT EARLY AGES IN THE CRACKING FRAME TEST

Foreword

The cracking frame is useful tool to assess thermal stresses in young concrete and to compare the cracking resistance of different concrete mixes in the laboratory. This Technical Recommendation was drafted by R. Breitenbücher, Germany.

10.3.1 Scope

To prevent thermal cracking in restrained concrete members at early ages not only the temperature increase but rather the stress development has to be considered. As well as the temperature development the development of the modulus of elasticity and the relaxation are of importance for the stress development. To judge the cracking tendency the development of tensile strength must also be considered. In an adiabatic or semi-adiabatic calorimeter only the heat of hydration can be measured. In order to measure thermal stresses and determine the cracking tendency tests in a restraining apparatus have to be performed. The cracking frame has proved to be a useful tool for this purpose. In this equipment a concrete beam hardens under semi-adiabatic conditions and under restraint. Such tests are carried out to judge the cracking tendency of various concretes.

10.3.2 Definitions

- 1st Zero stress temperature $T_{Z,1}$: Temperature at which under restrained conditions compressive stresses during the generation of hydration heat first occur due to the increasing modulus of elasticity. $T_{Z,1}$ describes the change from plastic to visco-elastic behaviour.
- Maximum compressive stress $\sigma_{c,max}$: Maximum compressive stress under restrained conditions during the temperature increase. Due to the high relaxation of the compressive stresses $\sigma_{c,max}$ will usually be reached before the maximum temperature has occured.
- Maximum temperature T_{max}: Maximum temperature of the concrete specimen during hardening under semi-adiabatic conditions.
- 2nd Zero stress temperature $T_{Z,2}$: Temperature at which during the cooling phase the compressive stresses have decreased completely to zero and tensile stresses start to develop. Due to the relatively high relaxation at very early ages and the high modulus of elasticity on cooling down, $T_{Z,2}$ normally is much higher than $T_{Z,1}$ and only a few degrees below T_{max}.
- Cracking temperature T_c: Temperature at which the restrained specimen cracks, i.e. the tensile stresses have exceeded the tensile strength. The cracking temperature characterizes the cracking tendency of the concrete being tested: The tendency to thermal cracking at early ages is higher for higher cracking temperatures.
- Tensile strength $ß_t$: Tensile stress at time of cracking.

10.3.3 Cracking frame*

10.3.3.1 Apparatus

The rigid cracking frame consists of a frame with two cross-heads and two massive steel bars (Fig. 10.8).

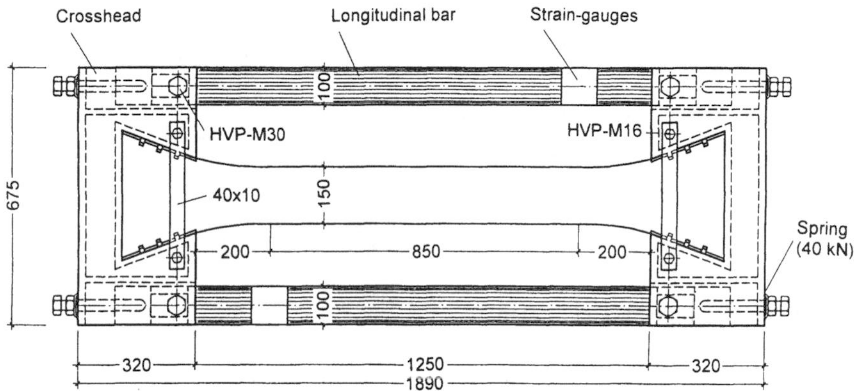

Fig. 10.8 Cracking frame

The longitudinal steel bars have a diameter of 100 mm. To guarantee minimal thermal deformation of the steel bars, a special steel with a low coefficient of thermal expansion has to be used. For this purpose steel containing 36% nickel is suitable. The cross-heads consist of normal steel. A dovetailed opening is located in the centre of each cross-head to fix the specimen when tensile stresses occur (Fig. 10.9a+9b).

The cross-heads should be enlarged to the top by an angle of about 2°, to enable easy removal of the specimen. The two cross-heads have to be fixed to the bars with prestressed screws of at least 30 mm in diameter. Additionally, the steel bars have to be prestressed against the cross-heads in the longitudinal direction by screws (16 mm in diameter) and springs; the prestressing force must be at least 40 kN. Thus a displacement between the bars and cross-heads during the change from compression to tension will be prevented.

The formwork is made using a thermally insulating material, e.g. polystyrene with a thickness of 50 mm and 24 mm thick wood as stabilizing element (Fig. 10.10).

---

* Addresses of suppliers of the cracking frame may be obtained at the Institute for Construction Materials, Technische Universität München, Baumbachstr. 7, D-81245 Munich, Germany, Fax: +49 89 289-27064, E-Mail: Baustoffinstitut@baustoffe.bauwesen.tu-muenchen.de

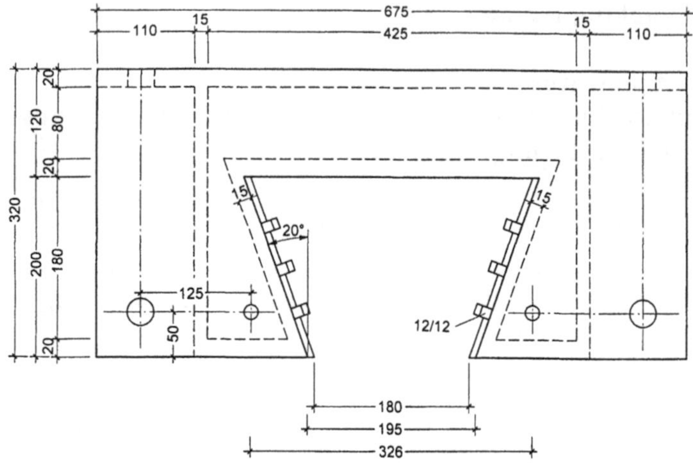

Fig. 10.9a Cross-head: Detail plan

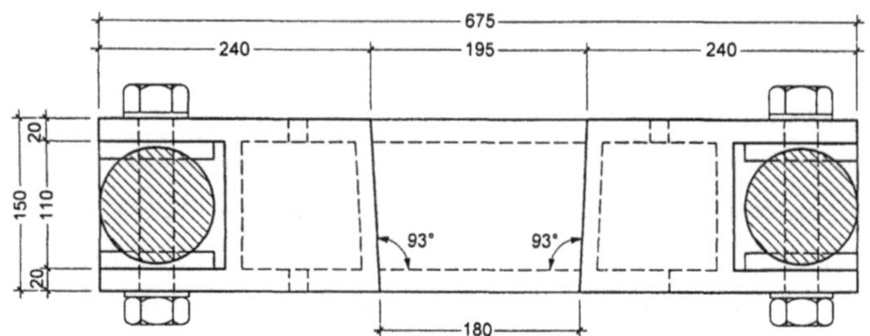

Fig. 10.9b Cross-head: Cross-section

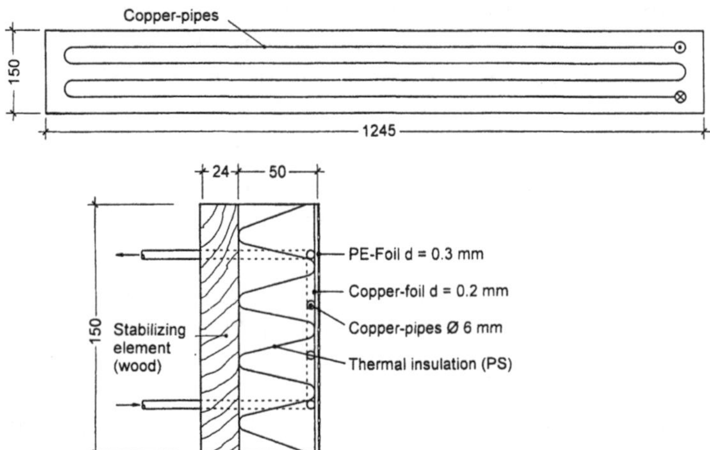

Fig. 10.10 Formwork

Inside the formwork (next to the concrete specimen) a 0.2 mm thick copper plate is fixed on the polystyrene. This causes the heating of the specimen to be similiar to that of a concrete member of about 50 cm in thickness and without any external heating. To enable additional heating or cooling within the thermal insulation, copper pipes (6 mm in diameter) are installed which are in contact with the copper plates. Water can be pumped through these pipes.

The 1.65 m long specimen is fixed within the dovetails of the cross-heads. To prevent notch stresses in the specimen near the cross-heads, the formwork must be formed in a sector of 200 mm in front of the cross-heads in such a way that the specimen is enlarged in a parabolic form continuously without any edge. Thus the specimen has a constant cross-section of 150 x 150 mm^2 over a length of 0.85 m.

To prevent the specimen from drying out during the test, the form-work is wrapped additionally with a 0.3 mm thick polyethylene foil.

10.3.3.2 Measurements

During the test, the temperature in the centre of the specimen and the stresses in the specimen have to be measured continuously.

For temperature measurement, thermometers (e.g. PT 100) are installed in the specimen soon after casting and finishing of the fresh concrete. This requires only a small hole (maximum 3 mm in diameter) for the insertion of the thermometer (otherwise an artificial failure will be located).

For the determination of the stress within the specimen, strain gauges have to be fixed on both of the longitudinal steel bars of the cracking frame in the form of a Wheatstone bridge. The elastic deformations of the steel bars due to the restraint force of the concrete specimen can be then measured. In view of the cross-section and the stiffness (modulus of elasticity) of the steel bars, the actual force (i. e. stress) in the specimen can be determined, because the force in the specimen is of the same value as that in the steel bars, however with opposite sign.

Before a test is carried out in the cracking frame, the strain gauges have to be calibrated. This can be done with, for example, two steel bars between the two cross-heads which are connected with a loading cell and a screw. With the latter the frame can be loaded under compression (i. e. tensile stress in the concrete beam).

10.3.4 Preparation of the concrete specimen

Before the concrete is placed into the cracking frame, the bottom and the two sides of the formwork have to be fixed at the frame. A joint of 2-3 mm between the cross-heads and the formwork must be present on each side otherwise the deformations of the specimen are restrained additionally in an uncontrolled way. These joints have to be sealed with a plastic material, e.g. silicone. In the same way the joints between the formwork elements have to be sealed. The fresh concrete is placed directly into the cracking frame. The temperature of the fresh concrete should be 20 ± 1°C. After compaction with an internal vibrator, the surface is finished in the conventional manner. The surface is covered by sticking a polyethylene sheet to the sides of the formwork and the cross-heads. The open ends of the dovetailed cross-heads are then connected with a steel plate (40 x 10 mm^2) each with screws on the upper surface. This prevents opening of the cross-heads due to tension in the specimen. The same plates have to be screwed onto the bottom of the cross-heads before the lower formwork is fixed to the frame.

The thermometer is then installed in the centre of the specimen. Finally, the thermally insulated upper formwork is fixed onto the specimen.

10.3.5 Storage conditions

The tests shall be performed at an ambient temperature of 20 ± 2°C and a relative humidity of 65 ± 5%.

10.3.6 Procedure for the standard test

After casting, the specimen for the standard test remains in the thermally insulated formwork under semi-adiabatic conditions for 96 hours without any artificial thermal treatment. Thus the temperature increase in the concrete specimen is only due to the heat of hydration in connection with the thermal insulation. During this period the temperature and the restraint stresses in the specimen have to be measured continuously.

After 96 hours the specimen has cooled down close to the ambient temperature. If at this time the concrete has not been yet cracked, the specimen must be cooled down artificially. The cooling rate should be 1 ± 0.1 K/h. For this purpose cooling water is pumped through the pipes within the formwork. When the specimen has been cracked (this can be established by the sudden drop of the measured stresses to nearly zero) the test can be terminated.

With this standard test the cracking tendency of various concretes can be compared under identical thermal and restraining conditions.

In order to remove the cracked specimen, the concrete is pulled out from the cross-heads using suitable hydraulic equipment. Care must be taken that the frame is not excessively bent or twisted when pulling out the specimen.

10.3.7 Concrete mix for standard test

Normally the cracking tendency of concretes has to be assessed at the beginning of a concrete project, when the exact mix composition is not yet known. Thus the cracking tendency of the individual components (e.g. various cements) has to be compared. For this purpose a constant mix composition must be used. Due to the present experience the following mix composition has proved to be appropriate for standard tests:

- Aggregate: A/B 32 mm according to DIN 1045
- Cement: 340 kg/m³ [1)]
- Water: 162 kg/m³
- Consistency: F 3 according to ENV 206 regulated by the addition of superplasticizers

[1)] The cement content has to be reduced appropriately when puzzolanic admixtures are used.

10.3.8 Temperature controlled test for mass concrete

In some special cases, e.g. for the simulation of the temperature in mass concrete members several meters thick, it may be necessary to regulate the temperature in the specimen. For this purpose, a separate calorimeter test has to be carried out beforehand. With these results the temperature behaviour within the simulated member can be calculated. During the cracking test, the calculated variation in temperature is applied to the specimen using a thermostat. Apart from this the test can be carried out as in the standard test. However, in this case the temperature and stress curves as well as the characteristic values, e.g. the cracking temperature, are different from the results obtained in standard tests.

For the comparison of different concrete mixes, the standard test gives a good indication even for mass concrete. This is because the development of elastic properties and of the relaxation on the first day is decisive for thermal stresses. During the first day, the temperature in both tests differs only slightly.

10.3.9 Test results

From the recorded data the following characteristic values should be extracted and documented in a test report.

- Temperature of the fresh concrete
- First zero stress temperature
- Age at which the first zero stress temperature has been reached
- Maximum in compressive stress during heating
- Age at which the maximum compression has been reached
- Maximum temperature
- Age at which the maximum temperature has been reached
- Second zero stress temperature
- Age at which second zero stress temperature has been reached
- Temperature at an age of 96 hours
- Stress at an age of 96 hours
- Maximum tensile stress at the time of cracking
- Temperature at the time of cracking
- Age at cracking

Additionally, it is recommended to plot the curves for stress and temperature development.

Keyword Index

Milton Keynes UK
Ingram Content Group UK Ltd.
UKHW021634071024
449327UK00020BA/1305